T0186398

LOW-ENERGY ELECTRON SCATTERING from MOLECULES, BIOMOLECULES and SURFACES

LOW-ENERGY ELECTRON SCATTERING from MOLECULES, BIOMOLECULES and SURFACES

EDITED BY
PETR ČÁRSKY • ROMAN ČURÍK

CRC Press
Taylor & Francis Group
Boca Raton London New York

CRC Press is an imprint of the
Taylor & Francis Group, an **informa** business

CRC Press
Taylor & Francis Group
6000 Broken Sound Parkway NW, Suite 300
Boca Raton, FL 33487-2742

First issued in paperback 2019

© 2012 by Taylor & Francis Group, LLC
CRC Press is an imprint of Taylor & Francis Group, an Informa business

No claim to original U.S. Government works

ISBN-13: 978-1-4398-3910-2 (hbk)
ISBN-13: 978-0-367-38180-6 (pbk)

This book contains information obtained from authentic and highly regarded sources. Reasonable efforts have been made to publish reliable data and information, but the author and publisher cannot assume responsibility for the validity of all materials or the consequences of their use. The authors and publishers have attempted to trace the copyright holders of all material reproduced in this publication and apologize to copyright holders if permission to publish in this form has not been obtained. If any copyright material has not been acknowledged please write and let us know so we may rectify in any future reprint.

Except as permitted under U.S. Copyright Law, no part of this book may be reprinted, reproduced, transmitted, or utilized in any form by any electronic, mechanical, or other means, now known or hereafter invented, including photocopying, microfilming, and recording, or in any information storage or retrieval system, without written permission from the publishers.

For permission to photocopy or use material electronically from this work, please access www.copyright.com (http://www.copyright.com/) or contact the Copyright Clearance Center, Inc. (CCC), 222 Rosewood Drive, Danvers, MA 01923, 978-750-8400. CCC is a not-for-profit organization that provides licenses and registration for a variety of users. For organizations that have been granted a photocopy license by the CCC, a separate system of payment has been arranged.

Trademark Notice: Product or corporate names may be trademarks or registered trademarks, and are used only for identification and explanation without intent to infringe.

Library of Congress Cataloging-in-Publication Data

Cárský, Petr
   Low-energy electron scattering from molecules, biomolecules, and surfaces / Petr Carsky and Roman Curik.
       p. cm.
    "A CRC title."
    Includes bibliographical references and index.
    ISBN 978-1-4398-3910-2 (alk. paper)
    1. Electrons--Scattering. 2. Chemistry, Physical and theoretical. 3. Surface chemistry.
I. Curik, Roman. II. Title.

QC793.5.E62C37 2012
541'.33--dc23   2011041490

Visit the Taylor & Francis Web site at
http://www.taylorandfrancis.com

and the CRC Press Web site at
http://www.crcpress.com

Contents

Preface

This book deals with the recent progress in the theory and experiment of electron–molecule collisions. The formal general scattering theory and description of the experimental setup will be suppressed to avoid repetition of reviews published thus far but retained to an extent necessary for a nonexpert reader to understand the importance of the recent development described for solving practical problems such as those met in plasma physics, microelectronics, nanolithography, DNA research, atmospheric chemistry, and astrochemistry. Accordingly, not much attention is paid to the gas-phase elastic electron scattering. The aim of this book is to present a comprehensive overview of the practical aspects of electron scattering, and it is focused primarily on topical problems of inelastic electron scattering. The expertise in this field achieved up to recently has been scattered in reviews on special topics, and the interested reader can find the respective references in this book.

Hardly in any other branch of science is the collaboration of theorists and experimentalists so topical, and this aspect will be emphasized in the book. It is hoped a nonexpert reader will learn from this book about topical problems and useful application of electron scattering and that theorists and experimentalists active in this field will find some useful information about the latest development.

This book is designed for international readership and it was therefore recommended by the publisher that all units should be provided in SI units. However, for practical reasons the use of non-SI units cannot be avoided completely. Strict requirement for the use of SI units can be still satisfied by viewing one atomic unit in energy (1 a.u.) as 1 E_h and one atomic unit in length (1 a.u.) as 1 a_0, where $E_h = 4.3598144 \times 10^{-18}$ J and $a_0 = 5.2917706 \times 10^{-11}$ m. Electron energy has been expressed traditionally in units of eV, for which 1 eV in the SI expression corresponds to 8065.5410 cm^{-1}.

Petr Čársky
Roman Čurík

Contributors

Michael Allan
Department of Chemistry
University of Fribourg
Fribourg, Switzerland

Roger Azria
Institut des Sciences Moléculaires
CNRS-Université Paris-Sud
Orsay, France

Laurent G. Caron
Groupe de Recherches en Sciences des
  Radiations
Université de Sherbrooke
Sherbrooke, Québec, Canada

Petr Čársky
J. Heyrovský Institute of Physical
  Chemistry
Academy of Sciences of the Czech
  Republic
Prague, Czech Republic

Martin Čížek
Institute of Theoretical Physics
Charles University
Prague, Czech Republic

Roman Čurík
J. Heyrovský Institute of Physical
  Chemistry
Academy of Sciences of the Czech
  Republic
Prague, Czech Republic

Jiří Horáček
Institute of Theoretical Physics
Charles University
Prague, Czech Republic

Karel Houfek
Institute of Theoretical Physics
Charles University
Prague, Czech Republic

Anne Lafosse
Institut des Sciences Moléculaires
CNRS-Université Paris-Sud
Orsay, France

Léon Sanche
Groupe de Recherches en Sciences
  des Radiations
Université de Sherbrooke
Sherbrooke, Québec, Canada

1 Electron Scattering as a Useful Tool for Research in Physics, Chemistry, and Biology
Overview and Introductory Remarks

Jiří Horáček

CONTENTS

1.1 PHYSICS OF LOW-ENERGY ELECTRON–MOLECULE COLLISIONS

The collision of electrons with molecules as a fundamental process is dealt with in many areas of physics, astrophysics, chemistry, technology, and even biology. This book has its purpose in describing some recent advances in experimental and theoretical studies of the processes of electron–molecule scattering. The scattering processes have been studied for many years and important results have been obtained. For an excellent introduction to the physics of electron–molecule collisions, we refer the readers to the book by Shimamura and Takayanagi (1984). A comprehensive description of the theory of electron scattering with polyatomic molecules can be found in Gianturco and Jain (1986). The problem of low-energy electron scattering, to which the present book is devoted, is, however, much more complicated than that at intermediate- and high-energy regions (Khare 2002). At high energies, various approximations (e.g., Born or Glauber approximations) can be used with success. At low energy, however, demanding *ab initio* calculations must be performed. Even at the experimental side the low-energy scattering process encounters many problems. Among others, it is difficult to provide an electron beam with the energy

1

resolution of several meV stable down to the meV range, which is needed to study fine resonance structures in the respective cross sections. On the theoretical side at low electron energies, processes like polarization of the target and electron correlation become essential and must be treated at an *ab initio* level. Most calculations on electron–molecule scattering have been performed with nuclear degrees of freedom frozen, the so-called fixed nuclei approximation. This approximation is the basic approximation in the electronic molecular structure theory. The nuclei are slow as compared to the swift motion of electrons and can, in first approximation, be treated as frozen. In any realistic scattering process, however, the nuclei move and the molecular degrees of freedom can be excited or the molecule can be disintegrated even at very low electron energies. It is obviously necessary to go beyond the fixed-nuclei approximation if we want to describe the processes in which the energy is transferred from light electrons to heavy nuclei. To describe processes of this type is one of the main purposes of this book.

As the first process, we mention here the process of *vibrational excitation by electron impact*. The calculation of vibrational excitation cross sections for polyatomic molecules is still a very demanding task (including the elastic scattering process $0 \rightarrow 0$). To treat the process of vibrational excitation, harmonic approximation for the molecule is usually assumed and the cross sections are calculated by some kind of adiabatic-type approximation starting from fixed-nuclei scattering data. This approach, however, fails at low energies and much more elaborated *ab initio* calculations must be performed. One of the modern methods for the calculation of vibrational excitation cross section, the so-called discrete momentum representation, will be discussed in detail in Chapter 8. Vibrational excitation is important, for example, as a prerequisite for *electron-driven processes in planetary atmospheres and comets*. The electron density of ionized layers in the Earth's atmosphere cannot be described correctly without inclusion of the main electron-loss process—the dissociative electron recombination with molecular ions. Another important electron-driven process is the vibrational excitation of N_2 giving rise to auroral infrared emissions. The NO molecule is known to be a major emitter of infrared radiation from the upper atmosphere. If vibrationally excited N_2 collides with O^+, NO^+ is produced in the reaction

$$O^+ + N_2(v) \rightarrow NO^+ + N$$

The rate of this reaction is much higher for excited vibrational levels of N_2 (i.e., $v > 0$). This enhancement leads to a significant decrease of the electron density (Campbell and Brunger 2009).

The process of *dissociative electron attachment* has received much attention in the last few decades and was found to be of utmost importance in many fields from biology (radiation damage) to technological applications (plasma etching). In this process, the incoming electron is captured by the target and the molecule dissociates. This process which requires substantial rearrangement of the molecule may be very efficient even at zero kinetic energy of the colliding electron. The dissociative attachment process is, in essence, a resonant process (Kukulin et al. 1989). During the collision, a scattering complex (resonance) is formed whose lifetime is long compared to the time the electron needs to travel through the molecular space. Resonance processes of this type will be discussed in Chapters 4 and 5. In the life sciences, the role of electron-driven processes has been recognized as crucial for our understanding of *radiation damage of biological tissues.* The degradation has been commonly attributed to a direct impact of high-energy quanta or by complex radical chemistry. This explanation has, however, been questioned by the pioneering work of Sanche and coworkers who suggest that DNA lesion is induced by secondary low-energy electrons generated by the primary ionizing radiation. Sanche and coworkers have shown that: (1) low-energy electron irradiation directly induces both single- and double-strand breaks at energies below the ionization limit of DNA, and (2) the probability of strand breaks are one to two orders of magnitude larger for electrons than for photons with the corresponding energy. This development will be discussed in detail in Chapter 6. A very interesting study of the process of dissociative attachment was done in Innsbruck (Ptasinska et al. 2005). It provided insight into *bond- and site-selective loss of H atoms from nucleobases by very low-energy electrons.* Experiments on the isolated nucleobases (NB) thymine, cytosine, adenine, guanine, and uracil have demonstrated that they all effectively capture low-energy electrons below 3 eV. During the collision, a transient negative ion is formed which subsequently decomposes by the loss of a neutral hydrogen atom. The process can be expressed as

$$e + NB \rightarrow (NB)^{-*} \rightarrow (NB - H)^{-} + H$$

The reaction is effective already at energies below the threshold for electronic excitation. The loss of hydrogen is bond and site selective (Ptasinska et al. 2005).

The process of dissociative attachment has many other applications. For example, the reaction

$$e + H_2(v) \rightarrow H + H^{-}$$

serves as a source of negative atomic ions for the process of neutral heating in Tokamacs (Janev 1998, Celiberto et al. 2001). It is worth mentioning that the low-energy dissociative electron attachment cross section to hydrogen is extremely sensitive to the vibrational state of the target molecule and may increase by several orders of magnitude when the target is originally vibrationally excited. For details of this behavior, see Horáček et al. (2004, 2006).

The process of *associative detachment* is inverse to the process of dissociative attachment. Negative ions collide with neutral species and new molecules are formed with the subsequent release of electrons. This process is of fundamental importance, for example, in astrophysics. A decisive aspect of the formation of first stars and galaxies is *the cooling of the hot material.* At some stages, efficient cooling is provided only by vibrational and rotational excitation of molecules. Its efficiency is clearly determined by the number of molecules available. At the early stages of development of the Universe, the only molecule available was molecular hydrogen. It is believed that the hydrogen molecule was formed via a two-step process (Lepp et al. 2002): by radiative electron capture

$$e + H \rightarrow H^{-} + \hbar\omega$$

and in the second step by associative electron detachment

$$H + H^- \rightarrow H_2(v) + e$$

Once the molecule is formed the energy can be released, for example, by the process of vibrational or rotational excitation

$$H + H_2(v_1) \rightarrow H + H_2(v_2)$$

The process of formation of the hydrogen molecule can be precisely studied theoretically by the so-called nonlocal resonance model (Domcke 1991), which will be discussed in Chapters 4 and 5. The calculated rates of formation of the hydrogen molecule were found to be in excellent agreement with the results of a very sophisticated experiment performed recently at the Columbia University, New York (Kreckel et al. 2010). The rate of molecular formation determines, for example, the mass of emerging stars, their central density, and temperature. In addition, the nonlocal resonance model has led to the prediction of temporary negative molecular ion states with extremely long lifetimes caused by the high rotational excitation of the molecular target. The existence of these mysterious states was later fully confirmed experimentally and the calculated lifetimes were found to be in good agreement with the measured ones (Golser et al. 2005).

Thus far, we considered collisions of electrons with single molecules (gas phase scattering). However, there exist other experimental settings when the electron beam is scattered by molecules adsorbed on a surface (*processes at surfaces*). When the molecule is located near a surface, its characteristics change and also the scattering process is influenced by the presence of the surface. This fact opens new exciting possibilities for applications as well as fundamental research. An appealing feature of this technique for chemists is its ability to determine the structure of species adsorbed on the surface (Conrad and Kordesh 2010). This subject will be discussed in Chapter 7.

There are many other important processes like dissociative recombination, dissociative excitation, and dissociative ionization. These will not be treated here. In addition to its fundamental nature, the study of electron–molecule collisions is also motivated by its relation to other areas of physics, chemistry, and technology. The applications are numerous. See, for example, Becker (1998). Let us mention just one of them—*molecular bridges*. It is well known that electronic devices that incorporate single molecules as active elements are considered to be promising alternatives to semiconductor-based electronics. Recent advantages in experimental techniques have allowed the study of the conductance properties of nanoscale molecular junctions, where single molecules are chemically bound to metal electrodes. It is possible to show that the current–voltage characteristics of single-molecule junctions resemble those of basic electronic devices, such as transistors. An important element for the design of molecular memory or logic devices is a molecular switch. A molecular junction may be used as a nanoswitch, if the molecule can exist in two or more differently conducting states that are sufficiently stable and can be reversibly transferred into each other (Benesch et al. 2009).

1.2 ORGANIZATION OF THE BOOK

This book is organized as follows:

Chapter 2. Introduction to Numerical Methods. This chapter represents a brief description of selected numerical techniques used for the calculation of electron–molecule scattering cross sections. Also some typical results are provided. For more detailed information on this subject, we refer the readers to the book by Huo and Gianturco (1995) and to the references therein. The selection of methods in this book is personal and no claim of completeness is made. Because of a large variety of methods, some of them are not included either by oversight or by practical limitations. The objective was to include primarily methods that are currently used, are of general use, form a perspective for future development, and contributed greatly to the progress in this field.

Most computational methods presented in Chapter 2 are general and can be used for different applications. They can be used for sophisticated rigorous applications to describe fine details of experimental data. However, complexity and need for a great computational power limit such applications to diatomics or exceptionally to small polyatomic molecules. The theories and methods designed primarily for this purpose are dealt with in Chapters 4 and 5. For applications to polyatomic molecules, however, it is unavoidable to sacrifice some rigor and to accept some approximations. Chapter 8 presents the theory and a method for treatments of vibrational excitation of polyatomic molecules with several first-row atoms. Considerably larger molecular systems are the subject of Chapter 6. It demonstrates that theory can provide relevant results even for systems of biochemical interest.

Chapter 3. Measurement of Absolute Cross Sections of Electron Scattering by Isolated Molecules. In this chapter, experimental techniques for the determination of absolute scattering cross sections are discussed for electron scattering, dissociative electron attachment, and neutral dissociation. Representative results obtained at the University of Fribourg are presented for carbon monoxide, hydrogen halides, and other molecules. These data are probably the most accurate data available in the literature.

Chapter 4. Nonlocal Theory of Resonance Electron–Molecule Scattering. This chapter discusses the most elaborate and accurate modern theory capable of treating inelastic scattering processes such as vibrational excitation, dissociative attachment, and associative detachment from the first principles. This approach combines the exact description of the nuclear dynamics with a correct account of threshold features. In this way, even the smallest details of experimental cross sections can be recovered. At present, the method is applicable only to diatomics or polyatomic molecules which can, in first approximation, be described as molecules with essentially one degree of freedom. Nevertheless, much understanding can be reached on one-dimensional problems. For example, full understanding of threshold features as threshold peaks, Wigner cusps, and oscillating patterns of some cross sections can be obtained. Significant progress was achieved in extending the applicability of the nonlocal resonance model to molecules with two degrees of freedom.

Chapter 5. Applications of the Nonlocal Resonance Theory to Diatomic Molecules. In this chapter, the most important results obtained by the nonlocal resonance model in the last two decades are summarized and discussed. Vibrational excitation, dissociative attachment, and associative detachment processes are described for HF, HCl, HBr, HI, H_2, and other molecules.

Chapter 6. Theoretical Studies of Electron Interactions with DNA and Its Subunits: From Tetrahydrofuran to Plasmid DNA. This chapter focuses on the interaction of low-energy electrons with DNA and RNA, their constituent subunits and related molecules. The authors present an extensive review of the problem. In addition, they propose their own model of scattering on DNA based on multiple scattering having as the first objective elucidation of the role of stacking periodicity of bases on the interaction of low-energy electrons with the DNA structure.

Chapter 7. Low-Energy Electron Scattering at Surfaces. In contrast to the previous chapters, this chapter describes electron scattering by molecules adsorbed on surfaces. The presence of the surface brings new exciting possibilities for the study. Using high-resolution electron energy loss spectrometry, one can probe the vibrational pattern of surfaces or interfaces, including adsorbate vibrations as well as lattice vibrations. Electron elastic reflectivity may give insight into the sample conduction-band density of states, which modulates the probability of inelastic scattering for electrons coming close to the substrate. As an illustration of the technique, hydrogenated or deuterated synthetic polycrystalline diamond films are discussed in this chapter.

Chapter 8. Vibrational Excitations of Polyatomic Molecules. In this chapter, the so-called discrete momentum representation method developed by the editors is described and applied to the calculation of vibrational excitation cross sections for polyatomic molecules. This method goes beyond the adiabatic approximation. As examples, methane, cyclopropane, and diacetylene are discussed.

References are given at the end of each chapter.

ACKNOWLEDGMENT

This work was supported by Záměr MSM0021620860 of the Ministry of Education, Youth, and Sports of the Czech Republic.

REFERENCES

Becker, K. H. 1998. *Novel Aspects of Electron–Molecule Collisions.* Singapore: World Scientific.
Benesch, C., M. F. Rode, M. Čížek, R. Härtle, O. Rubio-Pons, M. Thoss, and A. L. Sobolewski 2009. Switching the conductance of a single molecule by photoinduced hydrogen transfer. *J. Phys. Chem. C 113*, 10315–10318.
Campbell, L. and M. J. Brunger 2009. On the role of electron-driven processes in planetary atmospheres and comets. *Phys. Scripta 80*(5), 058101.
Celiberto, R., R. K. Janev, A. Laricchiuta, M. Capitelli, J. M. Wadehra, and D. E. Atems 2001. Cross section data for electron-impact inelastic processes of vibrationally excited molecules of hydrogen and its isotopes. *Atomic Data Nucl. Data Tables 77*(2), 161–213.
Conrad, H. and M. E. Kordesh 2010. High resolution electron energy loss spectroscopy applications. In *Encyclopedia of Spectroscopy and Spectrometry*, J.C. Lindon, G.E. Tranter, and D. Koppenaal (Eds.), pp. 865–876. London: Academic Press. Second edition.

Domcke, W. 1991. Theory of resonance and threshold effects in electron-molecule collisions: The projection-operator approach. *Phys. Rep. 208*(2), 97–188.

Gianturco, F. A. and A. Jain 1986. The theory of electron scattering from polyatomic molecules. *Phys. Rep. 143*(6), 347–425.

Golser, R., H. Gnaser, W. Kutschera, A. Priller, P. Steier, A. Wallner, M. Čížek, J. Horáček, and W. Domcke 2005. Experimental and theoretical evidence for long-lived molecular hydrogen anions H_2^- and D_2^-. *Phys. Rev. Lett. 94*(22), 223003.

Horáček, J., M. Čížek, K. Houfek, P. Kolorenč, and W. Domcke 2004. Dissociative electron attachment and vibrational excitation of H_2 by low-energy electrons: Calculations based on an improved nonlocal resonance model. *Phys. Rev. A 70*(5), 052712.

Horáček, J., M. Čížek, K. Houfek, P. Kolorenč, and W. Domcke 2006. Dissociative electron attachment and vibrational excitation of H_2 by low-energy electrons: Calculations based on an improved nonlocal resonance model. II. Vibrational excitation. *Phys. Rev. A 73*(2), 022701.

Huo, W. M. and F. A. Gianturco (Eds.) 1995. *Computational Methods for Electron-Molecule Collisions*. New York: Plenum Press.

Janev, R. K. 1998. Atomic and molecular processes in SOL/divertor plasmas. *Contrib. Plasma Phys. 38*(1–2), 307–318.

Khare, S. P. 2002. *Introduction to the Theory of Collisions of Electrons with Atoms and Molecules*. New York: Kluwer Academic/Plenum Publishers.

Kreckel, H., H. Bruhns, M. Čížek, S. C. O. Glover, K. A. Miller, X. Urbain, and D. W. Savin 2010. Experimental results for H_2 formation from H^- and H and implications for first star formation. *Science 329*, 69–71.

Kukulin, V. I., V. M. Krasnopolsky, and J. Horáček 1989. *Theory of Resonances*. Boston: Kluwer Academic Publishers.

Lepp, S., P. C. Stancil, and A. Dalgarno 2002. Atomic and molecular processes in the early Universe. *J. Phys. B 35*(10), R57.

Ptasinska, S., S. Denifl, P. Scheier, E. Illenberger, and T. D. Märk 2005. Bond- and site-selective loss of H atoms from nucleobases by very-low-energy electrons (<3 eV). *Agnew. Chem. Int. Ed. 44*, 6941–6943.

Shimamura, I. and K. Takayanagi (Eds.) 1984. *Electron–Molecule Collisions*. New York: Plenum Press.

2 Introduction to Numerical Methods

Roman Čurík and Jiří Horáček

CONTENTS

This chapter presents an overview of numerical methods for calculations of cross sections for electron scattering by molecules. Since a large variety of methods reported in the literature precludes any attempt for a complete coverage, we decided for a selection of methods which have contributed greatly to the progress in this field, can be applied to molecular targets of general structure, and are intensively used at present. Obviously, the choice of methods, as well as our remarks on merits and limitations of methods, is affected by our personal opinion and taste.

2.1 SINGLE-CENTER EXPANSION

The core idea of the single-center expansion (SCE) method lies in an angular expansion of continuum electron's degrees of freedom around a chosen center (typically the center of mass of molecule). SCE was also employed in the pioneering times of *ab initio* bound-state calculations of diatomic systems with basis sets formed by Slater-type orbitals (STO) (Allen and Karo 1960). Later, for calculations of polyatomic molecules, the SCE approach was abandoned, when merits of Gaussian basis sets were recognized and thus the evaluation of three- and four-center integrals ceased to be a problem.

In the case of collisional calculations, the application of the SCE approach to diatomic molecules was viewed as a natural extension of the techniques developed in electron–atom scattering models. First, a systematic study employing SCE on H_2^+ ion was carried out by Temking and Vasavada (1967) and later in an extended version (Temkin et al. 1969). This original work was followed by a vast number of calculations applied to diatomic molecules such as H_2 (Henry and Lane 1969, Chang and Temkin 1969), N_2 (Burke and Sinfailam 1970, Burke and Chandra 1972), CO (Chandra 1977, Jain and Norcross 1992), and some others, summarized in excellent reviews (Lane 1980, Brunger and Buckman 2002).

The SCE approach belongs to a class of methods employing an optical potential. The optical potential is a one-electron, nonlocal interaction obtained by projecting out internal target's degrees of freedom (those of bound electrons, of vibrations and rotations) in the scattering equations. Some simple forms of the optical potential will be described in Sections 2.2 and 2.3. At this moment, we postpone discussion of the optical potential to those sections and for the moment we assume the interaction between the scattered electron and the molecule is described by a one-electron interaction potential V. In general, V may contain local and nonlocal (exchange or correlation) components:

$$V(\mathbf{r},\mathbf{r}') = V_l(\mathbf{r})\delta(\mathbf{r} - \mathbf{r}') + W(\mathbf{r},\mathbf{r}').\tag{2.1}$$

The three-dimensional Schrödinger equation for the impinging electron's wave function $\psi(\mathbf{r})$ may then be written as

$$\left[-\frac{1}{2}\nabla^2 + V_l(\mathbf{r}) - E\right]\psi(\mathbf{r}) = -\int d\mathbf{r}'\, W(\mathbf{r},\mathbf{r}')\psi(\mathbf{r}').\tag{2.2}$$

For simplicity, we also omit nuclear degrees of freedom and a discussion on frame of reference. We assume molecular frame of reference (also called body-frame).

The SCE defines a numerical implementation for a solution of the above equation. In order to reduce Equation 2.2 into a set of coupled radial equations, we expand the solution $\psi(\mathbf{r})$ into a sum of orthonormal angular functions with coefficients dependent on the radial distance r:

$$\psi(\mathbf{r}) = \frac{1}{r}\sum_{lm} \psi_{lm}(r)Y_{lm}(\hat{\mathbf{r}}).\tag{2.3}$$

By projecting the Schrödinger Equation 2.2 onto the orthonormal angular basis of Y_{lm}, we arrive at a set of coupled second-order integro-differential equations (positive collision energy $E = k^2/2$):

$$\left[\frac{d^2}{dr^2} - \frac{l(l+1)}{r^2} + k^2\right]\psi_{lm}(r) = 2\sum_{l'm'}\left[V_{lm,l'm'}(r)\psi_{l'm'}(r) + \int dr' W_{lm,l'm'}(r,r')\psi_{l'm'}(r')\right].\tag{2.4}$$

The equations above are coupled via anisotropy of the optical potential V. The radially dependent coupling elements $V_{lm,l'm'}(r)$ and $W_{lm,l'm'}(r,r')$ may be expressed as follows:

$$V_{lm,l'm'}(r) = \langle Y_{lm}(\hat{\mathbf{r}}) \mid V_l(\mathbf{r}) \mid Y_{l'm'}(\hat{\mathbf{r}}) \rangle, \tag{2.5}$$

$$W_{lm,l'm'}(r,r') = \langle Y_{lm}(\hat{\mathbf{r}}) \mid W(\mathbf{r},\mathbf{r}') \mid Y_{l'm'}(\hat{\mathbf{r}}') \rangle, \tag{2.6}$$

where the scalar products are carried out only on the angular space $\hat{\mathbf{r}}$ leaving the dependence on the radial coordinate r.

The number of coupled equations N in Equation 2.4 is solely defined by the cutoff l_{max} value we chose to limit the angular space to. However, the above set of second-order differential equations with one boundary condition fixed at the origin ($\psi_{lm}(0) = 0$) has N independent solutions and therefore one has to introduce a second pair of indices to the wave function, that is, $\psi_{lm,l_0m_0}(r)$. In this way, the first channel index $i \equiv lm$ defines the angular shape of the wave function $\psi(\mathbf{r})$ via Equation 2.3 while the second channel index $j \equiv l_0m_0$ counts all the independent solutions. A linear combination of these solutions is again a solution of Equation 2.4 and thus the matrix of solutions ψ_{ij} can be right-multiplied by a nonsingular matrix to form a different matrix of independent solutions. Asymptotically, in a zero-interaction region, the set of solutions $\psi_{ij}(r)$ is essentially just a linear combination of two independent free solutions ($f^0(r)$ and $g^0(r)$) as follows:

$$\psi_{ij}(r) \xrightarrow{r \to \infty} f_{ij}^0(r)\delta_{ij} - g_{ij}^0(r)K_{ij}. \tag{2.7}$$

The K-matrix on the right hand side allows us to calculate all the necessary scattering parameters (S-matrix, T-matrix, phase shifts or cross sections, as explained in the literature (Newton 2002, Gianturco et al. 1995, Aymar et al. 1996)).

The angular functions chosen earlier are the spherical harmonics $Y_{lm}(\hat{\mathbf{r}})$ that are typical in implementation by Itikawa and collaborators (Nishimura and Itikawa 1996, Takekawa and Itikawa 1999). Such a choice leads to a set of complex radial coefficients $\psi_{lm}(r)$, although the wave function $\psi(\mathbf{r})$ may be chosen as real for real (and hermitian) potentials $V(\mathbf{r},\mathbf{r}')$. Therefore, some authors (Gianturco et al. 1995) use real spherical harmonics created by a combination of Y_{lm} and Y_{l-m}. In case of molecules with higher symmetry, one can combine (real) spherical harmonics in subspace with fixed l-value in a way such that they belong to irreducible representations of the molecular point group (Altmann and Cracknell 1965, Gianturco et al. 1995). Different irreducible representations then do not mix and the set of coupled Equations 2.4 may be decoupled into independent groups according to corresponding irreducible representations. Reduced dimensions of coupled equations make the SCE method very efficient for molecules that exhibit symmetry, starting with C_{2v} triatomics (Takekawa and Itikawa 1999, Nishimura and Itikawa 1996, Gianturco and Stoecklin 1997, Curik et al. 1999, 2001). Larger polyatomic molecules were explored computationally as well. Among many, we mention those belonging to the following point groups:

T_d: CH_4 (Gianturco and Jain 1986, Gianturco et al. 1987, Gianturco and Scialla 1987), CF_4 (Gianturco et al. 1994, Curik et al. 2000), CCl_4 (Curik et al.

2000), SiH_4 (Cascella et al. 2001), GeH_4 (Jain et al. 1991, Cascella et al. 2001)

D_{3h}: cyclopropane C_3H_6 (Čurík and Gianturco 2002a,b)

D_{6h}: benzene C_6H_6 (Gianturco and Lucchese 1998, 2000)

We conclude this section with four paragraphs containing several practical notes.

The nonlocal part $W(\mathbf{r}, \mathbf{r}')$ of interaction (2.1) makes a numerical solution of the coupled Equations 2.4 rather cumbersome as the integral on the r.h.s. of Equation 2.4 includes the unknown set of the radial scattering functions $\psi_{lm}(r)$. Some authors have solved this issue via iterative scheme (Collins et al. 1980, Gianturco et al. 1995) while most of the applications employing the SCE scheme work exclusively with some local models of the nonlocal parts (exchange and correlation) of the optical potential (Riley and Truhlar 1975, Gianturco and Scialla 1987, Curik et al. 2000).

SCE is a natural method for propagation of solutions ψ_{ij} at larger distances from the molecule. The interaction is then fairly simple and local, usually described by a few multipoles and polarizability of the molecule. Such interaction leads to a weak coupling between the partial waves in Equation 2.4 and no local potential approximations are necessary. The SCE approach is also heavily exploited in cold-electron regime (under collision energies of 1 eV), where very few partial waves penetrate the centrifugal barrier (Isaacs and Morrison 1992, Fabrikant 1984).

The major numerical obstacle in using the SCE method is a strong short-range coupling of partial waves in the core region of the molecule. Curik et al. (2000) used all the partial waves up to $l = 50$ to achieve the convergence of cross sections for e–CCl_4 collisions. The high angular anisotropy of the interaction V is caused by angular expansion of nuclear cusps. In Figure 2.1, we illustrate this by performing the angular expansion of two components of the electrostatic part V_s of the interaction V for N_2 molecule. These components are nuclear attraction V_n and a repulsion with bound electrons V_e:

$$V_s = V_n + V_e. \tag{2.8}$$

Their angular expansions in case of the homo-nuclear diatomic may be written as follows:

$$V_n(\mathbf{r}) = V_n(r, \theta, \phi) = \sum_L N_L(r) P_L(\cos\theta),$$

$$V_e(\mathbf{r}) = V_e(r, \theta, \phi) = \sum_L E_L(r) P_L(\cos\theta). \tag{2.9}$$

Figure 2.1 shows that all the nuclear components peak at -14.0 (total nuclear charge of the molecule) with the higher partial contributions exhibiting a narrower peak structure. On the other hand, the electronic partial contributions behave well and their higher partial contributions quickly diminish in amplitude.

The SCE method naturally leads to a more advanced laboratory-frame formulation of the collision theory. Calculations based on its implementation successfully explained fine details in ro-vibrationally inelastic scattering by diatomic molecules

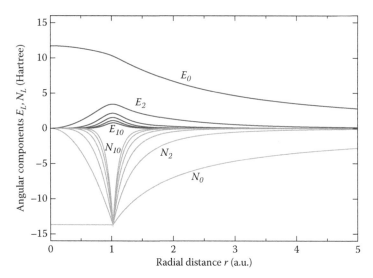

FIGURE 2.1 (**See color insert.**) The angular expansion (2.9) of the nuclear attractive potential (red lines) and the electronic repulsive part (blue lines). Their sum forms the electrostatic interaction between scattered electron and N_2 molecule. The partial contributions are shown for even L-values because the odd L-values do not contribute in case of the homonuclear diatomic molecule. The contributions up to $L = 10$ are displayed. The target electron density was obtained by the Hartree–Fock method.

at collision energies down to a few meV (Sun et al. 1995, Morrison et al. 1997, Telega et al. 2004, Telega and Gianturco 2006). However, we feel that in the long run, it may be expected that the SCE approach will be abandoned for more rigorous methods, as it happened in the electronic structure theory.

2.2 STATIC-EXCHANGE APPROXIMATION

Static-exchange (SE) approximation is the first level of approximation describing an interaction between a continuum electron and a target molecule during a collisional event. The SE model disregards a distortion of the bound electron density in the presence of a field of the incoming particle. The SE approximation also excludes possible electronic excitations of the molecule and tends to underestimate the strength of the interaction at resonant energies (Gianturco et al. 1996, Curik et al. 1999). In order to discuss the form of the SE approximation, we restrict our attention to a closed-shell target molecule in the ground state. We also exclude vibrational and rotational degrees of freedom as they do not play any role in the following discussion. Here, we shall not attempt to present all details necessary to derive the SE equations. Instead we just outline the initial model and present the final interaction components as their meaning can be understood well intuitively. More with regard to the procedure can be found in remarkable reviews (Lane 1980, p. 47 and Gianturco et al. 1995, p. 78).

We start with the electronic target states $\Phi_\alpha(x_1 \ldots x_N)$ that satisfy the Schrödinger equation

$$[H_m(\mathbf{x}_1 \ldots \mathbf{x}_N) - E_\alpha]\, \Phi_\alpha(\mathbf{x}_1 \ldots \mathbf{x}_N) = 0. \tag{2.10}$$

The molecule under investigation has N electrons, their space coordinates will be denoted as $\mathbf{r}_i$ while the coordinates $\mathbf{x}_i$ stand for both $\mathbf{r}_i$ and spin. The electronic Hamiltonian of the target molecule is H_m. The electronic part $\Psi(x_1 \ldots x_{N+1})$ of the total wave function describing the system of the target molecule and the scattered electron then satisfies

$$\left[H(\mathbf{x}_1 \ldots \mathbf{x}_{N+1}) - E \right] \Psi(\mathbf{x}_1 \ldots \mathbf{x}_{N+1}) = 0, \tag{2.11}$$

where the full Hamiltonian

$$H(\mathbf{x}_1 \ldots \mathbf{x}_{N+1}) = H_m(\mathbf{x}_1 \ldots \mathbf{x}_N) - \frac{1}{2}\nabla_{N+1}^2 + V(\mathbf{x}_1 \ldots \mathbf{x}_{N+1}). \tag{2.12}$$

The interaction potential V contains the Coulombic interaction of the continuum electron with N target-bound electrons and M nuclei of the molecule:

$$V(\mathbf{x}_1 \ldots \mathbf{x}_{N+1}) = -\sum_i^N \frac{1}{|\mathbf{r}_{N+1} - \mathbf{r}_i|} + \sum_j^M \frac{Z_j}{|\mathbf{r}_{N+1} - \mathbf{R}_j|}. \tag{2.13}$$

The vectors $\mathbf{R}_j$ are positions of the nuclei. The total electronic wave function Ψ may then be expanded in terms of complete set of the target eigenfunctions

$$\Psi(\mathbf{x}_1 \ldots \mathbf{x}_{N+1}) = A \sum_\alpha \Phi_\alpha(\mathbf{x}_1 \ldots \mathbf{x}_N)\psi_\alpha(\mathbf{x}_{N+1}). \tag{2.14}$$

The SE approximation may then be defined by retaining only the ground state in the above equation, that is,

$$\Psi_{SE}(\mathbf{x}_1 \ldots \mathbf{x}_{N+1}) = A \Phi_0(\mathbf{x}_1 \ldots \mathbf{x}_N)\psi(\mathbf{x}_{N+1}). \tag{2.15}$$

The operator A antisymmetrizes the product of the target state wave function Φ_0 and the continuum wave function ψ for the scattered electron.

A one-electron SE equation for ψ then can be obtained by using approximate wave function (2.15) in Equation 2.11 and then projecting Equation 2.11 onto the ground target state Φ_0 (Lane 1980, Gianturco et al. 1995). The resulting equation has a structure of the one-electron Schrödinger equation for the continuum electron:

$$\left[-\frac{1}{2}\nabla_r^2 + V_s + V_x - \frac{k^2}{2} \right]\psi(\mathbf{r}) = 0, \tag{2.16}$$

where V_s is a sum of an attraction with the nuclei and a mean-field repulsion with the bound electrons

$$V_s(\mathbf{r}) = -\sum_j^M \frac{Z_j}{|\mathbf{r} - \mathbf{R}_j|} + 2\sum_i^{N/2} \int d\mathbf{r}' \frac{|\phi_i(\mathbf{r}')|^2}{|\mathbf{r} - \mathbf{r}'|}, \tag{2.17}$$

with $\varphi_i(\mathbf{r})$ being doubly occupied Hartree–Fock orbitals of the target molecule. An action of V_x on the continuum function $\psi(\mathbf{r})$ may be written as

$$\langle \mathbf{r} \mid V_x \mid \psi \rangle = -\sum_i^{N/2} \int d\mathbf{r}' \frac{\phi_i(\mathbf{r}')\psi(\mathbf{r}')}{\mid \mathbf{r} - \mathbf{r}' \mid} \phi_i(\mathbf{r}). \tag{2.18}$$

In derivation of the above results (Lane 1980, Gianturco et al. 1995), the continuum function ψ was taken to be orthogonal to all the bound Hartree–Fock (HF) orbitals φ_i. This is substantiated because the resulting Equation 2.16 is an exact HF equation for ψ with the proper interaction components (2.17) and (2.18) that ensure orthogonality in the HF theory.

However, a complication arises when the nonlocal exchange interaction (2.18) is replaced by a local approximative version (Hara 1969, Morrison and Collins 1978, Riley and Truhlar 1975). Orthogonality is also broken when a third potential component is added into Equation 2.16. Such a potential often attempts to account for a correlation and polarization forces of the target electrons that are missing in the exact form of the SE approximation (Padial and Norcross 1984, Gianturco and Rodriguezruiz 1993, Tonzani and Greene 2006, Čurík and Šulc 2010).

2.3 CORRELATION–POLARIZATION POTENTIALS

The correlation and polarization forces are the interaction components that were neglected in the previous section when moving on from Equation 2.14 to Equation 2.15. The correlation–polarization effect originates in the virtual excitations of the target molecule as can be seen from Equation 2.14. Although the excitations in Equation 2.14 may be energetically unreachable, they contribute to a relaxation of the total wave function and are often referred as contributions of closed channels.

In many-electron methods, such as R-matrix implementations (Gillan et al. 1995, Pfingst et al. 1995, Morgan et al. 1998) or Schwinger multichannel method (Takatsuka and McKoy 1981, Winstead and McKoy 2000), a long-standing focus of research has been the selection of the most important excited states Φ_α that provide compact and computationally tractable representation of polarization effects. The Φ_α may be physical states of the target molecule (Pfingst et al. 1995, Brems et al. 2002, Winstead et al. 2005), but they need not be (Danby and Tennyson 1991, Tarana and Tennyson 2008).

In the many-body approaches, one of the simplest expansions in Equation 2.14 may be achieved by a target ground state wave function Φ_0 being a single determinant of HF theory, while the only excited states included are the single excitations from Φ_0. This level of correlation–polarization correction (Beyer et al. 2000, Winstead et al. 2005) is called static-exchange plus polarization (SEP) due to resemblance with bound-state theory where only single excitations of the molecule contribute to the dipole polarizability (Joachain 1983, p. 611).

Another, more model-oriented approach searches for an effective, local potential that mimics a change in the interaction due to the distortion of the target's electronic density and thus describing the polarization forces that come into play when the impinging electron is far from the target. However, ideally we also require this potential to describe the complicated dynamical correlation that is, missing in the SE wave function (2.15). A modification of the SEP by a local form of interaction accounting for correlation and even polarization of the target's bound orbitals is an old idea. The first models used the long-range asymptotic form

$$V_p = -\alpha_0 / 2r^4, \tag{2.19}$$

with a short-range cutoff function in order to remove the singular behavior at the origin (Morrison and Collins 1978). A single parameter of this model, the cutoff radius, was then determined by adjusting the calculated cross sections to some well-established feature of the results. For modeling the short-range correlation between the scattered electron and the target electron density, use of the density functional theory (DFT) was proposed (Connell and Lane 1983, Padial and Norcross 1984). They employed a local spin density (LSD) approximation emerging from the field of solid-state physics (Vosko et al. 1980, Perdew and Zunger 1981), both later refined by Perdew and Wang (1992). All these forms of the LSD correlation are based on Green's function Monte Carlo simulation for electrons in a finite volume, subject to periodic boundary conditions. The correlation energy per electron was then extrapolated to infinite volume (Ceperley and Alder 1980). The practical accuracy of all these LSD approximations is similar and they are a core component of all modern DFT functionals used in quantum chemistry. They are often referred to as LDA (local density approximation) in the literature of electron–molecule collisions.

A further way to improve LSD approximation came in the form of a generalized gradient approximation (GGA) by Perdew et al. (1992) (the functional is referred in the literature as PW91). A different way to incorporate the density gradient corrections was used by Lee et al. (1988), who turned the correlation-energy formula of Colle and Salvetti (1975) into a DFT functional form by use of the first and second gradients of electron density (LYP functional).

All the aforementioned correlation potentials decrease exponentially outside the molecule following the exponential decay of the bound electron density. This defect in DFT was noted a long time ago (Almbladh and von Barth 1985, Umrigar and Gonze 1994). Such an incorrect long-range behavior of the DFT potentials is a cause of many problems in the description of induced moments of delocalized charge densities. Accordingly, special treatment has been undertaken in calculations of charge transfers or dispersion forces (Antony and Grimme 2006, Zhao and Truhlar 2007). Rigorously, the long-range correlation potential should be taken as (Almbladh and von Barth 1985, Umrigar and Gonze 1994)

$$V_p(\mathbf{r}) = -\frac{1}{2r^6} \sum_{ij} \alpha_{ij} x_i x_j, \tag{2.20}$$

where the symmetric 3×3 matrix α_{ij} is a polarizability tensor of the remaining core (in our case the target molecule) and $r^2 = x_1^2 + x_2^2 + x_3^2$. A smooth connection of a general and anisotropic asymptotic form (2.20) to a short-range anisotropic potential is not uniquely defined. A plausible suggestion (Telega et al. 2004) was to find a crossing point r_c of a spherically symmetric components of the long- and short-range parts. Of course r_c is not a proper crossing point of the higher partial-wave components.

Overall smoothness is then achieved by adding the higher-order-induced multipoles leading to a modification of the asymptotic form (2.20), where faster-decaying terms are added.

An alternative approach may be performed by keeping the asymptotic form (2.20) while connecting smoothly to it the relevant short-range partial waves (Čurík and Šulc 2010). This procedure is based on an observation that symmetric tensor α_{ij} of the polarizabilities contains only $l = 0$ and $l = 2$ partial waves. These six components may then be smoothly connected to corresponding $l = 0$ and $l = 2$ components of the short-range correlation potential V_c while the other partial-wave components of V_c decay exponentially as mentioned earlier. Implementational details of this procedure are explained in Section 8.7 of this book.

The typical shape and magnitude of the correlation–polarization potential V_{cp} is shown in Figure 2.2. The example is made again for the nitrogen molecule. Figure 2.2a displays spherical components of all the three important optical potential contributions: The static part V_s defined in Equation 2.17, a local form of the exchange interaction V_x, and the correlation–polarization potential V_{cp}. Note that the free-electron gas exchange model V_x (Hara 1969) is not discussed here and only serves us to compare magnitudes of the contributing interactions. The short-range part of V_{cp} employs LSD approximation mentioned above. Figure 2.2b displays absolute values of all the three potentials in logarithmic scale.

Figure 2.2 confirms a general observation in quantum chemistry where the dominant energy contribution comes from the electrostatic energy and the exchange energy, both of similar magnitude. The correlation contribution is roughly 10% of the exchange part (Perdew and Zunger 1981) in the region of the molecular charge density; however, together with quadrupole component of V_s (not displayed in Figure 2.2) V_{cp} dominates the long-range part of the electron–molecule interaction.

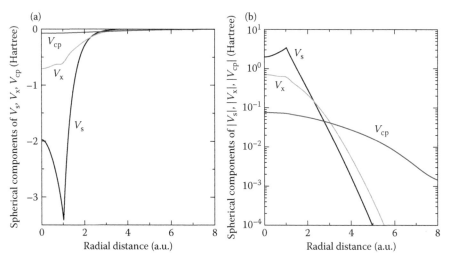

FIGURE 2.2 (**See color insert.**) (a) Spherical components of static V_s, exchange V_x and correlation–polarization V_{cp} (LSD model) potentials for nitrogen molecule. (b) Absolute values of the potentials on a logarithmic scale.

Much as in quantum chemistry, the use of the DFT vastly expanded the number of applications of the collision models for molecules of biochemical interest (Gianturco et al. 2008, Baccarelli et al. 2009) that are well summarized in Chapter 6 of this book. However, an extension of the DFT potentials for general-purpose calculations is not free of complications. An enormous effort has been devoted in quantum chemistry to put the DFT on more rigorous grounds. Therefore, we expect that this effort will bring a benefit to the scattering calculations in the near future.

2.4 *R*-MATRIX APPROACH

The derivative matrix or *R*-matrix is usually defined via the relationship between the matrix of radial channel solutions ψ_{ij}, see, for example, Equation 2.7, and their derivatives at some radial point r_0. Thus,

$$\psi_{ij}(r_0) = \sum_k R_{ik}\psi'_{kj}(r_0).$$ (2.21)

Therefore, in matrix form we may write

$$\mathbf{R} = \psi(r_0)[\psi'(r_0)]^{-1}.$$ (2.22)

It can be shown that the *R*-matrix as defined here is real and symmetric. It is also worth noting that $\mathbf{R}$ does not depend on the asymptotic normalization (2.7) as any effect of a nonsingular matrix that transforms the set of solutions ψ_{ij} into a different set of independent solutions will cancel in Equation 2.22.

In practice, the matrix $\mathbf{R}$ is evaluated at a distance r_0 beyond which the interaction of scattered electron with the molecule is simple and local (conveniently described by lowest multipoles and polarizability). In this case, *R*-matrix defines the starting condition for a numerical propagation in an outer region, utilizing a radial set of coupled Equations 2.4 as described in the literature (Burke and Robb 1972, 1975; Gillan et al. 1995).

Under the assumption that interaction can be neglected for $r > r_0$, a direct relationship between *K*- and *R*-matrices may be derived (Aymar et al. 1996, Equation 2.34):

$$\mathbf{K} = [\mathbf{f}^0 - \mathbf{f}^{0\prime}\mathbf{R}][\mathbf{g}^0 - \mathbf{g}^{0\prime}\mathbf{R}]^{-1},$$ (2.23)

where the diagonal matrices $\mathbf{f}^0$ and $\mathbf{g}^0$ contain regular and irregular free solutions in respective channels. As the *R*-matrix contains all the necessary information about the scattering wave function inside the sphere $r < r_0$, it is often exploited as a parameter (or a number of parameters) used to fit experimental data and understand the mechanism underlying the collisional process (Friedman et al. 2009, Gallup and Fabrikant 2011).

The *R*-matrix method was first introduced to nuclear physics to study resonance reactions by Wigner and Eisenbud 1947. Although the Wigner–Eisenbud form was not derived variationally, the final form of the *R*-matrix is equivalent to the variational approach of Kohn (1948) or to a noniterative reformulation (Greene 1983, Le Rouzo and Raseev 1984) of the Fano–Lee eigenchannel *R*-matrix method (Fano and Lee 1973). In our opinion, the essence of the *R*-matrix method can be most clearly understood in the following variational formulation (Kohn 1948, Nesbet 1980).

For simplicity, we focus on a single-channel case. The entire configuration space is first divided into two parts: $r > r_0$ and $r \leq r_0$. We attempt to minimize a functional

$$I = 2\int_0^\infty dr\, \psi_t(r)[H - E]\psi_t(r), \tag{2.24}$$

with the radial Hamiltonian

$$H = -\frac{1}{2}\frac{d^2}{dr^2} + V(r). \tag{2.25}$$

The trial wave function $\psi_t(r)$ is assumed to be exact ψ for $r > r_0$. Integrating by parts the kinetic energy term in Equation 2.24, we obtain (Kohn 1948, Nesbet 1980, Adhikari 1998)

$$I(E,\lambda) = \int_0^{r_0} dr\left[(\psi_t'(r))^2 + (2V(r) - k^2)\psi_t^2(r)\right] - \lambda\psi_t^2(r_0). \tag{2.26}$$

The parameter λ is the logarithmic derivative:

$$\lambda = \frac{\psi'(r_0)}{\psi(r_0)} = \frac{1}{R}, \tag{2.27}$$

fixed by presumably an exact solution ψ for $r > r_0$ and $E = k^2/2$.

There are two ways to look at Equation 2.26. For the known λ, Equation 2.26 offers a variational principle to calculate a stationary energy k^2. This fact is the basis of bound state energy calculations. However, one can equally well regard Equation 2.26 as variational principle to obtain a stationary expression for λ for a given fixed value of k^2. This second approach is more convenient for collision problems, in which the energy of the system is prescribed.

In order to address different implementations of this principle available in modern codes, we introduce a linearly independent basis set φ_i. The trial wave function ψ_t is then approximated by a finite expansion:

$$\psi_t(r) = \sum_i c_i \varphi_i(r). \tag{2.28}$$

Variational equations are then obtained by derivatives of $I(E, \lambda)$ in Equation 2.26 with respect to the coefficients c_i. The resulting matrix equation has the following form:

$$\sum_j A_{ij}c_j = \lambda\varphi_i(r_0)\psi_t(r_0), \tag{2.29}$$

where the matrix elements of the symmetrized Hamiltonian are

$$A_{ij} = \int_0^{r_0} dr\,[\varphi_i'(r)\varphi_j'(r) + \varphi_i(r)(2V - k^2)\varphi_j(r)]. \tag{2.30}$$

Does Equation 2.29 provide a means to calculate the stationary value λ uniquely? This question has been already answered positively by Kohn (1948). To prove this, one replaces $\psi_t(r_0)$ in Equation 2.29 by its expansion (2.28). The resulting set of linear equations

$$\sum_j [A_{ij} - \lambda \varphi_i(r_0)\varphi_j(r_0)]c_j = 0 \tag{2.31}$$

has a nontrivial solution only if the determinant vanishes:

$$\det[A_{ij} - \lambda \varphi_i(r_0)\varphi_j(r_0)] = \left[1 - \lambda \sum_{ij} \varphi_i(r_0)[A^{-1}]_{ij}\varphi_j(r_0)\right]\det[A_{ij}] = 0. \tag{2.32}$$

In this equation, we used the matrix determinant lemma (Harville 2008). In contrast to the usual secular equations, whose degree in the unknown energy is as high as the number of basis functions in Equation 2.28, the secular Equation 2.32 is a linear equation for λ.

Finally, the usual form of the R-matrix theory can be obtained as follows. We first formally solve the set of linear Equations 2.29 for the coefficients

$$c_i = \lambda \sum_j [A^{-1}]_{ij}\varphi_j(r_0)\psi_t(r_0), \tag{2.33}$$

and insert them back to expansion (2.28) evaluated at $r = r_0$ as

$$\psi_t(r_0) = \sum_i \varphi_i(r_0)c_i = \lambda \sum_{ij} \varphi_i(r_0)[A^{-1}]_{ij}\varphi_j(r_0)\psi_t(r_0). \tag{2.34}$$

The identity above directly gives the desired expression for the R-matrix:

$$R = \frac{1}{\lambda} = \sum_{ij} \varphi_i(r_0)[A^{-1}]_{ij}\varphi_j(r_0). \tag{2.35}$$

Some authors (Greene 1983, Le Rouzo and Raseev 1984, Aymar et al. 1996) directly use the form of Equation 2.35, that is, in order to obtain the scattering quantities they inverse the A-matrix for each collision energy $E = k^2/2$ in Equation 2.30. Multichannel implementation of this technique is called the Eigenchannel R-matrix method.

If one needs to calculate cross sections for multiple collision energy points, it may be convenient to replace the multiple inversions by a single diagonalization of the symmetrized Hamiltonian $\bar{H}$ in Equation 2.30. It was shown (Robicheaux 1991) that Equation 2.35 can be expressed in the Wigner–Eisenbud form for $R(E)$, involving a sum of poles. In his procedure, he first splits the elements of the matrix $\mathbf{A}$ as follows:

$$A_{ij} = \int_0^{r_0} dr[\varphi_i'(r)\varphi_j'(r) + 2\varphi_i(r)V(r)\varphi_j(r)] - \int_0^{r_0} dr\, \varphi_i(r)k^2\varphi_j(r) \tag{3.36}$$

$$= 2\bar{H}_{ij} - k^2 S_{ij}, \tag{3.37}$$

where S_{ij} is an overlap matrix of the basis set and the symmetrized Hamiltonian $\bar{H}$ may be expressed by adding a surface term to the original Hamiltonian (2.25):

$$\bar{H} = H + \frac{1}{2}\delta(r - r_0)\frac{d}{dr}. \tag{2.38}$$

The Hamiltonian H is not hermitian in finite space due to the kinetic energy term. The second term in Equation 2.38, the Bloch operator (Bloch 1957), effectively symmetrizes H. In order to carry out the inversion of r.h.s. of Equation 2.37, we need to diagonalize first $\bar{H}$ in a generally nonorthogonal basis $\varphi_i(r)$ (Robicheaux 1991). In matrix form

$$\bar{\mathbf{H}} = \mathbf{S}\mathbf{V}\varepsilon\mathbf{V}^+\mathbf{S}, \tag{2.39}$$

where the diagonal matrix ε contains the eigenvalues ε_α and the columns of $\mathbf{V}$ form corresponding eigenvectors of $\bar{\mathbf{H}}$. Simple algebraic manipulations then give

$$\mathbf{A}^{-1} = (2\bar{\mathbf{H}} - k^2\mathbf{S})^{-1} = \mathbf{V}(2\varepsilon - k^2)\mathbf{V}^+. \tag{2.40}$$

A substitution of the above result into Equation 2.35 directly leads to the Wigner–Eisenbud form, involving the sum of poles:

$$R(E) = \frac{1}{2}\sum_\alpha \frac{v_\alpha(r_0)v_\alpha(r_0)}{\varepsilon_\alpha - E}, \tag{2.41}$$

where $v_\alpha(r_0)$ are so-called surface amplitudes, that is, the eigenfunctions of $\bar{H}$ evaluated at $r = r_0$:

$$v_\alpha(r_0) = \sum_i V_{i\alpha}\varphi_i(r_0), \tag{2.42}$$

and ε_α are the variational energy eigenvalues. Equation 2.41 is very efficient for multiple R-matrix evaluations over a dense energy grid. The only time-consuming step is the single diagonalization of $\bar{\mathbf{H}}$ and the evaluation of surface amplitudes. After that, the scattering quantities can be computed via sum (2.41) for each total energy of the system E.

The Wigner–Eisenbud form of the R-matrix is a base for the UK Molecular R-matrix Scattering Package (Gillan et al. 1995, Morgan et al. 1998) and also for the Bonn R-matrix implementation (Pfingst et al. 1995). Both packages are designed to apply R-matrix approach to the scattering of electrons from polyatomic molecules and they are both based on quantum chemistry codes to solve an $N + 1$-electron problem inside the sphere $r < r_0$. In order to exploit immense development in quantum chemistry, both methods describe the continuum inside the sphere by a basis of Gaussian functions placed in the center of the sphere. Atomic and molecular repulsion integrals are then evaluated very efficiently (Tennyson 1996, Nestmann et al. 1991); however, such a basis may encounter severe linear dependencies as the size of the R-matrix volume (i.e., value of r_0) increases (Tarana and Tennyson 2008). Despite all the mentioned similarities of the two packages, it may be worthy to mention the implementations mainly differ in the way they form the total $N + 1$-electron wave function and therefore they both include correlation energy differently. The UK Molecular R-matrix Scattering Package also allows carrying the electron impact electronic excitation of the target molecule (Branchett et al. 1991, Gorfinkiel and Tennyson 2005). More details about the available polyatomic R-matrix implementations would exceed the scope of this chapter and we refer an interested reader to several excellent reviews (Gillan et al. 1995, Pfingst et al. 1995, Morgan et al. 1998).

Many-electron R-matrix implementations are rigorous in the treatment of the target electronic structure, together with the treatment of the correlation effects between the scattered electron and the bound electrons of the molecule. That makes these implementations one of the best choices for diatomics and small polyatomics. However, even with a rapid increase of the available computing power, we feel that the problem of the basis set linear dependence remains to be solved before the method can be applied to larger molecules.

2.5 SCHWINGER VARIATIONAL PRINCIPLE

Variational principles play an important role in the field of electron–molecule collisions. The most important of these are probably the Schwinger variational principle and the Kohn variational principle that are well described in many reviews (Callaway 1978, Abdel-Raouf 1982, 1984, Lucchese et al. 1986, Gerjuoy et al. 1983, Nesbet 1979, 1980). The Schwinger variational principle (SVP) was introduced by Julian Schwinger in his lectures at Harvard University in 1947. It is based on the integral form of the Schrödinger equation—the Lippmann–Schwinger equation (Lippmann and Schwinger 1950). Since there exist excellent reviews of SVP (for reviews related to atomic and molecular physics, see, for example, Lucchese et al. 1983, 1986, Winstead and McKoy 1996a, Watson 1989, and numerical methods are discussed in Rescigno et al. 1995), we restrict ourselves here to a brief description of the method. The essential quantity for the calculation of cross sections is the scattering T-matrix which is defined as (Newton 2002)

$$T_{\alpha\beta} = \langle \psi_{0\beta} \mid V \mid \psi_\alpha^{(+)} \rangle = \langle \psi_\beta^{(-)} \mid V \mid \psi_{0\alpha} \rangle, \tag{2.43}$$

where V is the interaction, $\mid \psi_{0\alpha} \rangle$ and $\langle \psi_{0\beta} \mid$ are the unperturbed initial states and $\mid \psi_\alpha^{(+)} \rangle$ is the solution of the Lippmann–Schwinger equation (Lippmann and Schwinger 1950)

$$\mid \psi_\alpha^{(+)} \rangle = \mid \psi_{0\alpha} \rangle + G_0^{(+)} V \mid \psi_\alpha^{(+)} \rangle, \tag{2.44}$$

where $G_0^{(+)}$ is the unperturbed Green function with outgoing scattering boundary conditions. For details, see Newton (2002).

Using Equation 2.44, the T-matrix can be expressed in two different forms. The first, the so-called bilinear form, reads

$$T_{\alpha\beta} = \langle \psi_{0\beta} \mid V \mid \psi_\alpha^{(+)} \rangle + \langle \psi_\beta^{(-)} \mid V \mid \psi_{0\alpha} \rangle - \langle \psi_\beta^{(-)} \mid V - V G_0^{(+)} V \mid \psi_\alpha^{(+)} \rangle. \tag{2.45}$$

It is easy to show that small variations of $\mid \psi_\alpha^{(+)} \rangle$ and $\langle \psi_\beta^{(-)} \mid$, respectively, around their exact values, lead to

$$\delta T_{\alpha\beta} = 0. \tag{2.46}$$

Expression 2.45 is thus stationary with respect to small variations of the solution and represents a variational principle. The second variant of SVP has the fractional form

$$T_{\alpha\beta} = \frac{\langle \psi_{0\beta} \mid V \mid \psi_\alpha^{(+)} \rangle \langle \psi_\beta^{(-)} \mid V \mid \psi_{0\alpha} \rangle}{\langle \psi_\beta^{(-)} \mid V - V G_0^{(+)} V \mid \psi_\alpha^{(+)} \rangle}. \qquad (2.47)$$

This expression is equivalent to Equation 2.45. It has the advantage over Equation 2.45 in being independent of the normalization of the unperturbed states $\langle \psi_{0\beta} \mid$ and $\mid \psi_{0\alpha} \rangle$. An important feature of SVP is that the trial functions $\langle \psi_\beta^{(-)} \mid$ and $\mid \psi_\alpha^{(+)} \rangle$ are always multiplied by the interaction V. If the interaction is of short range, we do not have to care about the asymptotic form of the trial functions and they can be well represented by a sum of square integrable functions (e.g., Gaussians, splines). Applications have shown that for molecular targets with no strong long-range potentials, for example, H_2, N_2^+, and CO_2 (Smith et al. 1984), the discrete basis set approach to SVP can be very effective. For strongly polar molecules, however, this approach must be modified. This was done by Smith et al. (1984). The authors developed an adaptation of SVP for long-range potentials in which the static component of the electron–molecule interaction is treated exactly and the exchange interaction is approximated by an expansion in short-range basis functions.

An obvious disadvantage of SVP is the occurrence of the Green function in Equations 2.45 and 2.47 and the necessity to calculate multidimensional integrals involving the Green function. This must be done, in general, numerically and this part of the SVP calculation represents a time-consuming step. For a detailed description of the numerical performance of SVP, see the quoted paper (Winstead and McKoy 2000). The first SVP calculations on electron scattering were carried out by Watson and McKoy (1979) using a separable form of the Green function (see also Lucchese and McKoy 1979, 1980, Watson et al. 1980, Lucchese et al. 1980). To account for polarization effect and multichannel coupling, Takatsuka and McKoy developed the so-called Multichannel Schwinger Method (Takatsuka and McKoy 1980). Some applications of the method to elastic and electronically inelastic electron scattering are reported in the quoted reviews (Winstead and McKoy 1995, 1996a).

There exist a large number of applications of SVP to electron–molecule scattering and photoionization. We selected a few recent applications to large molecules: furan (Bettega and Lima 2007), pyrrol (de Oliveira et al. 2010), pyrazine (McKoy and Winstead 2007a), and pyrimidine and purine bases and nucleosides of DNA (Winstead and McKoy 2006, Winstead et al. 2007). This impressive progress was made possible by availability of high computational power and massively parallel computational techniques (Winstead and Mckoy 1995, 1996b). In spite of the size of molecules, the SVP calculations bring the most accurate results for elastic scattering by polyatomics available in the literature. As an example, we mention the calculation of the elastic electron scattering by 3-hydroxytetrahydrofuran carried out using the multichannel Schwinger method in the SEP approximation (Vizcaino et al. 2008). In this paper, a set of angular differential cross section is provided for energies 6.5–20 eV and the energy dependence of the elastic cross section at a scattering angle of 120° was in very good agreement with the experiment as seen

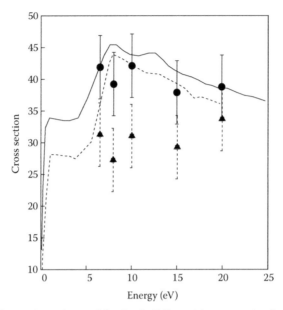

FIGURE 2.3 Energy dependence of the elastic differential cross section for 3-hydroxytetra-hydrofuran over an energy ranges from 3 to 20 eV at a fixed scattering angle of 120°—solid line. Cross sections are in units of 10^{-16} cm^2. Dashed line—the calculated cross sections for tetrahydrofuran. (Adapted from Vizcaino, V. et al. 2008. *New J. Phys. 10*, 053002.)

from Figure 2.3. As the last example, we discuss the application of SVP to the construction of a nonlocal resonance model for an e-HCl system. The model is based on the calculated fixed nuclei phase shifts (Fedor et al. 2010). The dissociative attachment cross sections and vibrational excitation cross sections calculated by the new model (Fedor et al. 2010) agreed much better with the measured data than that of the previous model.

2.5.1 HIGHER-ORDER EXPRESSIONS AND OTHER VARIANTS OF SVP

There are many possibilities regarding how to define other variational principles similar to functionals (2.45) and (2.47). For example, one can start from the iterated LS equation

$$| \psi_\alpha^{(+)} \rangle = | \psi_{0\alpha} \rangle + G_0^{(+)} V | \psi_{0\alpha} \rangle + G_0^{(+)} V | \psi_\alpha^{(+)} \rangle \tag{2.48}$$

instead of Equation 2.44 (see, e.g., Adhikari 1998). A series of variational functionals can be obtained in this way. Next, we mention one approach that goes beyond simple iterations of the LS equation namely the so-called Schwinger–Lanczos method (also termed as the method of continued fractions), which combines the variational principles of the Schwinger-type with the method of Lanczos of forming the tridiagonal representation of the T-matrix.

2.5.2 SCHWINGER–LANCZOS METHOD

The method of continued fractions (MCF) was proposed by Horáček and Sasakawa (1983, 1984, 1985), and reformulated by Meyer et al. (1991, Čížek et al. 2000) as the Schwinger–Lanczos (SL) method. For simplicity, we explain the MCF method here for the case of one-channel scattering. Multichannel generalization can be found in the paper by Znojil (1984). Let us consider the one-channel on-shell LS equation at a given energy E:

$$|\varphi\rangle = |u\rangle + G_0(E)V|\varphi\rangle. \tag{2.49}$$

The essence of the MCF method consists in the decomposition of the interaction V into two parts, one of which is separable:

$$V = V_1 + \frac{V|u\rangle\langle u|V}{\langle u|V|u\rangle}. \tag{2.50}$$

Making use of this definition, the problem of solving Equation 2.49 is reduced to the treatment of the following Lippmann–Schwinger-type equation for a new wave function $|\varphi_1\rangle$ (for details, see Horáček and Sasakawa 1983):

$$|\varphi_1\rangle = |u_1\rangle + G_0(E)V_1|\varphi_1\rangle, \tag{2.51}$$

where

$$|u_1\rangle = G_0(E)V|u\rangle. \tag{2.52}$$

From the definition of the potential V_1, Equation 2.50, we find that the free particle wave function $|u\rangle$ lies in the null space of V_1, that is,

$$V_1|u\rangle = 0, \langle u|V_1 = 0. \tag{2.53}$$

As a result, there is no scattering on this potential at energy E, that is, the energy-dependent potential V_1 is transparent. The equation

$$|\omega\rangle = |u\rangle + G_0(E)V_1|\omega\rangle \tag{2.54}$$

has as the unique solution the free particle wave function $|\omega\rangle = |u\rangle$.

By repeating analogical decompositions for V_1, V_2, and so on, the T-matrix takes the continued fraction form (Horáček and Sasakawa 1983)

$$T = \cfrac{\langle u|V|u\rangle^2}{\langle u|V|u\rangle - \langle u|V|u_1\rangle - \cfrac{\langle u_1|V|u_1\rangle^2}{\langle u_1|V_1|u_1\rangle - \langle u_1|V_1|u_2\rangle - \cdots}}, \tag{2.55}$$

where

$$|u_i\rangle = G(E)_0 V_{i-1}|u_{i-1}\rangle \tag{2.56}$$

$$V_{i-1} = V_{i-2} - \frac{V_{i-2}|u_{i-2}\rangle\langle u_{i-2}|V_{i-2}}{\langle u_{i-2}|V_{i-2}|u_{i-2}\rangle}, \tag{2.57}$$

$$V_0 = V, \tag{2.58}$$

$$|u_0\rangle = |u\rangle. \tag{2.59}$$

By virtue of definitions (2.56) and (2.57), the functions u_i satisfy the following orthogonality relations:

$$V_n |u_i\rangle = \langle u_i | V_n = 0, \quad i = 0,1,\ldots,n-1. \tag{2.60}$$

The first iteration of MCF yields the T-matrix in the form

$$T = \frac{\langle u | V | u \rangle\langle u | V | u \rangle}{\langle u | V - VG_0V | u \rangle}. \tag{2.61}$$

This is exactly what follows from SVP when the trial wave function is represented by the plane wave, $|\psi\rangle = |u\rangle$ and $\langle\psi| = \langle u|$. The second-order MCF expression for the T-matrix appears as follows:

$$T_2 = \frac{\langle u | V | u \rangle\langle u | V | u \rangle}{\langle u | V - VG_0V | u \rangle - \dfrac{\langle u_1 | V_1 | u_1 \rangle^2}{\langle u_1 | V_1 - V_1G_0V_1 | u_1 \rangle}}. \tag{2.62}$$

This formula can be interpreted in two ways: First as a variational expression resulting from the application of SVP to the "reduced" Lippmann–Schwinger Equation 2.51 or as a variational result following from a new variational principle:

$$[T_2] = \frac{\langle u | V | u \rangle\langle u | V | u \rangle}{\langle u | V - VG_0V | u \rangle - Z_2}, \tag{2.63}$$

where

$$[Z_2] = \frac{\langle \psi | V_1 | u_1 \rangle\langle u_1 | V_1 | \psi \rangle}{\langle \psi | V_1 - V_1G_0V_1 | \psi \rangle}. \tag{2.64}$$

This expression is manifestly stationary under variations of ψ around ϕ_1, where ϕ_1 is given by Equation 2.51. The T-matrix obtained in this way is clearly a variational approximation to the exact T-matrix given by Equation 2.47. The remainder Z_2 therefore represents a correction to the exact T-matrix and is in a sense minimized. On the other hand, side $[Z_2]$ is itself a variational expression. Hence, the new variational principle (2.63) is in some sense doubly variational (this fact partly explains the unusual stability of MCF). One can therefore expect that a crude approximation to the trial function ψ in Equation 2.64 should give very good results for the T-matrix T_2.

The method was adapted in 1995 by Lee et al. (1995b) to study low-energy electron scattering by neutral and ionic He, as well as positron scattering by atomic hydrogen. The method was also applied to study low-energy electron scattering by linear molecules (Lee et al. 1995a). Since then, the applicability of the method has been extended, for example, to the calculation of elastic and excitation cross section for electron scattering by H_2 in the low- and intermediate-energy range (Lee et al. 1997, Machado and Lee 1999). The MCF method has also been applied to electron–molecule scattering calculations for diatomics by Lee Mu-Tao and coworkers (Riberio et al. 2001, Machado and Lee 1999, Machado et al. 2001, Taviera et al. 2006, Lee and Mazon 2002, Mazon et al. 2001). Lately, the method was generalized to study the elastic electron scattering by polyatomic molecules of arbitrary symmetry, including those with no symmetry at all. Elastic scattering cross sections have been calculated for CH_4, H_2O, and NH_3. The obtained elastic differential cross sections agree very well with other calculations and experimental data (Riberio et al. 2001). The strong points of MCF are (1) there are no basis functions needed and (2) fast convergence. The weak points are (1) high memory demands; (2) high demands on computing time for the polyatomic version, since no symmetry properties have been thus far considered.

2.6 KOHN VARIATIONAL PRINCIPLE

The Kohn variational principle (KVP) (Kohn 1948) is a very popular method for solving electron molecule scattering problem and exists in several modifications. Here again, we present only a brief description of KVP for the case of potential scattering. The solution $\psi_l(r)$ of the Schrödinger equation for a partial wave l

$$-\frac{d^2\psi_l(r)}{dr^2} + \frac{l(l+1)}{r^2}\psi_l(r) + V(r)\psi_l(r) = k^2\psi_l(r) \tag{2.65}$$

must behave at large distances, where the interaction $V(r)$ is negligible, as

$$\psi_l(r) \approx \frac{1}{k}\sin\left(kr - \frac{l\pi}{2}\right) + K_l \cos\left(kr - \frac{l\pi}{2}\right), \tag{2.66}$$

where

$$\tan\delta_l\ (k) = kK_l \tag{2.67}$$

and ψ_l must be regular at the origin $\psi_l(0) = 0$. Let us consider the expression

$$I[\phi,\phi] = \int_0^\infty \phi_l(r)\left\{\frac{d^2}{dr^2} + k^2 - V(r) - \frac{l(l+1)}{r^2}\right\}\phi_l(r)dr, \tag{2.68}$$

which vanishes if ϕ_l is the correct solution of Equation 2.65. It is easy to show that the quantity $I + kK_l$ is stationary

$$\delta(I + kK_l) = 0, \tag{2.69}$$

when evaluated at some $\phi = \psi_l + \delta\psi_l$ satisfying the boundary condition Equation 2.66 and the regularity at the origin. Finally, we get

$$I + kK_l = k\tan\delta_l(k), \tag{2.70}$$

which represents a variational result for the phase shift $\delta_l(k)$. In contrast to the Schwinger variational principle, the asymptotic behavior of the trial wave function must be specified, see Equation 2.66. On the other hand, the Kohn variational principle requires calculation of the matrix elements of the Hamiltonian only which is a much easier task as compared to the calculation of the Green function matrix elements needed in applications of the Schwinger variational principle. In fact, both principles are closely related (Gil et al. 1994, McCurdy and Rescigno 1989, Meyer 1994, Rescigno et al. 1990, 1995, Schneider and Rescigno 1988). As noted by Schwartz (Nesbet 1980), when using real functions in the evaluation of Equation 2.69, one is faced with the problem of spurious singularities (Adhikari and Kowalski 1991). In the Kohn method, one has to invert the operator $(E-H)$, where H is the full Hamiltonian in the space of real functions. As H has a continuum spectrum at positive energies, this leads to the appearance of singularities. This problem can be eliminated by the use of complex basis functions. This method—the so-called complex Kohn variational problem—has found many applications (see, e.g., Gil et al. 1994, Rescigno et al. 1995, Schneider and Rescigno 1988, McCurdy and Rescigno 1989, Trevisan et al. 2003). For review, see Rescigno et al. (1995), where also applications to electronically inelastic electron scattering are reviewed. As two recent applications of the method, we mention the calculation of electron elastic scattering by tetrahydrofuran and formic acid by Trevisan et al. (2006a,b). In this application, the fixed-nuclei wave function is written as

$$\Psi(\mathbf{r}) = A[\Phi_0(\mathbf{r}_1,\ldots,\mathbf{r}_N)F(\mathbf{r}_{N+1})] + \sum_n d_n \Theta_n(\mathbf{r}_1,\ldots,\mathbf{r}_N), \qquad (2.71)$$

where Φ_0 is the wave function describing the ground state of the target molecule, A is the standard antisymmetrization operator and the sum contains square-integrable terms that describe dynamic polarization and other effects. In the present application of the complex Kohn method, the scattering wave function $F(\mathbf{r}_{N+1})$ is further expanded in a combined basis of square-integrable functions, φ_i and numerical continuum functions, j_l^+ and h_l^+, which incorporate the outgoing-wave boundary conditions

$$F(r) = \sum_i c_i \phi_i(r) + \sum_{lm} [j_l(kr)\delta_{ll_0}\delta_{mm_0} + T_{ll_0 mm_0}h_l^+(kr)]Y_m(\hat{r})/r, \qquad (2.72)$$

where $Y_{lm}(\hat{r})$ are spherical harmonics. This trial function is substituted into the stationary form of the Kohn variational principle written here for the T-matrix

$$T = T_{\text{trial}} - 2\int \Psi(H-E)\Psi. \qquad (2.73)$$

As a result, a set of linear equations for the coefficients c_i, d_n, and $T_{ll_0 mm_0}$ is obtained. In Figure 2.4, the momentum transfer cross section for electron–formic acid calculated by the use of the complex Kohn method is shown (Trevisan et al. 2006b). The calculated cross section compares well with the measured data with exception of the sharp resonance feature near 1.9 eV. The measured cross section does not show any evidence of a resonance. This is probably due to small branching ratio of the resonance to the vibrationally elastic channel.

2.7 COMPLEX ROTATION

Generally, the problem to calculate scattering wave functions and cross sections is much more difficult than the bound state problem. This is because the asymptotic

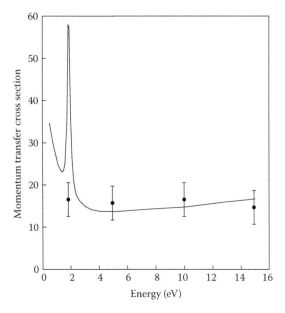

FIGURE 2.4 Electron scattering from formic acid: momentum transfer cross section. Cross sections are in units of 10^{-16} cm^2. Solid line—the complex Kohn calculation. (Adapted from Trevisan, C. S., A. E. Orel, and T. N. Rescigno. 2006a. *J. Phys. B 39*, L255–L260.)

boundary condition for the bound state wave function is very simple. At large distances, the bound state wave function exponentially decreases to zero whereas the scattering state wave functions are oscillating and nonzero even at infinity. There exists an approach, the so-called complex rotation method, which was developed to calculate resonance energies and widths. The idea is to replace the coordinate **r**, which is real, by a complex quantity obtained by rotating **r** by an angle θ into the complex plane:

$$\mathbf{r} \rightarrow \mathbf{r}e^{i\theta}. \tag{2.74}$$

The free particle wave function, $\psi(r) = e^{ikr}$, is thus transformed into

$$\psi(r) = e^{ikr} \rightarrow e^{ikr(\cos\theta + i\sin\theta)} = e^{ik\cos\theta}e^{-kr\sin\theta}. \tag{2.75}$$

If the angle θ is properly chosen, $(\sin \theta > 0)$, the magnitude of the scattering wave function (in the new coordinate) decreases exponentially with increasing distance and the wave function in essence resembles the bound state wave function. This idea is used in the so-called hybrid approach proposed by Rescigno et al. (2005). In this approach, the space is partitioned into several regions: An inner region $r < r_0$ which contains the nuclei, and an outer region which extends to some large distance r_{max}. The outer region is further divided into an intermediate region $r_0 < r < r_c$ and an asymptotic region $r_c, r < r_{max}$. In the hybrid approach, the inner region is spanned by Gaussians, in the intermediate region DVR basis functions are used. Only in the asymptotic region, the coordinate is complex scaled (the so-called exterior complex scaling). This approach obviates the need of slowly convergent SCEs, and allows one to study a variety of electron–molecule collision problems. This approach has been used successfully recently (Yip et al. 2008) to the study of photoionization of diatomic Li_2^+.

2.8 MULTIPLE SCATTERING METHODS

The multiple scattering method (MSM) represents an approximate approach of how to treat multicenter problems. Essentially, it is based on the division of the space into nonoverlapping regions, solving the Schrödinger equation separately in each of the regions and then constructing a global smooth and continuous solution over the whole region. The MSM was proposed originally by Korringa (1947) and by Kohn and Rostocker (1954) for calculating the electronic structure of solids. In solid-state physics, the method is known as the KKR method (Korringa 1947, Kohn and Rostocker 1954, Morse 1956). For a comprehensive review of the state of the art in solid-state physics, we refer the reader to the book by Gonis and Buttler (2000). The method was later extended to polyatomic molecules by Slater and Johnson (1972). In what follows, we will restrict our attention only to the electron–molecule scattering problem.

2.8.1 Continuum Multiple Scattering Method

For solving the electron–molecule scattering problem, in the early 1970s, Dehmer, Dill, and coworkers proposed the so-called continuum multiple scattering method (CMSM). The method had been developed earlier in nuclear physics (Agassi and Gal 1973) to compute the scattering cross sections.

The essence of the CMSM is to treat the multicenter electron–molecule problem as a multiple scattering problem. After penetrating into the molecule, the electron subsequently encounters the molecular constituents (atoms) and scatters from them. In the original proposal of CMSM, the atoms were represented by spherically symmetric potentials. Obviously, the method represents a rather rough approximation to a real problem but it has one big advantage: It can be applied to molecules consisting of a large number of constituents (up to several hundreds of atoms (Yang et al. 1995)).

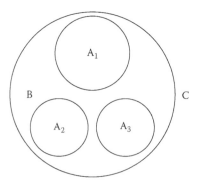

FIGURE 2.5 Separation of the molecular space in the CMS method: A_i—atomic regions, B—interatomic region, C—outer space.

In the Dehmer–Dill approach, the molecule is separated into several nonoverlapping regions: 1. Spherical atomic regions A_i with radii ρ_i surrounding each atom. 2. A molecular spherical region B comprising all atomic spheres, and 3. The outer space C. The radii of the spheres are chosen to achieve maximum packing and to reflect the atomic radii of the atoms. In the atomic regions, it is assumed that the electron interacts with each atom via a spherically symmetric potential V_i. The Schrödinger equation is then solved for a set of angular momentum l for each atomic region. In region B, it is assumed that the potential is constant (the muffin-tin approximation well known in solid-state physics). This, of course, is not realistic but this assumption greatly facilitates the calculations. In the outer region C, a spherically symmetric potential is assumed that may include a polarization term that is important for low-energy electron–molecule scattering. In region B, where the potential is assumed to be constant, the wave function is known analytically and the calculation reduces to finding the expansion coefficients of the wave functions so as to make the logarithmic derivative smooth at the atomic boundaries. Finally, fitting the solution to the asymptotic wave functions at the outer boundary C determines the scattering cross section. Since the theory contains no multicenter integrals, it is possible to carry out the calculations over very large energy regions and for arbitrary molecular configurations. Using this approach, Dehmer, Dill and coworkers calculated, for example, elastic scattering cross sections (Siegel et al. 1980, 1978, Dill and Dehmer 1974, 1977), shape resonance-enhanced vibrational excitation cross sections (Dehmer et al. 1979, 1980, Dill et al. 1979) for N_2, and photoionization cross sections for several molecules (Dill et al. 1978, Dehmer and Dill 1975, 1976, Dill and Dehmer 1974). The CMSM provides results which are approximate but quite realistic. This is demonstrated in Figure 2.6, where the calculated fixed-nuclei elastic scattering cross section of electrons with N_2 molecules is compared with the experimental data. The peak represents the effect of nuclear motion near the π_g resonance.

The number of examples of application of the CMSM method to electronic collisions decreased in the 1980s, probably because by then, more accurate *ab initio* methods had been developed. Nevertheless, the method has been in use. See, for example, Tossell and Davenport (1977) (data for CX_4 and SX_4 ($X = H$, F, Cl)), Bloor et al. (1981), Bloor and Sherrod (1986), and so on.

This continuum MSM has some potential for improvement. First, the potential in the interatomic region B may be taken as not constant but rather depending on the distance between the electron and the atom. Second, the regions A_i need not be spherical and may even overlap. Finally, the assumption that the electron–atom interaction is represented by a potential may be abandoned and the wave function calculated by other more realistic methods.

2.8.2 Multiple Scattering Approach to Elastic Electron Collisions with Molecular Clusters

Recently, the idea of multiple scattering has been extended (Bouchiha et al. 2008, Caprasecca et al. 2009) to handle electron scattering with molecular clusters. Instead of atomic constituents, the subunits are supposed to be molecules. The second

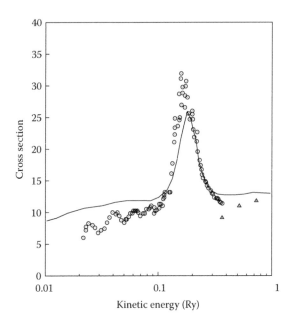

FIGURE 2.6 The adiabatically R-averaged elastic scattering e^-—N_2 cross section—solid line. Experimental data are represented by circles and triangles. (Adapted from Siegel, J., J. L. Dehmer, and D. Dill. 1980. *Phys. Rev. A 21*, 85–94.) Cross sections are in units of Å^2.

important departure from the method of Dill and Dehmer is that the subunits are not described by a potential but the R-matrix calculation is performed for each subunit: This makes the results more realistic. Details of the derivation of the MS equations are described in the literature (Caron et al. 2007, Bouchiha et al. 2008, Caprasecca et al. 2009). Here we present just the main equations of the method in a symbolic way. Let us denote as T^N the T-matrix of the N-th monomer as obtained by the R-matrix approach. Then the total T-matrix describing the scattering on the full cluster, T^{tot}, takes the form

$$T^{tot} = A \cdot T \cdot (1 - X \cdot T)^{-1} \cdot B, \tag{2.76}$$

where A and B are matrices that account for the transformation from the center of mass of the monomers to that of the cluster. T is a block diagonal matrix whose matrix elements are the monomer T-matrices T_N. The term $(1 - X T)^{-1}$ represents the multiple scattering between the scatterers. The formation of the cluster T-matrix thus requires calculations of the R-matrix for each monomer and the calculation of the transformation matrices. The method was applied in the treatment of a water cluster consisting of two water molecules (Caprasecca et al. 2009). Elastic cross sections were given in the electron energy range 0–10 eV. The calculated cross sections are compared with the more accurate R-matrix calculations treating the whole cluster as one unit in Figure 2.7. The agreement is good for energies above 1 eV, while below 1 eV the CMSM seems to overestimate the cross section. The method is computationally inexpensive once the R-matrix data are generated for the monomers (Caprasecca

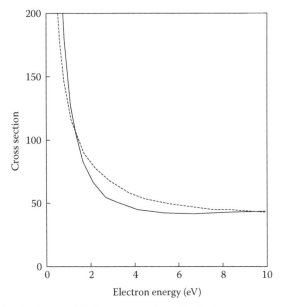

FIGURE 2.7   Elastic electron-$(H_2O)_2$ cross sections. Dashed line—R-matrix cross section, solid line—the CMS calculation. For details see Bouchiha et al. (2008). Cross sections are in units of a_0^2.

et al. 2009). Using this approach, clusters consisting of several monomers can be treated.

Recently, the CMSM has been used to deal with electron scattering of DNA constituents by Sanche and coworkers. The results are discussed in detail in Chapter 6. The CMSM approach has also been applied (Caron et al. 2007) to the low-energy electron scattering from a molecule embedded in a solid.

ACKNOWLEDGMENTS

JH acknowledges support of Záměr MSM0021620860 of the Ministry of Education, Youth and Sports of the Czech Republic. RČ would like to thank the Czech Ministry of Education (Grants OC09079 and OC10046) and by the Grant Agency of the Czech Republic (Grants 202/08/0631 and P208/11/0452).

REFERENCES

Abdel-Raouf, M. A. 1982. On the variational methods for bound-state and scattering problems. *Phys. Rep. 84*, 163–261.
Abdel-Raouf, M. A. 1984. On the variational methods for bound-state and scattering problems II. *Phys. Rep. 108*, 1–164.
Adhikari, S. K. 1998. *Variational Principles and the Numerical Solution of Scattering Problems*. New York: John Wiley & Sons, Inc.

Adhikari, S. K. and K. L. Kowalski. 1991. *Dynamical Collision Theory and its Applications*. New York: Academic Press.

Agassi, D. and A. Gal. 1973. Scattering from non-overlapping potentials. I. General formulation. *Ann. Phys. 75*, 56–76.

Allen, L. C. and A. M. Karo. 1960. Basis functions for *ab initio* calculations. *Rev. Mod. Phys. 32*, 275–285.

Almbladh, C. and U. von Barth. 1985. Exact results for the charge and spin-densities, exchange-correlation potentials, and density-functional eigenvalues. *Phys. Rev. B 31*, 3231–3244.

Altmann, S. L. and A. P. Cracknell. 1965. Lattice harmonics I. Cubic groups. *Rev. Mod. Phys. 37*, 19–32.

Antony, J. and S. Grimme. 2006. Density functional theory including dispersion corrections for intermolecular interactions in a large benchmark set of biologically relevant molecules. *Phys. Chem. Chem. Phys. 8*, 5287–5293.

Aymar, M., C. H. Greene, and E. Luc-Koenig. 1996. Multichannel Rydberg spectroscopy of complex atoms. *Rev. Mod. Phys. 68*, 1015–1123.

Baccarelli, I., F. Sebastianelli, F. A. Gianturco, and N. Sanna. 2009. Modelling dissociative dynamics of biosystems after metastable electron attachment: The sugar backbones. *51*, 131–136.

Bettega, M. H. F. and M. A. P. Lima. 2007. Electron collisions with furan. *J. Comp. Phys. 126*, 194317.

Beyer, T., B. Nestmann, and S. Peyerimhoff. 2000. Study of electron polarization and correlation effects in resonant and background electron scattering off CF_3Cl. *Chem. Phys. 255*, 1–14.

Bloch, C. 1957. Une formulation unifie de la théorie des réactions nucléaires. *Nucl. Phys. 4*, 503–528.

Bloor, J. E. and R. E. Sherrod. 1986. A continuum multiple scattering method for the treatment of elastic electron-molecule scattering. Total and DCS cross sections for argon and methane. *J. Phys. Chem. 90*, 5508–5518.

Bloor, J. E., R. E. Sherrod, and F. A. Grimm. 1981. An endogenous multiple scattering method for the calculation of the elastic scattering cross sections of electron shape resonances in polyatomic molecules. *Chem. Phys. Lett. 78*, 351–356.

Bouchiha, D., L. G. Caron, J. D. Gorfinkiel, and L. Sanche. 2008. Multiple scattering approach to elastic low-energy electron collisions with the water dimer. *J. Phys. B 41*, 045204.

Branchett, S., J. Tennyson, and L. Morgan. 1991. Differential cross-sections for electronic excitation of molecular-hydrogen using the *R*-matrix method. *J. Phys. B 24*, 3479–3490.

Brems, V., T. Beyer, B. Nestmann, H. Meyer, and L. Cederbaum. 2002. Ab initio study of the resonant electron attachment to the F_2 molecule. *J. Comp. Phys. 117*, 10635–10647.

Brunger, M. J. and S. J. Buckman. 2002. Electron-molecule scattering cross-sections. I. Experimental techniques and data for diatomic molecules. *Phys. Rep. 357*, 215–458.

Burke, P. and W. Robb. 1972. Elastic-scattering of electrons by hydrogen and helium-atoms. *J. Phys. B 5*, 44–54.

Burke, P. B. and N. Chandra. 1972. Electron-molecule interactions III. A pseudo-potential method for e^--N_2 scattering. *J. Phys. B 5*, 1696–1711.

Burke, P. B. and A. L. Sinfailam. 1970. Electron-molecule interactions II. Scattering by closed-shell diatomic molecules. *J. Phys. B 3*, 641–659.

Burke, P. G. and W. D. Robb. 1975. *R*-matrix theory of atomic processes. *Adv. At. Mol. Phys. 11*, 143.

Callaway, J. 1978. The variational method in atomic scattering. *Phys. Rep. 45*, 89–173.

Caprasecca, S., J. D. Gorfinkiel, D. Bouchiha, and L. Sanche. 2009. Multiple scattering approach to elastic electron collisions with molecular clusters. *J. Phys. B 42*, 095202–095212.

Caron, L., D. Bouchiha, J. D. Gorfinkiel, and L. Sanche. 2007. Adapting gas-phase electron scattering R-matrix calculations to a condensed-matter environment. *Phys. Rev. A 76*, 032716.

Cascella, M., R. Curik, F. Gianturco, and N. Sanna. 2001. Electron-impact vibrational excitation of polyatomic gases: Exploratory calculations. *J. Comp. Phys. 114*, 1989–2000.

Ceperley, D. and B. Alder. 1980. Ground-state of the electron-gas by a stochastic method. *Phys. Rev. Lett. 45*, 566–569.

Chandra, N. 1977. Low-energy electron scattering from CO. II. Ab initio study using the frame-transformation theory. *Phys. Rev. A 16*, 80–108.

Chang, E. S. and A. Temkin. 1969. Rotational excitation of diatomic molecules by electron impact. *Phys. Rev. Lett. 23*, 399–403.

Čížek, M., J. Horáček, and H.-D. Meyer. 2000. Schwinger-Lanczos algorithm for calculation of off-shell T-matrix elements and Wynn's epsilon algorithm. *Comput. Phys. Commun. 131*, 41–51.

Colle, R. and O. Salvetti. 1975. Approximate calculation of correlation energy for closed shells. *Theor. Chim. Acta 37*, 329–334.

Collins, L. A., W. D. Robb, and M. A. Morrison. 1980. Electron scattering by diatomic molecules: Iterative static-exchange techniques. *Phys. Rev. A 21*, 488–495.

Connell, O. K. and N. F. Lane. 1983. Nonadjustable exchange-correlation model for electron scattering from closed-shell atoms and molecules. *Phys. Rev. A 27*, 1893–1903.

Čurík, R. and F. Gianturco. 2002a. A computational analysis of low-energy electron scattering from gaseous cyclopropane. *J. Phys. B 35*, 717–732.

Čurík, R. and F. Gianturco. 2002b. Quantum calculations for resonant vibrational excitations of cyclopropane by electron impact. *J. Phys. B 35*, 1235–1250.

Curik, R., F. Gianturco, R. Lucchese, and N. Sanna. 2001. Low-energy electron scattering and resonant states of NO_2 ($\tilde{X}$ $^2A_1$). *J. Phys. B 34*, 59–79.

Curik, R., F. Gianturco, and N. Sanna. 1999. Resonant features in low-energy electron scattering from ozone. *J. Phys. B 32*, 4567–4580.

Curik, R., F. Gianturco, and N. Sanna. 2000. Electron and positron scattering from halogenated methanes: A comparison of elastic cross sections. *J. Phys. B 33*, 615–635.

Čurík, R. and M. Šulc. 2010. Towards efficient ab initio calculations of electron scattering by polyatomic molecules: III. modelling correlation-polarization interactions. *J. Phys. B 43*, 175205.

Danby, G. and J. Tennyson. 1991. R-matrix calculations of vibrationally resolved positron N_2 scattering cross-sections. *J. Phys. B 24*, 3517–3529.

Dehmer, J. L. and D. Dill. 1975. Shape resonances in k-shell photoionization of diatomic molecules. *Phys. Rev. Lett. 35*, 213–215.

Dehmer, J. L. and D. Dill. 1976. Angular distribution of Xe $5s \rightarrow \varepsilon p$ photoelectrons: Direct evidence for anisotropic final-state interaction. *Phys. Rev. Lett. 37*, 1049–1052.

Dehmer, J. L., D. Dill, and S. Wallace. 1979. Shape-resonance-enhanced nuclear-motion effects in molecular photoionization. *Phys. Rev. Lett. 43*, 1005–1008.

Dehmer, J. L., J. Siegel, J. Welch, and D. Dill. 1980. Origin of enhanced vibrational excitation in N_2 by electron impact in the 15–35 eV region. *Phys. Rev. A 21*, 101–104.

Dill, D. and J. L. Dehmer. 1974. Electron-molecule scattering and molecular photoionization using the multiple-scattering method. *J. Chem. Phys. 61*, 692–697.

Dill, D. and J. L. Dehmer. 1977. Total elastic electron scattering cross section for N_2 between 0 and 1000 eV. *Phys. Rev. A 16*, 1423–1431.

Dill, D., S. Wallace, J. Siegel, and J. L. Dehmer. 1978. Molecular-photoelectron angular distributions as a probe of dynamic symmetry breaking. *Phys. Rev. Lett. 41*, 1230–1233.

Dill, D., J. Welch, J. L. Dehmer, and J. Siegel. 1979. Shape-resonance-enhanced vibrational excitation at intermediate energies (10–40 ev) in electron-molecule scattering. *Phys. Rev. Lett. 43*, 1236–1239.

Fabrikant, I. 1984. Effective-range analysis of low-energy electron-scattering by non-polar molecules. *J. Phys. B 17*, 4223–4233.

Fano, U. and C. M. Lee. 1973. Variational calculation of *R*-matrices. Application to Ar photoabsorption. *Phys. Rev. Lett. 31*, 1573–1576.

Fedor, J., C. Winstead, V. McKoy, M. Čížek, K. Houfek, P. Kolorenč, and J. Horáček. 2010. Electron scattering in HCl: An improved nonlocal resonance model. *Phys. Rev. A 81*, 042702.

Friedman, J. F., T. M. Miller, L. C. Schaffer, A. A. Viggiano, and I. I. Fabrikant. 2009. Electron attachment to Cl-2 from 300 to 1100 K: Experiment and theory. *Phys. Rev. A 79*, 032707.

Gallup, G. A. and I. I. Fabrikant. 2011. Vibrational Feshbach resonances in dissociative electron attachment to uracil. *Phys. Rev. A 83*, 012706.

Gerjuoy, E., A. R. P. Rau, and L. Spruch. 1983. A unified formulation of the construction of variational principles. *Rev. Mod. Phys. 55*, 725–774.

Gianturco, F., A. Jain, and L. Pantano. 1987. Electron methane scattering via a parameter-free model interaction. *J. Phys. B 20*, 571–586.

Gianturco, F. and R. Lucchese. 1998. One-electron resonances and computed cross sections in electron scattering from the benzene molecule. *J. Comp. Phys. 108*, 6144–6159.

Gianturco, F. and R. Lucchese. 2000. Angular distributions of scattered electrons from gaseous benzene molecules. *J Comp. Phys. 113*, 10044–10050.

Gianturco, F., R. Lucchese, and N. Sanna. 1994. Calculation of low-energy elastic cross-sections for electron-CF4 scattering. *J. Chem. Phys. 100*, 6464–6471.

Gianturco, F., R. Lucchese, and N. Sanna. 1996. Low-energy electron scattering by halomethanes: Elastic and differential cross sections for CF4. *J. Comp. Phys. 104*, 6482–6490.

Gianturco, F. and J. Rodriguezruiz. 1993. Correlation forces in electron-scattering processes via density-functional theory—Electron collisions with closed-shell atoms. *Phys. Rev. A 47*, 1075–1086.

Gianturco, F. and S. Scialla. 1987. Local approximations of exchange interaction in electron molecule collisions—The methane molecule. *J. Phys. B 20*, 3171–3189.

Gianturco, F. A. and A. Jain. 1986. The theory of electron scattering from polyatomic molecules. *Phys. Rep. 143*, 347–425.

Gianturco, F. A., F. Sebastianelli, R. R. Lucchese, I. Baccarelli, and N. Sanna. 2008. Ring-breaking electron attachment to uracil: Following bond dissociations via evolving resonances. *J. Comp. Phys. 128*(17), 174302.

Gianturco, F. A. and T. Stoecklin. 1997. Calculation of rotationally inelastic processes in electron collisions with CO_2 molecules. *Phys. Rev. A 55*, 1937–1944.

Gianturco, F. A., D. G. Thompson, and A. Jain. 1995. Electron-scattering from polyatomic molecules using a single-center-expansion formulation. In W. M. Huo and F. A. Gianturco (Eds.), *Computational Methods for Electron-Molecule Collisions*, pp. 75–118. New York, NY: Plenum Press.

Gil, T. J., T. N. Rescigno, C. W. McCurdy, and B. H. Lengsfield III. 1994. *Ab initio* complex Kohn calculations of dissociative excitation of water. *Phys. Rev. A 49*, 2642–2650.

Gillan, C. J., J. Tennyson, and P. G. Burke. 1995. The UK molecular *R*-matrix scattering package: A computational perspectives. In W. M. Huo and F. A. Gianturco (Eds.), *Computational Methods for Electron-Molecule Collisions*, pp. 239–254. New York, NY: Plenum Press.

Gonis, A. and W. H. Butler. 2000. *Multiple Scattering in Solids*. Graduate texts in contemporary physics. New York, NY: Springer Verlag, Inc.

Gorfinkiel, J. and J. Tennyson. 2005. Electron impact ionization of small molecules at intermediate energies: The molecular *R*-matrix with pseudostates method. *J. Phys. B 38*, 1607–1622.

Greene, C. H. 1983. Atomic photoionization in a strong magnetic field. *Phys. Rev. A 28*, 2209–2216.

Hara, S. 1969. A 2-center approach in low energy electron-H2 scattering. *J. Phys. Soc. Jap. 27*, 1009.

Harville, D. A. 2008. *Matrix Algebra From a Statistician's Perspective* (Corrected ed.). New York: Springer.

Henry, R. J. W. and N. F. Lane. 1969. Polarization and exchange effects in low-energy electron-H_2 scattering. *Phys. Rev. 183*, 221–231.

Horáček, J. and T. Sasakawa. 1983. Method of continued fractions with application to atomic physics. *Phys. Rev. A 28*, 2151–2156.

Horáček, J. and T. Sasakawa. 1984. Method of continued fractions with application to atomic physics. II. *Phys. Rev. A 30*, 2274–2277.

Horáček, J. and T. Sasakawa. 1985. Method of continued fractions for on- and off-shell *T* matrix of local and nonlocal potentials. *Phys. Rev. C 32*, 70–75.

Isaacs, W. A. and M. A. Morrison. 1992. Modified effective range theory as an alternative to low-energy close-coupling calculations. *J. Phys. B 25*, 703–725.

Jain, A., K. Baluja, V. Dimartino, and F. Gianturco. 1991. Differential, integral and momentum-transfer cross-sections for electron-scattering with germane (GeH4) molecules at 1–100 eV. *Chem. Phys. Lett. 183*, 34–39.

Jain, A. and D. W. Norcross. 1992. Slow-electron collisions with CO molecules in an exact-exchange plus parameter-free polarization model. *Phys. Rev. A 45*, 1644–1656.

Joachain, C. J. 1983. *Quantum Collision Theory*. New York, NY: Elsevier Science Publishing Company Inc.

Kohn, W. 1948. Variational methods in nuclear collision problems. *Phys. Rev. 74*, 1763–1772.

Kohn, W. and N. Rostoker. 1954. Solution of the Schrödinger equation in periodic lattices with an application to metallic lithium. *Phys. Rev. 94*, 1111–1120.

Korringa, J. 1947. On the calculation of the energy of a Bloch wave in a metal. *Physica 13*, 392–400.

Lane, N. F. 1980. The theory of electron–molecule collisions. *Rev. Mod. Phys. 52*, 29–119.

Lee, C., W. Yang, and R. Parr. 1988. Development of the Colle–Salvetti correlation-energy formula into a functional of the electron-density. *Phys. Rev. B 37*, 785–789.

Lee, M.-T., M. M. Fujimoto, and I. Iga. 1997. Application of the method of continued fractions to low-energy electron scattering by the hydrogen molecule. *J. Mol. Structure: THEOCHEM 394*, 117–125.

Lee, M.-T., I. Iga, M. M. Fujimoto, and O. Lara. 1995a. Application of the method of continued fractions for electron scattering by linear molecules. *J. Phys. B 28*, 3325–3334.

Lee, M.-T., I. Iga, M. M. Fujimoto, and O. Lara. 1995b. The method of continued fractions for electron (positron)-atom scattering. *J. Phys. B 28*, L299.

Lee, M.-T. and K. T. Mazon. 2002. Electron scattering by vibrationally excited H_2 in the low-energy range. *Phys. Rev. A 65*, 042720.

Le Rouzo, H. and G. Raseev. 1984. Finite-volume variational method: First application to direct molecular photoionization. *Phys. Rev. A 29*, 1214–1223.

Lippmann, B. A. and J. Schwinger. 1950. Variational principles for scattering processes. I. *Phys. Rev. 79*, 469–480.

Lucchese, R. R. and V. McKoy. 1979. Application of the Schwinger variational principle to electron scattering. *J. Phys. B 12*, L421.

Lucchese, R. R. and V. McKoy. 1980. Application of the Schwinger variational principle to electron-ion scattering in the static-exchange approximation. *Phys. Rev. A 21*, 112–123.

Lucchese, R. R., K. Takatsuka, and V. McKoy. 1986. Application of the Schwinger variational principle to electron–molecule collision and molecular photoionization. *Phys. Rep. 131*, 147–221.

Lucchese, R. R., K. Takatsuka, D. K. Watson, and V. McKoy. 1983. *Electron–Atom and Electron–Molecule Collisions*, pp. 29–49. New York: Plenum Press.

Lucchese, R. R., D. K. Watson, and V. McKoy. 1980. Iterative approach to the Schwinger variational principle for electron–molecule collisions, *Phys. Rev. A 22*, 421–426.

Machado, A. M., M. M. Fujimoto, A. M. A. Taveira, L. M. Brescansin, and M.-T. Lee. 2001. Application of the method of continued fractions to multichannel studies on electronic excitation of H_2 by electron impact. *Phys. Rev. A 63*, 032707.

Machado, A. M. and M.-T. Lee. 1999. Application of the method of continued fractions for the distored-wave Green function in electron–molecule scattering. *J. Phys. B 32*, L81–L87.

Machado, A. M., A. M. A. Taveira, L. M. Brescansin, and M.-T. Lee. 2001. Application of the method of continued fractions to multichannel studies on electron-impact excitation of the $B^1\Sigma_u^+, C^1\Pi_u$ and $E(F)^1\Sigma_g^+$ states in H_2. *J. Mol. Structure: THEOCHEM 574*, 133–140.

Mazon, K. T., R. Fujiwara, and M.-T. Lee. 2001. Exact exchange effects on vibrational excitation of H_2 by electron impact. *Phys. Rev. A 64*, 042705.

McCurdy, C. W. and T. N. Rescigno. 1989. Collisions of electrons with polyatomic molecules: Electron-methane scattering by the complex Kohn variational method. *Phys. Rev. A 39*, 4487–4493.

McKoy, V. and C. Winstead. 2007a. Low-energy electron scattering with pyrazine. *Phys. Rev. A 76*, 012712.

Meyer, H.-D. 1994. The equivalence of the log derivative Kohn principle with the *R*-matrix method. *Chem. Phys. Lett. 223*, 465–468.

Meyer, H.-D., J. Horáček, and L. S. Cederbaum. 1991. Schwinger and anomaly-free Kohn variational principles and a generalized Lanczos algorithm for nonsymmetric operators. *Phys. Rev. A 43*, 3587–3596.

Morgan, L. A., J. Tennyson, and C. J. Gillan. 1998. The UK molecular *R*-matrix codes. *Comput. Phys. Comm. 114*, 120–128.

Morrison, M. and L. Collins. 1978. Exchange in low-energy electron-molecule scattering—Free-electron-gas model exchange potentials and applications to e-H_2 and e-N_2 collisions. *Phys. Rev. A 17*, 918–938.

Morrison, M. A., W. Sun, W. A. Isaacs, and W. K. Trail. 1997. Ultrasimple calculation of very-low-energy momentum transfer and rotational-excitation cross sections: e–N_2 scattering. *Phys. Rev. A 55*, 2786–2798.

Morse, P. M. 1956. Waves in a lattice of spherical scatterers. *Proc. Math. Aced. Sci. USA 42*(5), 276–286.

Nesbet, R. K. 1979. Low-energy electron scattering by complex atoms: Theory and calculations. *Adv. At. Mol. Phys. 13*, 315–382.

Nesbet, R. K. 1980. *Variational Methods in Electron-Atom Scattering Theory*. Physics of atoms and molecules. New York, NY: Plenum Press.

Nestmann, B., R. Nesbet, and S. Peyerimhoff. 1991. A concept for improving the efficiency of *R*-matrix calculations for electron molecule-scattering. *J. Phys. B 24*, 5133–5149.

Newton, R. G. 2002. *Scattering Theory of Waves and Particles* (2nd ed.). New York, NY: Dover Publications.

Nishimura, T. and Y. Itikawa. 1996. Vibrationally elastic and inelastic scattering of electrons by hydrogen sulphide molecules. *J. Phys. B 29*, 4213–4226.

de Oliveira, E. M., M. A. P. Lima, M. H. F. Bettega, S. d. Sanchez, R. F. da Costa, and M. T. d. N. Varella. 2010. Low-energy electron collisions with pyrrole. *J. Comp. Phys. 132*, 204301.

Padial, N. T. and D. W. Norcross. 1984. Parameter-free model of the correlation-polarization potential for electron-molecule collisions. *Phys. Rev. A 29*, 1742–1748.

Perdew, J., J. Chevary, S. Vosko, K. Jackson, M. Pederson, D. Singh, and C. Fiolhais. 1992. Atoms, molecules, solids, and surfaces—Applications of the generalized gradient approximation for exchange and correlation. *Phys. Rev. B 46*, 6671–6687.

Perdew, J. and Y. Wang. 1992. Accurate and simple analytic representation of the electron-gas correlation-energy. *Phys. Rev. B 45*, 13244–13249.

Perdew, J. and A. Zunger. 1981. Self-interaction correction to density-functional approximations for many-electron systems. *Phys. Rev. B 23*, 5048–5079.

Pfingst, K., B. M. Nestmann, and S. D. Peyerimhoff. 1995. Tailoring the *R*-matrix approach for application to polyatomic molecules. In W. M. Huo and F. A. Gianturco (Eds.), *Computational Methods for Electron-Molecule Collisions*, pp. 75–118. New York, NY: Plenum Press.

Rescigno, T. N., D. A. Horner, F. L. Yip, and C. W. McCurdy. 2005. Hybrid approach to molecular continuum processes combining Gaussian basis functions and the discrete variable representation. *Phys. Rev. A 72*, 052709.

Rescigno, T. N., B. H. Lengsfield, and C. W. McCurdy. 1990. Electronic excitation of formaldehyde by low-energy electrons: A theoretical study using the complex Kohn variational method. *Phys. Rev. A 41*, 2462–2467.

Rescigno, T. N., B. H. Lengsfield, and C. W. McCurdy. 1995a. The incorporation of modern electronic structure methods in electron-molecule collision problems: Variational calculations using the complex Kohn method. In D. R. Yarkony (Ed.), *Modern Electronic Structure Theory*, Vol. 1, pp. 501–588. Singapore: World Scientific.

Rescigno, T. N., C. W. McCurdy, A. E. Orel, and B. H. Lengsfield III. 1995b. The complex Kohn variational method. In W. M. Huo and F. A. Gianturco (Eds.), *Computational Methods for Electron-Molecule Collisions*, pp. 1–44. New York: Plenum Press.

Riberio, E. M. S., L. E. Machado, M.-T. Lee, and L. M. Brescansin. 2001. Application of the method of continued fractions to electron scattering by polyatomic molecules. *Comput. Phys. Comm. 136*, 117–125.

Riley, M. and D. Truhlar. 1975. Approximations for exchange potential in electron-scattering. *J. Comp. Phys. 63*, 2182–2191.

Robicheaux, F. 1991. Driving nuclei with resonant electrons: Ab initio study of $(e + H_2)^2 \Sigma_u^+$. *Phys. Rev. A 43*, 5946–5955.

Schneider, B. I. and T. N. Rescigno. 1988. Complex Kohn variational method: Application to low-energy electron-molecule collisions. *Phys. Rev. A 37*, 3749–3754.

Siegel, J., J. L. Dehmer, and D. Dill. 1980. Elastic-electron-scattering cross sections for N_2 from 0 to 1000 eV. energy-dependent exchange potentials. *Phys. Rev. A 21*, 85–94.

Siegel, J., D. Dill, and J. L. Dehmer. 1978. Differential elastic electron scattering cross sections for N_2 from 0 to 30 eV. *Phys. Rev. A 17*, 2106–2109.

Slater, J. C. and K. H. Johnson. 1972. Self-consistent-field X_α cluster method for polyatomic molecules and solids. *Phys. Rev. B 5*, 844–853.

Smith, M. E., R. R. Lucchese, and V. McKoy. 1984. Schwinger variational principle applied to long-range potentials. *Phys. Rev. A 29*, 1857–1864.

Sun, W., M. A. Morrison, W. A. Isaacs, W. K. Trail, D. T. Alle, R. J. Gulley, M. J. Brennan, and S. J. Buckman. 1995. Detail theoretical and experimental analysis of low-energy electron-N_2 scattering. *Phys. Rev. A 52*, 1229–1256.

Takatsuka, K. and V. McKoy. 1980. Schwinger variational principle for multichannel scattering. *Phys. Rev. Lett. 45*, 1396–1399.

Takatsuka, K. and V. McKoy. 1981. Extension of the Schwinger variational principle beyond the static-exchange approximation. *Phys. Rev. A 24*, 2473–2480.

Takekawa, M. and Y. Itikawa. 1999. Theoretical study of electron scattering from carbon dioxide: Excitation of bending vibration. *J. Phys. B 32*, 4209–4223.

Tarana, M. and J. Tennyson. 2008. Polarization effects in electron collisions with Li_2: Application of the molecular *R*-matrix method with pseudostates. *J. Phys. B 41*, 205204.

Taviera, A. M. A., L. M. Brescansin, A. M. Machado, and M.-T. Lee. 2006. Multi-channel coupling effects for electronic excitations leading to the $b^3 \sigma_u^+$, $a^3 \sigma_g^+$, and $c^3 \pi_u$ states of H_2. *Int. J. Quant. Chem. 106*, 2006–2013.

Telega, S., E. Bodo, and F. Gianturco. 2004. Rotationally inelastic collisions of electrons with H_2 and N_2 molecules: Converged space-frame calculations at low energies. *Eur. Phys. J. D 29*, 357–365.

Telega, S. and F. A. Gianturco. 2006. Electron-molecule scattering in gases at very low energies: A comparison of theory and experiment for the nitrogen ($^1\Sigma_g^+$) target. *Mol. Phys. 104*, 3147–3154.

Temkin, A. and K. V. Vasavada. 1967. Scattering of electrons from H_2^+: The method of polarized single-center orbitals. *Phys. Rev. 160*, 109–117.

Temkin, A., K. V. Vasavada, E. S. Chang, and A. Silver. 1969. Scattering of electrons from H_2^+. II. *Phys. Rev. 186*, 57–66.

Tennyson, J. 1996. A new algorithm for Hamiltonian matrix construction in electron-molecule collision calculations. *J. Phys. B 29*, 1817–1828.

Tonzani, S. and C. Greene. 2006. Low-energy electron scattering from DNA and RNA bases: Shape resonances and radiation damage. *J. Comp. Phys. 124*, 054312.

Tossell, J. A. and J. W. Davenport. 1977. MS-X_α calculation of the elastic electron scattering cross sections and X-ray absorption spectra of CX_4 and SiX_4 (X = H,F,Cl). *J. Chem. Phys. 80*, 813–821.

Trevisan, C. S., A. E. Orel, and T. N. Rescigno. 2003. *Ab initio* study of low-energy electron collisions with ethylene. *Phys. Rev. A 68*, 062707.

Trevisan, C. S., A. E. Orel, and T. N. Rescigno. 2006a. Elastic scattering of low-energy electrons by tetrahydrofuran. *J. Phys. B 39*, L255–L260.

Trevisan, C. S., A. E. Orel, and T. N. Rescigno. 2006b. Low-energy electron scattering by formic acid. *Phys. Rev. A 74*, 042716.

Umrigar, C. and X. Gonze. 1994. Accurate exchange-correlation potentials and total-energy components for the helium isoelectronic series. *Phys. Rev. A 50*, 3827–3837.

Vizcaino, V., J. Roberts, J. P. Sullivan, M. J. Brunger, S. J. Buckman, C. Winstead, and V. McKoy. 2008. Elastic electron scattering from 3-hydroxytetrahydrofuran: Experimental and theoretical studies. *New J. Phys. 10*, 053002.

Vosko, S., L. Wilk, and M. Nusair. 1980. Accurate spin-dependent electron liquid correlation energies for local spin-density calculations—A critical analysis. *Can. J. Phys. 58*, 1200–1211.

Watson, D. K. 1989. Schwinger variational methods. Volume 25 of *Advances in Atomic and Molecular Physics,* pp. 221–250. London: Academic Press.

Watson, D. K., R. R. Lucchese, V. McKoy, and T. N. Rescigno. 1980. Schwinger variational principle for electron-molecule scattering: Application to electron-hydrogen scattering. *Phys. Rev. A 21*, 738–744.

Watson, D. K. and V. McKoy. 1979. Discrete-basis-function approach to electron-molecule scattering. *Phys. Rev. A 20*, 1474–1483.

Wigner, E. P. and L. Eisenbud. 1947. Higher angular momenta and long range interaction in resonance reactions. *Phys. Rev. 72*, 29–41.

Winstead, C. and V. McKoy. 1995. Studies of electron-molecule collisions on massively parallel computers. In D. R. Yarkony (Ed.), *Modern Electronic Structure Theory. Part II,* pp. 1375–1462. Singapore: World Scientific.

Winstead, C. and V. McKoy. 1996a. Electron scattering by small molecules. Volume XCVI of *Advances in Chemical Physics.* New York: John Wiley.

Winstead, C. and V. McKoy. 1996b. Highly parallel computational techniques for electron-molecule collisions. *Ad. Atom. Mol. Optic. Phys. 36*, 183–219.

Winstead, C. and V. McKoy. 2000. Parallel computational studies of electron molecule collisions. *Comput. Phys. Comm. 128*, 386–398.

Winstead, C. and V. Mckoy. 2006. Low-energy electron scattering by deoxyribose and related molecules. *J. Comp. Phys. 125*, 074302.

Winstead, C., V. McKoy, and M. H. F. Bettega. 2005. Elastic electron scattering by ethylene, C_2H_4. *Phys. Rev. A 72*, 042721.

Winstead, C., V. Mckoy, and S. d. Sanchez. 2007. Interaction of low-energy electrons with the pyrimidine bases and nucleosides of DNA. *J. Comp. Phys. 127*, 085105.

Yang, W., G. M. Stocks, W. A. Shelton, D. M. C. Nicholson, Z. Szotek, and W. M. Temmerman. 1995. Order-N multiple scattering approach to electronic structure calculations. *Phys. Rev. Lett. 75*, 2867–2870.

Yip, F. L., C. W. McCurdy, and T. N. Rescigno. 2008. Hybrid Gaussian–discrete-variable representation approach to molecular continuum processes: Application to photoionization of diatomic Li_2^+. *Phys. Rev. A 78*, 023405.

Zhao, Y. and D. G. Truhlar. 2007. Density functionals for noncovalent interaction energies of biological importance. *J. Chem. Theory Comput. 3*, 289–300.

Znojil, M. 1984. Two continued-fractional treatments of multichannel scattering. *Phys. Rev. A 30*, 2080–2081.

3 Measurement of Absolute Cross Sections of Electron Scattering by Isolated Molecules

Michael Allan

CONTENTS

3.1 INTRODUCTION

There is twofold interest in the measurement of the various quantitative cross sections described in this chapter. The first is direct use of the measured quantities to understand and to model systems where electron–molecule (atom) collisions occur, that is, technological plasmas, upper partially ionized layers of planetary atmospheres, comets, but also in flames. The second is providing data for testing various theories. The two areas are related. Experiments cannot measure all the quantities required for the applications, it is very hard to measure, for example, cross sections for reactive intermediates and for electronically excited atoms and molecules and these quantities must be provided by theory. On the other hand, development of a theory is not possible without experimental data for comparison.

The electron scattering research is not new, pioneering studies date from the 1930s and have been revived by the discovery of resonances in the late 1960s (Schulz 1973). But the field goes through a phase of intense activity, caused by novel areas of application (radiation therapy, nanofabrication), improvement of instrumentation permitting observations not available in the past, and by improvement of theory and computational power leading to demand for new experimental data for comparison.

Recently, there has been increasing emphasis on the processes which cause chemical change and thus promise to be useful in plasma chemistry, material science, or radiation therapy. They include the dissociative electron attachment (DEA), electronic excitation followed by dissociation, and dissociative ionization.

The measurements of DEA and vibrational excitation (VE), on which this chapter concentrates, are complementary, and ideally both should be measured for each compound. This is because the important electron-driven processes proceed via resonances (temporary negative ions) and the various processes described here are often competing decay channels of the same resonances and thus a combined knowledge provides more detailed information on them. In particular, bands in VE cross sections provide valuable information on resonances, helpful for understanding DEA.

Resonant electron-induced processes start with a vertical attachment of an electron to a molecule—the anion has initially the geometry of the target. The anion then starts to relax, it distorts as a consequence of modifications of bonding caused by the extra electron. The distortion may be trivial, for example, the C≡C or C=C bond lengthens because the π* orbital into which the extra electron is accommodated is antibonding along the bond, or less trivial, like the symmetry lowering (Renner–Teller effect) related to vibronic coupling (Estrada et al. 1986). In the process of this relaxation, autodetachment proceeds at a rate given by the width Γ of the resonance at each geometry, leading generally to a vibrationally excited state of the neutral molecule. The selectivity into which vibrational mode the resonance decays, and how many quanta are excited, are indicative of what geometry change occurred in the process of the relaxation, as outlined in the pioneering study of Walker et al. (1978). DEA is competing with the VE; it is due to those anions which survived, did not autodetach, and whose geometry was distorted beyond the stabilization point— the crossing of the anion and neutral potential surfaces. The information about the path of the relaxation, derived from the VE, can thus help in understanding DEA.

This chapter starts by presenting some of the experimental techniques used to measure various absolute cross sections, focusing on the instrumentation used in Fribourg. It will then present a selection of examples, highlighting some of the interesting areas, such as shape and core excited resonances and the use of VE, in particular the selectivity of VE, to obtain detailed information on the resonances, threshold phenomena, in particular, vibrational Feshbach resonances (VFR), and the selectivity and the mechanisms of dissociative electron attachment.

3.2 EXPERIMENTAL METHODS

3.2.1 ELECTRON SCATTERING

The basic quantity of electron scattering is the grand total cross section σ_T, which can be derived from the attenuation of an electron beam traversing a chamber of a given length and with a known sample pressure in a transmission experiment. Pioneering studies were performed by Ramsauer and Kollath who reached surprisingly low energies and whose results were proven correct by later experiments. They applied their method to rare gases (Ramsauer and Kollath 1929), where they discovered the Ramsauer–Townsend minimum, and to small molecules (Ramsauer and Kollath 1930). The more modern version of this instrument uses an axial magnetic field and the publications of the Gdańsk group are representative for this type of work (see, e.g., Szmytkowski et al. 1996). The principal problem of this method is the incomplete discrimination against electrons scattered into nearly forward direction. Apart from this, the data are very reliable because of the simplicity of the experiment and are valuable as a verification of the partial and differential cross sections. The integration of differential cross sections over all angles to yield the integral partial cross section, and their sum for all accessible processes yields the grand total cross section—and this must agree with the result of the transmission experiment. This important validation of the differential cross section sets will be exemplified on CO below.

The transmission method has been dramatically improved with the development of the "Surko" trap system for positron scattering (Murphy and Surko 1992, Gilbert et al. 1997, Sullivan et al. 2008) and has become remarkably powerful, both because cross sections can be measured with few incident particles, and because partial cross sections (elastic, vibrationally inelastic) can be measured using a retarding field analyzer. Resolution has been greatly improved by cooling the incident electrons in a trap with a buffer gas cooled by liquid nitrogen. The method is equally applicable to the measurement of electron collision cross section, and could gain more popularity for this application in the future since it is potentially simpler than the double-hemispherical system described below, and has the advantage of measuring integral cross sections which are often sufficient for the application in plasma simulation, bypassing the step of first measuring many differential cross sections and having to integrate them.

The magnetically collimated electron spectrometer (Allan 1982) using "trochoidal" electron energy analyzers (Stamatovic and Schulz 1968) can in some cases also be used to measure absolute inelastic cross sections, when an assumption about the angular distributions can be made and when suitable absolute data for normalization are available; an example is the work on H_2 (Poparić et al. 2010). This instrument has certain advantages—very high sensitivity and low-energy capacity, and the capacity to measure the 0°/180° cross section ratio.

The standard instrument to measure the differential cross sections is a spectrometer with electrostatic analyzers. Although the principle is not new (see, e.g., Pavlovic et al. 1972), it has been substantially improved over the past years, in terms of low-energy capacity, extending the angular capacity to the full angular range 0°–180°, improving the correction for the instrumental response function, and the resolution. This chapter emphasizes on this type of instrument, in particular, the version developed in Fribourg.

3.2.1.1 Double-Hemispherical Instrument

The Fribourg spectrometer with hemispherical analyzers, shown in Figure 3.1, had already been constructed in the late 1980s, but has been continuously improved (Allan 1992, 2005, 2007a, 2010). The energy resolution is typically about 15 meV in the energy-loss mode, corresponding to about 10 meV in the incident electron beam, at a beam current of around 400 pA, although a beam current of 2 nA can be reached at a resolution of 25 meV, and a resolution of 7 meV was reached with a current of 40 pA (Allan 2001). The energy of the incident beam is calibrated on the 19.365 eV (Gopalan et al. 2003) ^2S resonance in helium and is accurate to within ± 10 meV. The instrumental response function was determined on elastic scattering in helium and all spectra were corrected as described earlier (Allan 2005).

Absolute values of the elastic cross sections are determined by the relative flow technique as described by Nickel et al. (1989) using the theoretical helium elastic cross sections of Nesbet (1979) as a reference. The confidence limit is about ±20% for the elastic cross sections and ±25% for the inelastic cross sections (two standard deviations). The sample and helium pressures in the gas inlet line during the absolute measurements were kept low, typically 0.1 and 0.2 mbar, respectively. Background is determined by recording signal with gas flowing into the main chamber via a bypass line and not the nozzle. This background is generally negligible except in the more forward scattering and at low energies. The excitation functions and the angular distributions are measured at higher pressures and background is subtracted only when it is significant. Absolute inelastic cross sections are derived by integrating the areas under the elastic and inelastic peaks in energy-loss spectra recorded at constant

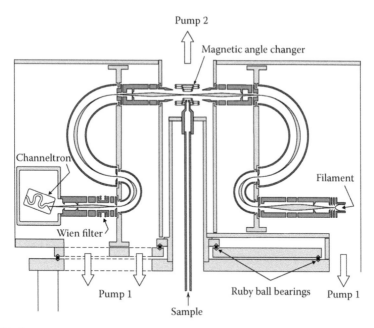

FIGURE 3.1 Schematic diagram of the Fribourg double-hemispherical spectrometer. (Reprinted with permission from Allan, M. 2004. *Phys. Scripta T110*(3), 161–165. Copyright 2004 IOP Publishing Ltd.)

incident energies, corrected for instrumental response functions, and normalizing to the elastic cross section measured by the relative flow method.

The instrument incorporates the magnetic angle changer (MAC) invented by Read and coworkers (Read and Channing 1996, Zubek et al. 1996), which permits measurements of scattering into the backward hemisphere. The particular form of the MAC realized in Fribourg (Allan 2000, 2004) is made of few windings of a thin (0.63 mm diameter) copper tubing, cooled by water. This design minimizes the obstruction of the gas flow, the local pumping speed in the collision region is nearly unaffected by the presence of the MAC. This reduces beam attenuation by background gas, is important for absolute measurements which rely on a definite gas flow, and improves resolution by reducing thermal Doppler broadening encountered in scattering by the background gas. The design further simplifies the power supplies since it has only one current loop for both the inner and the outer solenoids. The same computer controls the digital-to-analog convertors determining the incident and the scattered electron energies (E_i and E_r) and the MAC current, which is automatically adjusted to provide the desired deflection angle every time E_i and/or E_r are changed. The scattering angle was calibrated by guiding the incident beam into a rotatable Faraday cup and is accurate within ±3°.

The angular distributions are measured using combined mechanical setting of the analyzer and magnetic deflection with the magnetic angle changer (Read and Channing 1996, Zubek et al. 2000, Cho et al. 2004), correcting the curves for the instrumental response function (determined on helium and, for angles near 0°, on the $v = 1$ excitation cross section in N_2), and fitting them to the discrete absolute values measured at

$45°$, $90°$, $135°$, and $180°$. Details of the procedure were described by Allan (2005). The magnetic deflection is typically incremented in steps of $1°$ or $2.5°$.

Peak widths often depend on energy and scattering angle. One reason is the Doppler broadening which makes peaks of light targets wider at large scattering angles and higher energies. Another frequent reason is rotational excitation which is, particularly in a resonance region, strong (see, e.g., Jung et al. 1982), and leads to a substantial broadening of the energy-loss bands. Consequently, care has to be taken to derive all cross sections from the areas under the energy-loss bands and not from their heights. For angular distributions, and for the absolute measurements, an energy-loss peak is recorded at each energy and angle, which is then numerically integrated. Two methods are used for the excitation functions. One consists of recording a number of excitation functions at energy losses around the nominal energy loss of the process being recorded (elastic or a given vibrational level), thus covering the entire width of rotational excitations, and then making the sum. The other consists of recording a number of energy-loss spectra in the energy range of interest, then deriving numerically their heights and areas and constructing a "height-to-area correction function," used to correct an excitation function recorded at the top of the energy-loss peak. Both methods gave consistent results.

3.2.2 DISSOCIATIVE ELECTRON ATTACHMENT

3.2.2.1 Total Ion Collection Instruments

The importance of knowing absolute cross section for dissociative electron attachment has been realized early. To measure them, an early generation of total ion current collection tubes has been developed, for example, by Rapp and Briglia (1965) and Azria et al. (1974). The importance of absolute dissociative electron attachment (DEA) cross sections motivated the group in Lincoln (Aflatooni et al. 2006) and in Fribourg (Fedor et al. 2008) to construct modernized versions of this instrument. A slightly simplified version was constructed in Bologna (Modelli 2005). The principal weakness of this instrument is the background of scattered electrons which, at finite pressure, diffuse across the axial magnetic field which is confining them. This background limits the applications to relatively large cross sections. The great advantage of this method is that it is fundamentally an absolute measurement; it does not rely on normalization to other data.

3.2.2.2 Classical Mass Spectrometer

Early DEA spectra were measured at low resolution with modified commercial mass spectrometers (Dorman 1966). These instruments were very sensitive, but did not yield absolute DEA cross-section values. The workhorse of the majority of research groups studying DEA are quadrupole mass spectrometers. The resolution of the electron beam is generally improved by using a trochoidal electron monochromator (Stamatovic and Schulz 1968), sometimes electrostatic analyzers (Vizcaino et al. 2010). These instruments yielded a wealth of most interesting information, recently particularly related to biologically relevant molecules (Ptasinska et al. 2005). The majority of these instruments do not, however, measure the absolute cross sections.

3.2.2.3 Velocity Imaging Spectrometers

An exciting new development in the instrumentation for DEA is the velocity imaging technique. They are based on the COLd Target Recoil Ion Momentum Spectroscopy

(COLTRIMS) technique developed originally for photodissociation experiments and employ a pulsed electron beam to prevent its perturbation by the ion extraction field and to permit time-of-flight (TOF) analysis of the ions. Two versions of this instrument are currently in operation, the velocity slice technique in Mumbai (Nandi et al. 2011) and the COLTRIMS technique in Berkeley (Adaniya et al. 2009).

3.2.2.4 Quantitative Time-of-Flight Spectrometer

This section describes the instrument constructed in Fribourg, combining TOF with the total ion collection technique. More detailed descriptions are given by Fedor et al. (2008) and May et al. (2009). An alternative scheme, a TOF spectrometer measuring absolute DEA cross section by the relative flow technique, is operated in Mumbai (Prabhudesaia et al. 2005).

Schematic diagram of the instrument is shown in Figure 3.2. The instrument uses axial magnetic field to collimate the electrons and a trochoidal electron monochromator to reduce the electron energy spread. The collision chamber is equipped with an exit slit through which ions enter a TOF tube. A short (200 ns) pulse of electrons is sent through the interaction region while the ion repellers are on the potential of the chamber. A 4 µs long pulse with amplitude between −300 and −450 V is applied to the repellers about 200 ns later, after the electrons have left the collision chamber. The experiment is repeated at a rate of between 10 and 100 kHz.

Measurements of absolute cross sections require that the spectrometer is "quantitative," that is, the ion collection efficiency does not depend on their mass and their initial kinetic energy. The dependence on mass is given primarily by the fact that the ions have to traverse the magnetic field required for collimation of the electron beam. The curving of their trajectories is compensated by electrostatic deflectors in the first stage of the ion drift tube. The latter criterion is fulfilled by using a relatively high repeller voltage, accelerating the ions to energies much higher than their initial kinetic energy, and by using a slit 10 mm wide in the direction perpendicular to the direction of the electron beam. The yield of ions is further controlled by the focusing voltage on the center element of the ion drift tube, functioning as a zoom lens.

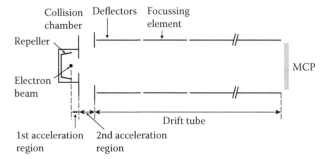

FIGURE 3.2   Schematic diagram of the Fribourg TOF spectrometer for absolute DEA cross sections. The orientation of electron beam is perpendicular to the paper plane. (Reprinted with permission from May, O., J. Fedor, and M. Allan. *Phys. Rev. A 80*, 012706, 2009. Copyright 2009 by the American Physical Society.)

Another potential problem stems from the fact that ions formed in the early stages of the electron beam pulse have time to travel a certain distance before the repeller pulse arrives, reducing their chance of being collected. This problem can be reduced, at the expense of signal intensity, by reducing the electron beam pulse duration. Ion trajectory calculations indicate that the design has a constant collection efficiency up to an initial ion kinetic energy of 2.5 eV, and retains 90% of it up to about 4 eV.

The ions which passed the TOF tube are detected with a microchannel plate (MCP), counted, and their arrival times are analyzed using the delayed coincidence scheme, with a time-to-amplitude converter (TAC) and a pulse-height analysis.

An additional potential problem with respect to the requirement of a "quantitative" mass spectrometer stems from the possible dependence of the detector efficiency on the type of ion. This problem is bypassed by accelerating the ions to very high energies, where the quantum yields for all ions concerned saturate. We have found that the quantum yield of the detector saturates already at 3 kV for all ions investigated (O^-, C_2H^-, C_2^-, H^-).

The pressure in the collision chamber is measured by a capacitance manometer and is kept typically between 0.3×10^{-3} and 0.6×10^{-3} Torr during the measurement. The cross sections were normalized on the 4.4 eV, O^-/CO_2 DEA band. The measurements of this cross section have been reviewed by Orient and Srivastava (1983) and by Itikawa (2002). We take a value of 14 pm^2, which is the average of the values listed by Orient and Srivastava, together with our own value from the total ion collection experiment. The entire TOF scheme was verified by measuring the cross sections for the O^-/N_2O and O^-/O_2 processes with the correct results.

The thermionic cathode warms the target chamber to about 60°C (measured by a Pt100 probe) during operation, making the Knudsen correction (Knudsen 1910) necessary in the total ion collection mode. This correction is redundant in the case of TOF measurements because it applies in the same way to both the measured gas and the calibrating gas CO_2.

3.2.3 NEUTRAL DISSOCIATION

Neutral dissociation, fragmentation which leads only to neutral fragments, is an important process in plasmas and other applications. It has been neglected experimentally because of the difficulty to detect neutral fragments, as opposed to charged fragments, which can be detected with high sensitivity using mass filters and electron multipliers. A more comprehensive account of the techniques, results, and associated problems has been described in a recent review (Moore et al. 2010).

There were few efforts to detect the neutral fragments by ionizing them by electron impact and then using a quadrupole mass spectrometer. The method suffers from the fact that the concentration of radicals to be ionized and detected is typically orders of magnitude less than that of the parent gas and that electron-impact ionization of the radicals yields the same masses as dissociative ionization of the parent. The electron ionizer of the mass spectrometer must therefore operate with low electron energies, just above the ionization energy (IE) of the radical to be detected, but below the IE of the parent. This results in low sensitivity because ionization efficiency is low near the threshold. This scheme has been realized by Sugai and collaborators (Tanaka et al. 1998).

A more powerful and nearly universal technique for the quantitative analysis of radicals has been developed by Motlagh and Moore (1998) at the University of Maryland. The technique, based on the method by which radicals in the gas phase were first identified, relies upon the efficient reaction of radicals with tellurium to yield volatile and stable organotellurides. A beam of electrons passes through a target gas in a collision cell that has a tellurium mirror on its inner surface. Radicals from electron-impact fragmentation react with tellurium within their first few encounters with the wall to produce volatile tellurides. The telluride partial pressure is measured mass spectrometrically and related to the radical production rate. The technique is specific for radicals since a target gas of stable (closed-shell) molecules does not react at the tellurium surface. In addition, the portion of the mass spectrum under observation is displaced by more than 128 amu (the nominal tellurium mass) from the region displaying peaks characteristic of the parent gas—bypassing the problem of having the same masses as fragments of the parent molecule.

This method has been employed to measure *partial* cross sections for the production of neutrals by electron impact on CH_4, CH_3F, CH_2F_2, CHF_3, CF_4, C_2F_6, and C_3F_8 (Motlagh and Moore 1998). The instrument is, however, no longer operational and there is worldwide no laboratory having this important technique.

3.3   NITROGEN

N_2 is perhaps the most studied molecule as far as electron collisions are concerned, both experimentally and theoretically, and is suitable for verifying new methods, both experimental and theoretical. It will therefore be taken as the example on which the experimental procedures will be explained. This chapter reviews the work of the article of Allan (2005), which also lists references to earlier work. Particular mention deserves the paper by Zubek et al. (2000), who used the MAC to extend the elastic and the $v = 0 \rightarrow 1$ differential measurements to 180°.

3.3.1   PROCEDURES OF MEASURING ABSOLUTE CROSS SECTIONS

3.3.1.1   Tuning

It is essential to obtain an optimal beam focus, and an optimal overlap of the three entities incident beam, analyzer acceptance cone (the "scattered beam"), and the gas beam, when the electron energies E_i and E_r and the scattering angle θ are changed. The six voltages which control these parameters, that is the two voltages on the central cylinders of the zoom lenses at the monochromator exit and the analyzer entrance, and the x and y deflector voltages of these two lenses, are empirically optimized for maximum signal at a number of energies and angles (the "pivotal points"), the values are stored, and the instrument then automatically sets the values interpolated between the pivotal points. This procedure is called "tuning" and a "tuned" instrument can be used as a "black box," that is, the incident and residual energies and the scattering angle, may be scanned and the optimal overlap is maintained automatically. The performance needs to be frequently verified on He, and the instrument re-tuned, to offset the surface potential drifts.

There is a subtle point in this procedure: Not only must the incident and scattered beams overlap optimally for all energies and angles, but also the distance of the overlap point from the gas nozzle must remain constant as E_i and E_r are changed. If this is not the case, then the instrument yields optimum signal for elastic scattering, but the signal for inelastic scattering can be reduced dramatically. This desired condition is reached by adjusting the analyzer x and y deflector voltages for low energies (E_r = 0–5 eV) on deeply inelastic scattering (2^3S and 2^3P excitation) from He. While doing so, the incident beam's energy is in the range of 20–25 eV, that is, it is high and changes only by a relatively small percentage—the distance of the incident beam from the nozzle is nearly constant within this range of the incident energies. The analyzer tuned to the incident beam in this way has consequently also a nearly constant distance from the nozzle even when E_r is changed. The monochromator is then tuned in the E_i = 0–5 eV range on elastic scattering, that is, to the scattered beam position, which has been fixed in the previous step. Elastic signal cannot be measured below θ = 10° and inelastic signal (e.g., $v = 0 \rightarrow 1$ in N_2) must be used for tuning at 0°.

3.3.1.2 Response Function for Energy Scans

The response function for elastic signal is relatively easy to determine, by measuring the He elastic signal as a function of energy and dividing it point-by-point by the theoretical cross section given by Nesbet (1979). In practice, a problem arises because thermal Doppler broadening and shift of the elastic peak due to momentum transfer are not negligible in He as analyzed in detail by Read (1975) (see Figure 3.3). The translational excitation is about 9 meV at 20 eV and 135°. The thermal Doppler broadening increases with θ and with the electron energy, but it also depends on θ for apparative reasons. It is smallest at 90° where the incident and scattered electron beams intersect

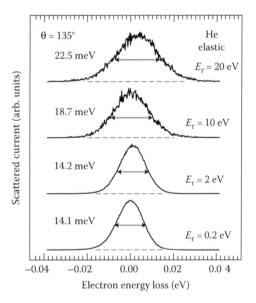

FIGURE 3.3 Energy-loss profiles of the He elastic line, illustrating the Doppler-broadening at higher energies.

in a very small volume (approximately a sphere of about 0.25 mm diameter) in the present instrument, in the front of the nozzle, and the molecules in this effective collision volume move all essentially in one direction. The apparative aspect is worst near $0°$ and $180°$ where the incident and scattered electron beams are nearly collinear, probe the effusive gas beam over a large length, with a substantial range of directions of the thermal velocities of the target gas. As a consequence, the elastic peak becomes broader and shifts to higher energy loss ΔE with increasing scattering angle and increasing incident electron energy E_i. The width of the He elastic peak was 14 meV at 0.4 eV and 22.5 meV at 20 eV at $\theta = 135°$ (Figure 3.3) during the N_2 measurements, indicating a Doppler broadening of 17.6 meV. This is still substantially less than the 45 meV calculated for a stationary sample gas using the expression given by Read (1975). The Doppler broadening is particularly pronounced around $180°$, where the apparative and inherent effects combine and the width of the elastic peak in helium may reach 50 meV. The consequence is that the area under the elastic or inelastic energy-loss peak must be taken in all measurements, not the peak height. In practice, this can be achieved either by recording the peak signal and correcting it by a smooth function expressing the area/height ratio measured at a few discrete energies, or by measuring many excitation functions at energy losses spanning the range of about $(-50\ \text{meV}; +50\ \text{meV})$, that is, covering the entire elastic peak, and then taking the sum. The latter method was used in the illustrative example shown in Figure 3.4.

In theory, the ideal response function should behave as $1/E_r$, because the ideal incident beam is constant and the analyzer acceptance angle should increase with decreasing E_r. The hemispheres operate at a constant pass energy (1.4–5 eV for this instrument, 3 eV for the N_2 measurements in this chapter), that is, the pencil angle, defined by the size of the pupil apertures, is constant between the hemispheres.

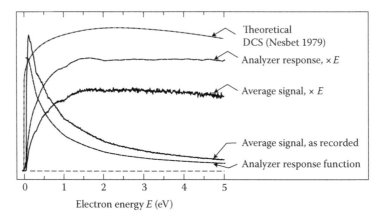

Electron energy E (eV)

FIGURE 3.4   Figure illustrating how the variation of response function with electron energy is derived. Some of the curves are shown multiplied by the electron energy E to improve the visibility of the deviation from the "ideal" $1/E$ behavior. The scattering angle is $\theta = 135°$. The vertical scale is linear and shows the signal intensity, response function, and differential He cross section in arbitrary units. (Reprinted with permission from Allan, M. 2005. *J. Phys. B: At. Mol. Opt. Phys. 38*, 3655–3672. Copyright 2005 IOP Publishing Ltd.)

The analyzer entrance pencil angle then increases as $1/\sqrt{E_r}$, the solid entrance angle as $1/E_r$, with decreasing E_r. This is true provided that the magnification of the analyzer entrance lens does not vary significantly with energy. Trajectory calculations (using the program of Read and Bowring 2005) indicate that this assumption is approximately true. At low energies, the pencil angle given by this relation exceeds the useful physical dimension of the entrance lens (i.e., the filling factor exceeds 50%), and the acceptance angle cannot grow any further. The response then increases slower than $1/E_r$ with decreasing energy, and becomes constant at very low energies. At extremely low energies, the electrons are lost because of stray fields. The incident electron beam current is constant down to low energies, but the pencil angle increases with decreasing energy until it reaches the useful diameter of the monochromator exit lens. Below this energy, the beam current starts to decrease. At low energies, the incident beam may also lose efficiency because it becomes diffuse or distorted by stray fields.

Figure 3.4 illustrates how the response function is derived. The signal generally drops rapidly with increasing energy and some of the curves are shown multiplied by E_r to facilitate visual judgment of the deviation from the ideal $1/E_r$ behavior. The signal integrated over all ΔE in the range (-50 meV; $+50$ meV) to bypass the effect of Doppler broadening is shown both as recorded and multiplied by E_r. The integrated signal divided by the theoretical DCS yields the response function. The curve representing the response function, multiplied by E_r, reveals that the sensitivity of the instrument behaves approximately ideally (as $1/E_r$) at energies $E_r > 1$ eV. At $E_r < 1$ eV the sensitivity still rises with decreasing E_r, but less rapidly than $1/E_r$, following the expectation outlined above. Finally, at $E_r < 100$ meV, the sensitivity drops rapidly, both because the pencil angle of the incident beam exceeds the useful diameter of the monochromator exit lens and because of stray fields. At an energy range of about 50–100 meV, this drop may still be taken into account for correcting the elastic cross sections by making a response function which follows this drop, but the result is less reliable.

The flat He ionization continuum has been used to derive the response function by a number of groups (Pichou et al. 1978, Brunger et al. 1989, Allan 1992). This method is, however, not suitable for routine response function determination, both because the continuum is flat only for θ not far from 90° (Pichou et al. 1978, Asmis and Allan 1997a) and because the continuum is weak and long accumulation is required during which the instrument drifts.

3.3.1.3 Response Function for Angular Scans

Experience shows that the present instrument cannot be optimized over the entire angular range with a single set of tuning voltages because of drifts. This problem is circumvented by repetitive magnetic scanning ±45° around one of the three mechanical analyzer positions $\theta = 45°$, 90°, and 135°. The steps involved in determining the response function's dependence on scattering angle are illustrated in Figure 3.5. A series of short energy-loss spectra around the elastic peak, one at each angle, is recorded. The series is labeled "composite spectrum" in Figure 3.5. The areas under the elastic peaks are then taken and divided by the theoretical DCS to yield the response function. The instrument cannot distinguish between nearly forward-scattered and forward-unscattered electrons, resulting in a large background for elastic spectra at low angle. The lowest attainable angle is about 15° at 1 eV and slightly below 10° at higher energies.

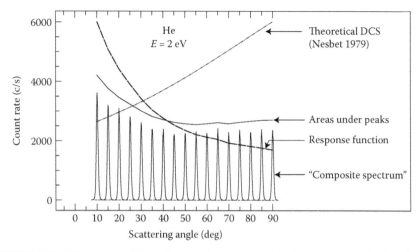

FIGURE 3.5 Illustration of how the angular response function is generated. (Reprinted with permission from Allan, M. 2005. *J. Phys. B: At. Mol. Opt. Phys. 38*, 3655–3672. Copyright 2005 IOP Publishing Ltd.)

The resulting response function has a minimum at 90° and increases below and above as the spatial overlap of the incident and scattered beams increases. It levels off near 0° and 180° because the overlap with the target gas beam becomes limiting. For a well-tuned instrument, the angular response function depends only weakly on energy.

3.3.1.4 Normalization to Absolute Values

Absolute elastic cross sections were determined at a number of discrete energies and angles by comparison with the theoretical helium elastic cross section of Nesbet (1979), using the relative flow method, described in detail by Nickel et al. (1989). They are accurate within about ±15%. The energy and angular scans of the elastic signal, corrected for the response functions, are then normalized to the discrete values, with a high degree of redundancy, that is, more discrete values are measured than required, providing a test of consistency of the shapes of the excitation functions.

Absolute inelastic cross sections are obtained by recording energy-loss spectra with constant incident energies at which the elastic cross section has been measured, correcting them for the instrumental response function and deriving the inelastic cross sections from the ratios of areas under the peaks. An example of an energy-loss spectrum is shown in Figure 3.6—it is useful that the background is low, about five orders of magnitude below the elastic peak.

3.3.2 Representative Results

3.3.2.1 Energy Scans

The two representative energy scans of the elastic cross section shown in Figure 3.7 illustrate how the shape of the resonant contribution varies with angle as a consequence

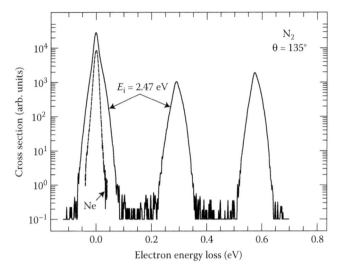

FIGURE 3.6 Electron energy-loss spectrum used to determine the absolute inelastic cross sections. The elastic peak recorded with neon, representative of the apparatus profile, is also shown. (Reprinted with permission from Allan, M. 2005. *J. Phys. B: At. Mol. Opt. Phys. 38*, 3655–3672. Copyright 2005 IOP Publishing Ltd.)

of the interference with the background scattering. As pointed out by Shi et al. (1993) and Sun et al. (1995), this leads to variations of the positions of the resonant peaks with scattering angle and complicates the comparison of angular distributions obtained by various experiments and by theory. The peak positions found here agree well with those of Sun et al. (1995). The peaks are wide and have a relatively flat top, making determination of the peak position better than about ±15 meV difficult even

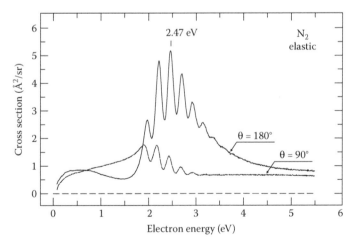

FIGURE 3.7 Rotationally summed elastic cross sections at 90° and 180°. (Data from Allan, M. 2005. *J. Phys. B: At. Mol. Opt. Phys. 38*, 3655–3672.)

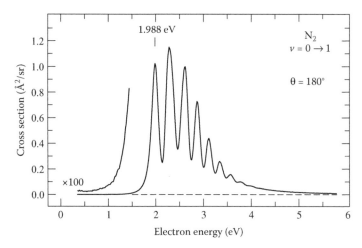

FIGURE 3.8 The rotationally summed $v = 0 \rightarrow 1$ cross section at 180°. (Reprinted with permission from Allan, M. 2005. *J. Phys. B: At. Mol. Opt. Phys. 38*, 3655–3672. Copyright 2005 IOP Publishing Ltd.)

with precise energy-scale calibration. A representative energy scan of the cross section for $v = 1$ excitation is shown in Figure 3.8.

3.3.2.2 Angular Scans

The results for $v = 1$ excitation at the first boomerang peak of the $^2\Pi_g$ resonance are shown in Figure 3.9 as a function of θ. The data are in excellent agreement with the earlier experimental work of Sun et al. (1995) and with the theoretical results of Morrison and coworkers (Feng et al. 2003, Feng, H., W. Sun, and M. A. Morrison (2005) *private communication.*). They use the body-frame vibrational close-coupling theory described by Sun et al. (1995).

3.4 CARBON MONOXIDE

This section is based on the recent study of Allan (2010), which was motivated by the data needed for simulations of the upper atmospheres of Venus and Mars and cometary comae (Campbell and Brunger 2008, 2009a,b). Electron collisions with CO have been studied many times and references to earlier work were given by Allan (2010). This section concentrates on the methods to obtain integral cross sections from the differential cross sections. An important test of the reliability of the relative flow method to measure absolute cross section is obtained by comparing the grand total cross section obtained from the present partial differential cross sections with the grand total cross section measured directly by the transmission method.

Whereas the elastic cross sections of Allan (2010) are in excellent agreement with those of Gibson et al. (1996), the $v = 1$ cross sections, shown in Figure 3.10, are about 20% higher in the resonance region. The difference becomes larger below about 30° at 1.94 eV. The data of Jung et al. (1982) (open triangles in Figure 3.10) agree very well with the present data in shape, but are slightly higher in magnitude.

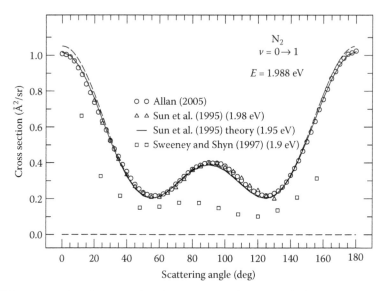

FIGURE 3.9 Rotationally summed $v = 0 \rightarrow 1$ cross section at $E = 1.988$ eV. The results of Sun et al. (1995), Sweeney and Shyn (1997) and the d_π wave distribution (dashed line, normalized to the experiment at 90°) are shown for comparison. (Reprinted with permission from Allan, M. 2005. *J. Phys. B: At. Mol. Opt. Phys. 38*, 3655–3672. Copyright 2005 IOP Publishing Ltd.)

The angular distribution in the resonance region, at 1.94 eV in Figure 3.10, is nearly symmetric around 90°, with the exception of a peak at 0°. It is dramatically different from that of N_2, as shown in Figure 3.9.

Poparić et al. (2004) measured the ratio of the forward and backward cross sections in the resonance region to be 1.00 ± 0.06, in an apparent contradiction to the

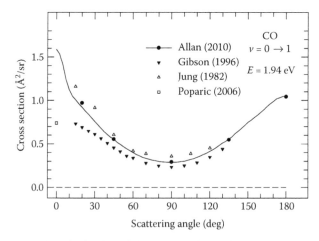

FIGURE 3.10 Angular distribution of electrons having excited the $v = 1$ level of CO at the peak of the $^2\Pi$ resonance. The data of Gibson et al. (1996), Jung et al. (1982), and of Poparić et al. (2006) are compared with that of Allan (2010). (Redrawn from Allan, M. 2010. *Phys. Rev. A 81*, 042706/1–042706/9.)

data in Figure 3.10, which has a forward peak. The two measurements are probably not contradictory; however, the difference could be a consequence of the fact that the magnetically collimated spectrometers such as those used by Poparić et al. have a wide acceptance angle at low energy (Asmis and Allan 1997b) and are thus insensitive to a narrow peak at $\theta = 0°$.

Angular distributions such as those in Figure 3.10 were multiplied by $2\pi\sin\theta$ and integrated to obtain integral cross sections (ICS). Examples of the ICSs obtained in this way are given, as empty circles, in Figure 3.11 for the elastic scattering and in Figure 3.12 for $v = 1$ excitation. The detailed shape of the cross section as a function of energy is then obtained as a weighted sum of several differential cross sections measured at several angles. These sums are shown as solid lines in Figures 3.11 and 3.12. They are in excellent agreement with the integral cross sections of Gibson et al. (1996) for the elastic cross section and are about 20% higher for the $v = 1$ cross section.

In the resonance region, there is an excellent agreement with the calculated cross section of Morgan (1991), both in terms of shape and in terms of absolute value. Below the resonance region, the present cross section is in reasonable agreement with that of Sohn et al. (1985). It is, however, nearly twice as high as the cross section derived from transport phenomena by Hake and Phelps (1967). This discrepancy between swarm and beam data were already noted by Schulz (1973) and Sohn et al. (1985).

Integral cross sections up to $v = 11$ were obtained in a similar way and used to construct the grand total cross section as the sum of the partial integral cross sections, as shown in Figure 3.13. This test is important because the grand total cross

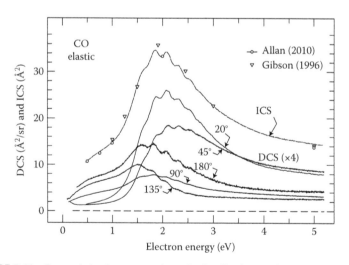

FIGURE 3.11 Integral elastic cross sections obtained by integration under the angular distributions are shown as circles (Allan 2010). The top (dashed) curve shows the shape of the ICS, obtained as a weighted sum of the differential cross sections, shown for comparison below the ICS (4 × vertically expanded). The data of Gibson et al. (1996) are shown as triangles. (Reprinted with permission from Allan, M. *Phys. Rev. A. 81*, 042706/1–042706/9, 2010. Copyright 2010 by the American Physical Society.)

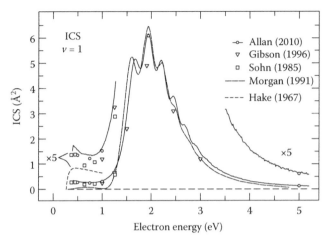

FIGURE 3.12 As Figure 3.11, but for excitation of $v = 1$. Also shown is the data of Gibson et al. (1996), Sohn et al. (1985), Hake and Phelps (1967) (from drift velocity measurements, shown 5 × vertically expanded), and Morgan (1991) (theory). (Reprinted with permission from Allan, M. *Phys. Rev. A. 81*, 042706/1–042706/9, 2010. Copyright 2010 by the American Physical Society.)

sections were measured in transmission-type experiments, which do not have many of the problems encountered in the measurement of partial differential cross sections, in particular the correction for instrumental response functions, or use of the relative flow method. The result is very satisfactory, the present sum agrees very well with the grand total cross sections of Szmytkowski et al. (1996) and of Buckman and Lohmann (1986). The experiment of Kwan et al. (1983), optimized for positron scattering, yielded cross sections which are larger at energies below 2 eV, but the problem is presumably only a small offset of the energy scale, by about 0.15 eV. The accuracy and the low-energy capacity of the 1930 data of Ramsauer and Kollath (1930) are remarkable in view of the simplicity of the equipment available at the time.

3.5 HYDROGEN HALIDES

The two molecules discussed in the preceding sections, N_2 and CO, had π^* resonances coupled to d or p waves, resulting in substantial centrifugal barriers and autodetachment widths Γ smaller but comparable to the vibrational spacing. This leads to the well-known boomerang structure (Birtwistle and Herzenberg 1971, Dubé and Herzenberg 1979, Schulz 1973).

Oscillatory structures were found in a number of resonances where the autodetachment width is much larger than the vibrational spacing, however, and Figure 3.14 shows several examples. The hydrogen halides show a second interesting feature—threshold peaks in the excitation of certain vibrational states. The threshold peaks are absent in H_2 and are related to the dipole moment (and polarizability) of the hydrogen halides.

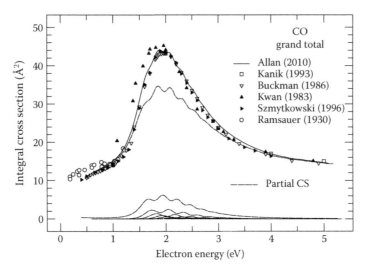

FIGURE 3.13 Grand total cross section. The results of Ramsauer and Kollath (1930), Buckman and Lohman (1986), Kwan et al. (1983), Szmytkowski et al. (1996), as well as the "recommended" data of Kanik et al. (1993) are shown for comparison. The present elastic and vibrational excitation integral cross sections are also shown. (Reprinted with permission from Allan, M. *Phys. Rev. A. 81*, 042706/1-042706/9, 2010. Copyright 2010 by the American Physical Society.)

Both features were studied a number of times. The oscillations in the excitation of high vibrational levels of H_2 were predicted by Mündel et al. (1985) and experimentally confirmed by Allan (1985). The oscillations in the $v = 1$ cross section of HCl were reported experimentally by Cvejanović and Jureta (1989) and predicted independently by Domcke (1989). The experiment was confirmed by Schafer and Allan (1991) and a high-resolution experimental study accompanied by nonlocal resonance model calculations was presented by Allan et al. (2000). The structures in HF were important in

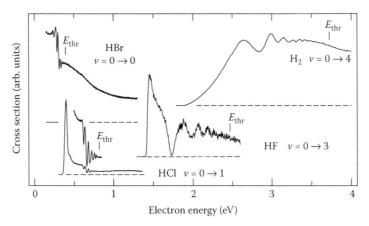

FIGURE 3.14 Selected cross sections of halogen halides and H_2 illustrating the common features, in particular oscillatory structures converging to the DEA threshold (marked as E_{thr}).

that they were the first positive experimental identification of vibrational Feshbach resonances (Knoth et al. 1989), called "nuclear-excited resonances" in older literature. They were studied up to high vibrational levels and with high resolution, accompanied by nonlocal resonance theory, by Sergenton et al. (2000) and Čížek et al. (2003). A high-resolution experimental study and nonlocal model calculations for HBr and DBr were reported by Čížek et al. (2001). The subject of threshold peaks was reviewed by Cvejanović (1993). The discovery of pronounced threshold peaks in the vibrational excitation (VE) cross sections of HF, HCl, and HBr by Rohr and Linder (1975) and Rohr (1978) has initiated intense experimental and theoretical research on low-energy collisions with hydrogen halides. A survey of the experimental developments has been given by Cvejanović (1993). The theoretical developments have been reviewed by Fabrikant (1990), Domcke (1991), and Horáček (2000). Various experimental and theoretical aspects of threshold phenomena were reviewed by Hotop et al. (2003).

The understanding of the observed phenomena, which emerged from the nonlocal resonance theory, can be rationalized using the schematic potential curves shown in Figure 3.15. At large internuclear distances R, the electron is captured in an antibonding valence σ* orbital. This reduces the bond order to 0.5 and the curve's minimum is shallower and at a larger R than that of the neutral molecule. A second potential curve arises in the hydrogen halides (but not H_2) from the binding of the incident electron by the dipole moment of the molecule. This binding decreases with decreasing R as the dipole moment decreases and the dipole-bound potential curve disappears below a critical R. An avoided crossing smoothly joins the valence and the dipole-bound curves to yield the adiabatic potential curve. The dipole-bound branch of the potential curve supports vibrational Feshbach resonances (VFRs) because, classically speaking, as the nuclei oscillate in the presence of the incident electron, they allow the electron cloud to leave at short R, but before it can leave completely, it is partly recaptured when the nuclei

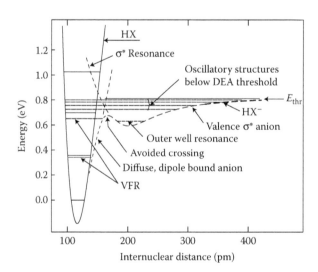

FIGURE 3.15 Generic potential curves illustrating the phenomena encountered in halogen halides. They are based in the theoretical findings of the nonlocal resonance theory, see, for example, Čížek et al. (2002) and references therein, and Figure 5.5 of Chapter 5 of this book.

return to larger R where the dipole moment is larger (Gauyacq 1987). The VFRs are narrower and only slightly below the parent vibrational state for low v, and become broader and further below the parent state for larger v. At higher v is the outgoing nuclear wave packet reflected not by the dipole-bound branch of the potential curve, but by the valence part, at a much larger R. The corresponding resonant states are consequently much closer spaced and a relation to specific parent vibrational levels is no longer meaningful—boomerang oscillations converging to the DEA threshold are observed.

The influence of a VFR, situated below its parent vibrational level, extends even above the parent level and increases the cross section for its excitation—it causes a threshold peak. Depending on the position of the avoided crossing and the minimum of the valence curve for a given molecule, the adiabatic potential curve may or may not have outer well. No outer well was found for HF, but it was found for HCl, where it gives rise to narrow "outer well resonances" (Allan et al. 2000). A (virtual state) cusp in the cross section is observed in the cross sections when the dipole (and polarizability) binding is not strong enough to support a VFR. Fabrikant (Hotop et al. 2003) nicely demonstrated how the appearance of the cross sections depends on the molecular properties, on a model case with adjustable dipole moment and polarizability—going from a virtual-state cusp to a sharp VFR as the polarizability increases.

The various resulting experimental features, and in particular the coupling between the various channels, are illustrated in Figure 3.16 on the example of HBr.

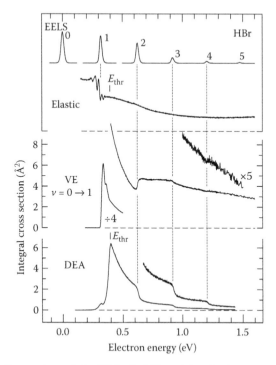

FIGURE 3.16 Comparison of the DEA, the VE, and the elastic cross sections in HBr. An energy-loss spectrum (recorded at $\theta = 90°$ and $E_r = 0.5$ eV) is shown on the top to indicate the vibrational thresholds. (Data from Čížek, M. et al. 2001. *Phys. Rev. A 63*, 062710.)

The energy-loss spectrum shown on the top visualizes the vibrational thresholds. The elastic cross section has a downward step (a cusp) at the $v = 1$ energy and, superimposed on it, boomerang oscillatory structure converging to the DEA threshold (labeled E_{thr}). (The downward step and the boomerang oscillations become separated in energy, and thus visible as two different entities, in DBr (Čížek et al. 2001)). The $v = 0 \rightarrow 1$ cross section has a threshold peak, with barely visible oscillations superimposed on it (their visibility is better in DBr), and then step-like structures at higher vibrational thresholds, alternatively up and down. Finally, the DEA cross section has a peak at threshold and step-wise drops at vibrational thresholds, that is, when new VE channels open. These steps were first reported in HCl by Abouaf and Teillet-Billy (1977) (see also Teillet-Billy and Gauyacq 1984) and in HBr by Abouaf and Teillet-Billy (1980).

The capacity of the nonlocal theory to reproduce the details of the shape of the DEA cross sections is illustrated in Figure 3.17 (Čížek et al. 2001). High-resolution absolute DEA spectra were obtained by recording their shapes under high resolution on the instrument with hemispherical analyzers, where electron and ion signals were separated using the built-in Wien filter in front of the detector, and normalizing their magnitudes to absolute values measured later with the total ion collection instrument (Fedor et al. 2008). The agreement is very good, in particular in view of the fact that not only the shapes, but also the absolute magnitudes are compared. A similar agreement of magnitudes was recently obtained also for HCl, after the parameters of the nonlocal resonance model were improved by fitting into recent *ab initio* scattering calculation results (Fedor et al. 2010).

Figure 3.18 compares experiment and theory for VE. The shape of the VE cross section was measured at 90° (Čížek et al. 2001). The absolute values were determined by the relative flow method at several angles, the angular distribution was measured at 2 eV and integrated to yield ICS (the authors unpublished results). Absolute differential VE cross section was also measured at 90° and 1 eV and the average of these two normalizations is shown in Figure 3.18. The agreement in the magnitude and the details of the shape is excellent. Theory predicts very narrow

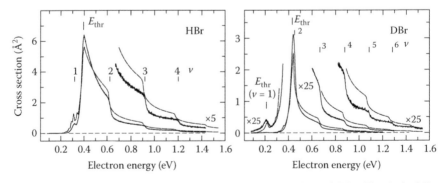

FIGURE 3.17 Comparison of the DEA cross sections for HBr and DBr. The dashed line shows the result of nonlocal resonant theory. (The theoretical data are from Čížek, M. et al. 2001. *Phys. Rev. A 63*, 062710; experimental data from Fedor, J., O. May, and M. Allan 2008. *Phys. Rev. A 78*, 032701.)

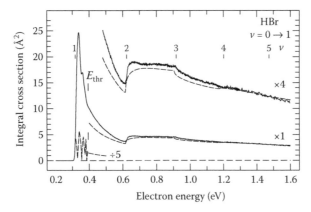

FIGURE 3.18 Integral experimental (solid line) and theoretical (dashed line) cross section for the excitation of $v = 1$ in HBr. The DEA threshold E_{thr} and the vibrational energies of neutral HBr are marked. (Data from Čížek, M. et al. 2001. *Phys. Rev. A 63*, 062710, but the experiment was normalized to absolute value in this work.)

oscillatory structure between the VE and the DEA thresholds, for which some evidence is seen in the experiment, as a shoulder. The visibility of this structure is better in DBr where the gap between the VE and the DEA thresholds is larger. Very good agreement of the shapes and structures in the VE and elastic cross sections (CSs) observed experimentally and calculated by the nonlocal resonance theory was obtained also for HCl (Allan et al. 2000, Čížek et al. 2002), although absolute values of the CSs are not yet available.

From the above discussion, it follows that no threshold peaks and oscillations should occur above the DEA threshold and, in line with this expectation, no threshold peaks and oscillations were found in HI (Sergenton and Allan 2000), and in HCl for excitation of vibrational levels above the DEA threshold, $v \geq 3$ (Čížek et al. 2002).

3.6 METHYL HALIDES

Carbon has an electronegativity similar to that of hydrogen and one may thus expect a similarity of behavior when the hydrogen in hydrogen halides is replaced by a methyl. This class of compounds has been extensively studied—see for example, Hotop et al. (2003), and only one example will be shown here.

3.6.1 CH₃I

Figure 3.19 compares the VE cross sections (both experimental and theoretical) for hydrogen iodide and methyl iodide. DEA is exothermic for both HI and CH_3I and this represents a major difference compared to the preceding cases of HCl and HBr. The potential curves of CH_3I were given by Hotop et al. (2003).

The shapes of the experimental curves are well reproduced by the nonlocal resonance theory (HI, Sergenton and Allan 2000) and the effective range R-matrix

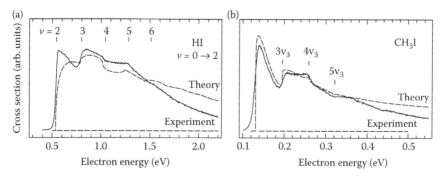

FIGURE 3.19 Cross sections for the excitation of two quanta of the H–I resp. H_3C–I vibrations in HI (a) and CH_3I (b). Theory for HI is from Horáček et al. (1997). ((a) Reprinted from *Chem. Phys. Lett. 319*(1–2), Sergenton, A.-C. and Allan, M., Excitation of vibrational levels of HI up to $v = 8$ by electron impact. 179–183, Copyright (2000), with permission from Elsevier. (b) Reprinted with permission from Allan, and M. Fabrikant, I. I. 2002. *J. Phys. B: At. Mol. Opt. Phys. 35*, 1025. Copyright 2002 IOP Publishing Ltd.)

theory (CH_3I, Allan and Fabrikant 2002). The essential physics is clearly seen to be the same for both molecules, bearing out their chemical similarities. Neither curve has a clear threshold peak, because the entire spectra are above the DEA threshold. There are clear similarities between HI and CH_3I, and the shape of the part of the HBr cross section which lies above the DEA threshold (Figure 3.18), and with the HCl and HBr cross sections for exciting higher vibrational levels, those which lie above the DEA threshold. All these curves are characterized by steps at vibrational thresholds, alternately up and down (with the up and down behavior alternating also when the final vibrational quantum is incremented). The DEA cross sections for CH_3I were measured under ultrahigh resolution and calculated by Schramm et al. (1999) (see also the review by Hotop et al. 2003). The DEA cross sections for CH_3Cl and CH_3Br were calculated by Wilde et al. (2000) and Gallup and Fabrikant (2007). These studies revealed further resemblance with HCl and HBr—the cross sections are characterized by the downward steps at vibrational thresholds caused by interchannel coupling. All these results confirm the prototype role played by the hydrogen halides.

3.7 VIBRATIONAL FESHBACH RESONANCES WITHOUT A PERMANENT DIPOLE: THE CASE OF CO_2

Carbon dioxide has been studied many times and only representative references can be given here. CO_2 has no permanent dipole moment and one would therefore expect that it, unlike the hydrogen halides, does not have VFRs. It has long been known that intense VE and DEA are caused by a $^2\Pi_g$ shape resonance centered at 3.8 eV (Boness and Schulz 1974). Large cross sections were found for certain vibrations even near threshold, and this cross section enhancement was ascribed to a virtual state (Morgan 1998, Estrada and Domcke 1985). The threshold peak did not have any structure, however, consistent with the notion that a virtual state does not involve any delay in the motion of electron. High-resolution measurement revealed that the virtual state excites only the upper member of the $\{(1, 0, 0), (0, 2, 0)\}$ Fermi dyad (Allan 2001).

A subsequent study with higher sensitivity such as that shown in Figure 3.20 revealed that only the topmost member of each Fermi polyad is excited by the virtual state at threshold, and narrow structure appeared on the threshold peaks. Similar to the case of H_2 and many other VE cross sections, the boomerang and other structures appear more pronounced in the cross sections for higher-lying vibrational states. The threshold regions of the cross sections for three higher-lying Fermi-coupled states are shown in Figure 3.21. Many qualitative aspects of the structures are strikingly similar to those found in HF, shown for comparison in Figure 3.21b. HF is the textbook case for VFRs, and the structures in CO_2 can thus also be assigned to VFRs. In both cases, the structures are shallower and narrower in the cross sections for the lower-lying final vibrational states, and become deeper and wider for the higher-lying final vibrational states. For each final vibrational state, the structures are narrower at lower energy and become wider with rising energy. There is an alternation in how the resonant contribution interferes with the nonresonant background—if a given VFR appears as a peak in the cross section for one final vibrational level, it will appear as a dip in final vibrational levels higher or lower by one. This relation is indicated by vertical dashed lines in the figure. The threshold structures in CO_2 were qualitatively reproduced by Vanroose et al. (2004).

Qualitative potential curves in Figure 3.22 rationalize and provide a physical picture for the observations. The curves for HF, adopted from the work of Čížek et al. (2003), are similar to the curves in Figure 3.15, except that the outer well is not developed in HF. At low energies, the nuclear wave packet is reflected on the dipole-bound outer wall of the potential, giving rise to the VFRs, which are deeper and below the respective parent vibrational levels with rising v, as the potential well

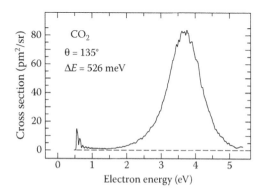

FIGURE 3.20 Cross section for exciting the topmost member of the Fermi polyad $\{(3, 0, 0), (2, 2, 0), \text{etc.}\}$, recorded at $\theta = 135°$ (Adapted from Allan, M. 2002. *J. Phys. B: At Mol. Opt. Phys. 35*, L387.).

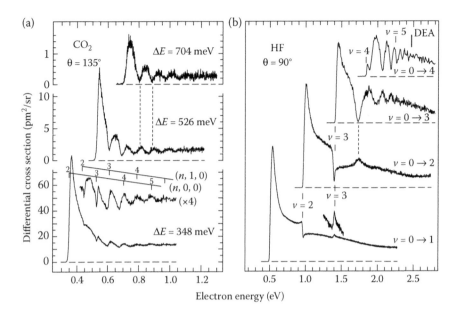

FIGURE 3.21 Comparison of the VE cross sections of HF (b) and, in each case, of the topmost member of the Fermi polyad involving the bending and symmetric stretch vibrations of CO_2 (a). (The label $(n, 0, 0)$ is meant to indicate, for example, for $n = 2$, the topmost member of the Fermi polyad $\{(2, 0, 0), (1, 2, 0), (0, 4, 0)\}$.) ((a) Reprinted with permission from Allan, M. 2002. *J. Phys. B: At. Mol. Opt. Phys. 35*, L387. Copyright 2002 IOP Publishing Ltd. (b) Reprinted with permission from Čížek, M. et al. 2003. *J. Phys. B: At. Mol. Opt. Phys. 36*, 2837–2849. Copyright 2003 IOP Publishing Ltd.)

widens. At higher energies, the wave packet is reflected on the valence-type outer wall, has to travel a long distance, the spacings of the oscillations become much narrower than the vibrational spacing of the neutral HF, and the concept of the parent vibrational state is no longer meaningful. The situation in CO_2 is similar, but with

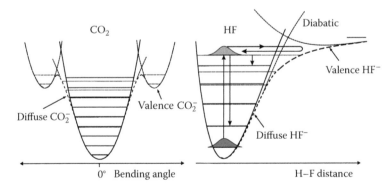

FIGURE 3.22 Qualitative potential curves for CO_2 and HF. The HF potential curves are based on those calculated for the nonlocal resonance model of Čížek et al. (2003), the CO_2 potential curves are hypothetical. (Figure on the left is reprinted with permission from Allan, M. 2002. *J. Phys. B: At. Mol. Opt. Phys. 35*, L387. Copyright 2002 IOP Publishing Ltd.)

respect to the bending motion. The potential of (the lower branch of) the $^2\Pi_u$ shape resonance responsible for the 3.7 eV peak in Figure 3.20 is known to descend, cross the potential of neutral CO_2, and be responsible for the "outer well," supporting a metastable CO_2^-. This potential bends down before crossing the neutral, because of dipole and polarizability binding, similar to the case of hydrogen halides. This view was supported by the high-level electronic structure calculations of Sommerfeld (2003) and Sommerfeld et al. (2003), who have shown that the potential energy surface of the anion bends down before reaching the potential energy surface of the neutral CO_2.

Very sharp VFRs were observed in the ultrahigh-resolution experiments on CO_2 clusters by Barsotti et al. (2002) (see also Denifl et al. 2010). These VFRs were identified also in thin films, where they make a substantial contribution to electron trapping (Michaud et al. 2007).

3.8 VIBRATIONAL FESHBACH RESONANCES AS DOORWAY STATES TO DEA: THE CASE OF N_2O

Early studies of N_2O (e.g., that of Azria et al. 1975) revealed a structureless $^2\Pi$ shape resonance at 2.3 eV, which has also been studied theoretically, for example, by Dubé and Herzenberg (1975) and Bettega et al. (2006). This resonance leads to O^- formation at 2.3 eV, with a shoulder at low energies and the cross section was measured by Rapp and Briglia (1965), Chaney and Christophorou (1969), Krishnakumar and Srivastava (1990), and by May et al. (2008) with a good mutual agreement. Chantry (1969) and Chaney and Christophorou (1969) reported that the shoulder increases dramatically with temperature. Weber et al. (1999) discovered very narrow VFRs in N_2O clusters using the laser photoelectron attachment method, later studied over a wider energy range by Vizcaino et al. (2010). The low-energy processes described here may also play a role in electron trapping by thin films of N_2O, found to occur at energies down to 0 eV (Michaud et al. 1997).

A study with higher resolution and sensitivity revealed threshold peaks in the VE cross sections, and very narrow structures were superimposed on these threshold peaks in the cross sections for excitation of higher vibrational levels (Allan and Skalický 2003). As an example, the excitation of two and four quanta of the N–O stretch vibration v_1 is shown in the bottom part of Figure 3.23. The structures were interpreted to have the same origin as in CO_2 (Figure 3.21), to be due to VFR supported by a branch of the potential surface which bends down before crossing the potential of neutral N_2O because of dipole and polarizability binding of the incoming electron. The peaks form two progressions in the bending vibration v_2, ($0n0$) and ($1n0$). The peaks appear about 5 meV below the vibrational thresholds, but this is within the error limit of the energy-scale calibration, so that, within the error limit, the structures appear at the energies of the parent vibrational states. This is in contrast to the prototype case of VFR, HF, shown in Figure 3.21, where the higher VFRs are clearly seen to be below the parent vibrational levels.

A consequence of the weak N–O bond is that the threshold for DEA is only $E_{thr} = 0.21$ eV. Since Figure 3.21 shows that VFRs are supported by the bending

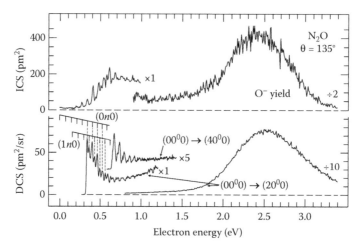

FIGURE 3.23 High-resolution DEA (top) and vibrational excitation curves for N_2O. (Data from Allan, M. and T. Skalický. 2003. *J. Phys. B 36*, 3397.)

motion up to an energy of about 0.9 eV, much higher than E_{thr}, N_2O is a suitable case where one can study whether VFRs lead to dissociation, that is, whether they act as doorway states to DEA. The O^- yield was therefore measured under high resolution with the same instrument as the VE cross sections, the result was normalized to the absolute value of May et al. (2008), and is shown in the upper part of Figure 3.23. Sharp structures are observed at the low-energy end of the spectrum, corresponding to the same progressions as those in the (20°) excitation. These peaks represent a direct indication of the VFR acting as doorway states. The positions of the peaks are about 10 meV under the peaks in the VE cross sections, but this is within the error limit of the energy-scale calibration.

N_2O is a suitable prototype case for the relaxation of VFRs into the dissociation channel, because the small size of the molecule makes the vibrational structure simple and well resolved, and because the VFR are observed in the two competing channels, VE and DEA. VFRs were discussed as intermediates in DEA in several cases, in particular for biomolecules (Abouaf and Dunet 2005, Sommerfeld 2007, 2008, Burrow et al. 2006, Scheer et al. 2005, Fabrikant 2009, Gallup and Fabrikant 2011). It is remarkable that VFRs are dominant intermediates in positron annihilation (Gribakin et al. 2010).

3.9 π* SHAPE RESONANCES AS DOORWAY STATES TO DEA

3.9.1 CHLOROBENZENE

DEA to chlorobenzene (referred to as PhCl) has been studied many times. In this section, we report on the insight gained by the comparison of VE cross sections (in particular their selectivity) with a high-resolution DEA spectrum, as presented by Skalický et al. (2002) (who also listed references to earlier DEA work).

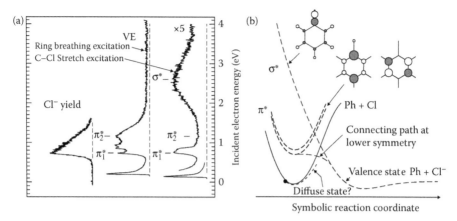

FIGURE 3.24 (a) Cross sections for VE and DEA in chlorobenzene. (b) Schematic potential curves and temporarily occupied orbitals. (Data from Skalický, T. et al. 2002. *Phys. Chem. Chem. Phys. 4*, 3583.)

Excitation of the ring breathing mode, shown in Figure 3.24a, reveals a broad-structured band in the 0.7–1.5 eV range which was assigned to two π* shape resonances with temporary occupation of the orbitals shown in Figure 3.24b. The narrow structure on the band indicates an autodetachment width of below 30 meV. The large width of the entire band must thus be due to Franck–Condon width and the fact that the Cl substituent lifted the degeneracy of the two π* orbitals of benzene, as indicated by the labels π_1^* and π_1^*. The theoretical treatment of Skalický et al. (2002), using the very simple method of estimating energies of shape resonances by empirical scaling of self-consistent field (SCF) virtual energies, revealed dramatic vibronic coupling in the chlorobenzene anion so that the π_1^* and π_2^* resonances are highly mixed, with very complex potential surfaces.

Excitation of the C–Cl stretch mode reveals an additional broad unstructured band at 2.6 eV. The selectivity of the VE indicates that it is a σ* resonance with temporary occupation of the σ* orbital shown in Figure 3.24b (note that the π* orbitals have a node in the plane of the paper, σ* orbital does not). This assignment agrees with that of Stricklett et al. (1986), based on electron transmission spectroscopy (ETS).

Excitation of both vibrational modes has an intense threshold peak, but its origin is not certain. It could be a threshold peak due to dipole and polarizability binding, analogous to the threshold peaks in HBr discussed in the preceding section. But it could simply be due to direct excitation of IR-active vibrations. This peak does not have any structure resembling that of the hydrogen halides or CH_3I.

The DEA spectrum was recorded with high resolution (about 10 meV) with the same instrument with hemispherical analyzers as the VE cross sections, using the Wien filter to separate ion and electron signals. An important observation is that the onset of the Cl^- signal is nearly vertical, about as steep as the VE cross section, and peaks at the same energy as the sharp $v = 0$ level of the π* shape resonance, 0.73 eV. This indicates that the π* resonance acts as a doorway state for DEA and dissociates without activation energy. The Cl^- formation can certainly not be a consequence of direct vertical electron capture into the σ^*_{C-Cl} resonance, which peaks at much higher energy. In fact, direct vertical electron capture into the σ^*_{C-Cl} orbital does not yield any Cl⁻ signal—the autodetachment width is too large. The Cl⁻ formation can also not follow the mechanism for which the hydrogen halides are a prototype—this mechanism leads to a vertical fragment ion onset at the DEA threshold, and the Cl⁻ signal onset is 0.22 eV above threshold (the threshold energy is 0.5 eV, Bulliard et al. 1994).

This indicates that the presence of a π electron system has a profound influence on DEA by offering a doorway state which dominates the process. The absence of an activation barrier is not trivial because the dissociation in C_{2v} symmetry is forbidden. It is due to symmetry-lowering, a consequence of vibronic coupling. The work of Skalický et al. (2002) indicates that the symmetry-lowering involves the expected out-of-plane bending of the C–Cl bond, but also a less expected out-of-plane bending of the hydrogen atoms.

3.9.2 ACETYLENE

Another example where a π* shape resonance acts as a doorway state is acetylene. A relative DEA spectrum was given, for example, by Dressler and Allan (1987). The absolute cross sections were measured by Azria and Fiquet-Fayard (1972), Abouaf et al. (1981), and May et al. (2008). The isotope effect was found to be large, the DEA cross section of the deuterated compound being 15× smaller, as shown in Figure 3.25 (May et al. 2009). VE was studied by Andrić and Hall (1988).

The theory of Chourou and Orel (2008) and of Chourou and Orel (2009a), which was refined to take into account the finite temperature of the sample, essential even at room temperature (Chourou and Orel 2009c), well reproduced the isotope effect, and the absolute magnitude was reproduced better than within a factor of two—a good agreement in view of the steep dependence on the primary results of the calculation, the resonance energy, and width. An important achievement of the theory (which is validated by the agreement with the experiment) is the insight which it provides into

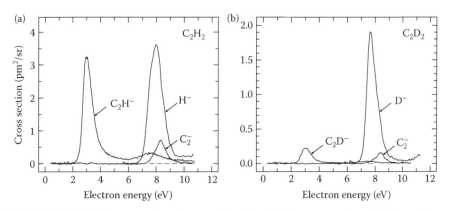

FIGURE 3.25 DEA cross sections for acetylene (a) and dideuteroacetylene (b). (Data from May, O., J. Fedor, and M. Allan. 2009. *Phys. Rev. A 80*, 012706.)

the mechanism of the dissociation. It shows that the nuclear wave packet, to bypass an energy barrier, needs first to move along the bending coordinate before it can dissociate (Chourou and Orel 2008). The fundamental principle is thus the same as in the case of chlorobenzene in the preceding section.

In this respect, it is interesting to observe that the same theoretical treatment, applied to the related case of HCN and DCN, revealed that, although the potential surface has a barrier in the dissociation path which can be bypassed by bending, like in the acetylene case, the system prefers to tunnel through the barrier instead of going around it (Chourou and Orel 2009b). The agreement with experiment (May et al. 2010) was not as good as it was in the acetylene case, but still qualitatively correct.

3.10 MOLECULES CONTAINING THE OH AND SH FUNCTIONAL GROUPS

3.10.1 FORMIC ACID

This section reviews the VE cross sections of the formic acid monomer (Allan 2006) and dimer (Allan 2007b) and discusses their implications on the resonant structure, nuclear dynamics, and DEA.

Targets with low and high concentrations of the dimer were obtained by expanding from either low-pressure (through a 250 μm dia nozzle) or high-pressure (through a 30 μm dia nozzle) gas. Pure dimer and monomer spectra were then obtained by minor spectra subtraction. The validity of the procedure was verified by the energy-loss spectra shown in Figure 3.26. The very-low-frequency in-plane rock vibration identifies the dimer, and its absence indicates a pure monomer.

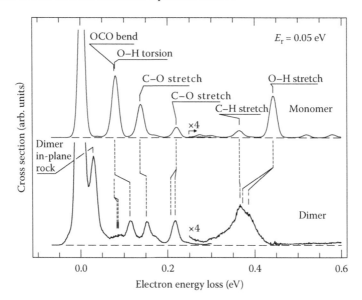

FIGURE 3.26 Electron energy-loss spectra of the monomer and the dimer of formic acid. (Reprinted with permission from Allan, M. *Phys. Rev. Lett.* **98**, 123201, 2007b. Copyright 2007 by the American Physical Society.)

The cross sections for the excitation of the C=O stretch vibrations, shown in Figure 3.27, identify the shape resonances. The 1.9 eV π^* resonance (studied theoretically by Gianturco and Lucchese 2004) of the monomer is split by 0.56 eV in the dimer and its center shifts slightly down. The splitting is quite large in view of the fact that the two C=O bonds are not directly conjugated in the dimer and the interaction happens over a large distance. It is an experimental indication of the large spatial extent of the temporarily occupied orbitals. The cross section is slightly smaller in the dimer, presumably because the temporarily captured electron is distributed over many bonds, and affects each of them less. The two π^* resonances of the dimer were calculated qualitatively correctly by Gianturco et al. (2005), although both the absolute energies and their splitting were larger than the experimental values. The threshold peak of the dimer is smaller than that of the monomer, presumably because of the lack of a permanent dipole moment.

The π^* resonance seen in the C=O and C–H stretch excitation in Figure 3.28 has a boomerang structure associated with the OCO-bending motion. This structure is barely visible in the O–H stretch excitation. The C=O stretch cross section has a high narrow threshold peak, but without structure. The peak is much lower in the C–H stretch cross section, presumably because this bond is much less polar.

An interesting question is whether there is resemblance between the O–H stretch cross sections of formic acid in Figure 3.28 and the VE cross sections of hydrogen halides in Figures 3.14, 3.16, 3.18, and 3.21. This may be expected because the hydrogen atom is bound to an electronegative atom in both cases, and both classes of compounds are acidic. Inspection of the spectra reveals a clear similarity in terms of a

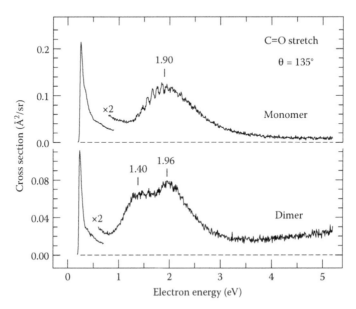

FIGURE 3.27   Cross sections for the excitation of the C=O stretch vibrations for the monomer and the dimer of formic acid. (Reprinted with permission from Allan, M. *Phys. Rev. Lett.* *98*, 123201, 2007b. Copyright 2007 by the American Physical Society.)

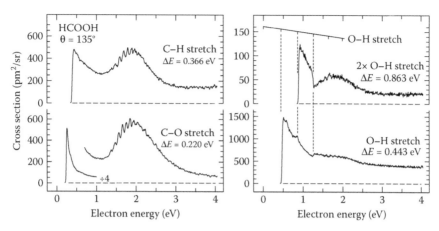

FIGURE 3.28   Cross sections for vibrational excitation of formic acid monomer. (Data from Allan, M. 2006. *J. Phys. B: At. Mol. Opt. Phys.* 39, 2939.)

threshold peak and step-wise drops and rises of the cross section at vibrational thresholds. There is no oscillatory structure converging to the DEA threshold (at 1.37 eV), however.

In this sense, the proposition of Gallup et al. (2009a) (see also Fabrikant 2010) that also DEA follows the same mechanism is justified. Their nonlocal *R*-matrix calculation was very successful in reproducing step-wise drops in the DEA cross section observed in the related molecule of glycine in the high-resolution study of Abouaf (2008), which are interpreted as having the same origin as those shown in Figure 3.17 for HBr. The description of this important feature, due to interchannel coupling, requires a nonlocal theory which is not yet available for more than one dimension, unfortunately. The possible role of the π* resonance, shown by Rescigno et al. (2006, 2007) to spontaneously distort out of planarity to permit dissociation in a way similar to chlorobenzene and acetylene presented earlier, thus needs further investigation. The two approaches led to a lively discussion in the literature (Rescigno et al. 2009, Gallup et al. 2009b).

Step-wise structure was also reported in formic acid by Pelc et al. (2002) albeit not as pronounced as in glycine. Other DEA studies of formic acid were reported by Pelc et al. (2002, 2003, 2005), Martin et al. (2005), and Prabhudesaia et al. (2005).

Intramolecular proton transfer driven by the attachment of an electron into the π* resonance carries its experimental signature as a yield of slow electrons, visible in Figure 3.29 (Allan 2007b). The species resulting from it, an ion–dipole complex of the formate anion and the protonated formic acid radical, and the reaction leading to it were studied theoretically by Bachorz et al. (2005). Proton-transferred complexes of formic acid with various biomolecules have also been generated in a high-voltage discharge ion source, mass selected, and studied by anion photoelectron spectroscopy—see, for example, the study on 1-methylcytosine (Ko et al. 2010).

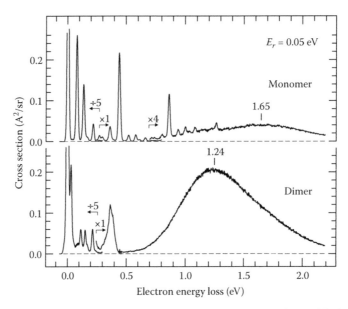

FIGURE 3.29 Energy loss spectra, recorded at a constant and very low residual energy, of formic acid monomer and dimer. The yield of low-energy electrons from the dimer is about 20× larger than that from the monomer (note that the pertinent section of the monomer spectrum is shown multiplied by 4). The slow electrons are interpreted as a consequence of an intradimer proton transfer. (Reprinted with permission from Allan, M. *Phys. Rev. Lett. 98*, 123201, 2007b. Copyright 2007 by the American Physical Society.)

3.10.2 Ethanol and Thioethanol

Alcohols and thioalcohols are chemically related to the hydrogen halides in the sense that hydrogen is bonded to an electronegative atom in both. Alcohols and thioalcohols are also acidic, but in contrast to the hydrogen halides only very weakly so. An interesting question is whether alcohols and thioalcohols show some of the phenomena described earlier for hydrogen halides and formic acid.

Two (relative) VE cross sections were compared with DEA spectra by Ibănescu et al. (2007) and are reproduced in Figure 3.30. The C–H stretch excitation cross section has a shape typical for VE cross sections in saturated hydrocarbons—with a peak around 8 eV and low values below about 4 eV (Allan and Andrić 1996). The O–H stretch excitation cross section on the other hand is very different. It peaks at threshold and then decreases slowly over a large energy interval. This cross section enhancement is too broad to be due to direct dipole excitation—it must be due to a resonant process and could be due to a σ* resonance combined with long-range attraction by the dipole moment (1.68 D). There are no structures on the cross section, however, in contrast to hydrogen halides. The cross section for the excitation of two quanta of the O–H vibration is hard to measure because the cross section drops very rapidly with increasing quantum—an indication of a large autodetachment width of the resonance. The DEA cross section (not shown here) has a nearly vertical

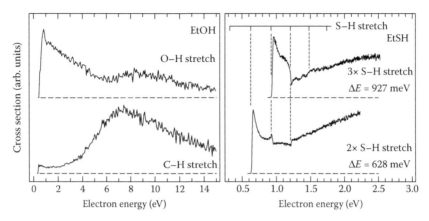

FIGURE 3.30   Cross sections for vibrational excitation of ethanol and thioethanol.

onset at the threshold energy, peaking at 2.75 eV (Ibǎnescu et al. 2007). The large isotope effect (4×) is another indication of a large autodetachment width.

Thioethanol resembles the hydrogen halides more, the excitation of high overtones of the S–H stretch vibration is pronounced and the spectra in Figure 3.30 exhibit step-wise structures reminiscent of those in HBr or HF, taken to be consequence of long-range attraction. The principal difference to ethanol must be the higher polariz-ability, because the dipole moment is nearly the same (1.58 D). DEA, measured by Ibanescu and Allan (2009), yields two fragments at low energies. The $(M–H)^-$ frag-ment has a vertical onset at the threshold energy of 1.8 eV. The $(M–H_2)^-$ fragment has a vertical rise at the same energy of 1.8 eV, but in addition a puzzling structureless band with an onset near 0 eV, peaking at 0.6 eV.

3.11   OTHER POLYATOMIC MOLECULES

3.11.1   NITROMETHANE

Nitromethane is an interesting prototype case. Compton et al. (1996) studied nitro-methane using the anion photoelectron spectra and Rydberg charge exchange. They recognized that Rydberg charge exchange leads to a dipole-bound state with an electron affinity of 12 ± 3 meV, which serves as a doorway to a valence state of the anion with a substantially different structure and an adiabatic electron affinity of 0.26 ± 0.08 eV. The interconversion of the two states became evident in the study of Ar-solvated nitromethane anion of Lecomte et al. (2000). Dessent et al. (2000) observed the VFR in nitromethane in photoabsorption and photofragment action spectra of the $X^- \cdot CH_3NO_2$ ($X^- = I^-$ and Br^-) complexes, with peaks at the onsets of the $-NO_2$ wag-ging, scissoring, and stretch modes. Lunt et al. (2001) reported absolute total integral and total backward scattering cross sections from 30 meV to 1 eV and found them to increase rapidly with decreasing energy, but observed no structures. Absolute DEA cross sections were estimated experimentally by Sailer et al. (2002), who found NO_2^- to be the by far dominant fragment, with a peak cross section of 620 pm^2 at 0.62 eV.

This low-energy band appears to be absent in the ion desorption spectra from thin films (Bazin et al. 2009, 2010).

Schematic potential curves of the two states were presented by Compton et al. (1996), and the experimental electron affinities were well reproduced by the high level *ab initio* calculations of Gutsev and Bartlett (1996). Sommerfeld (2002) (see also Sommerfeld 2008) presented high-level *ab initio* calculations of both states and calculated the coupling matrix element between them to be 30 meV.

Schematic potential curves based on the theoretical work are presented in Figure 3.31. Nitromethane has a supercritical dipole moment (3.46 D) and the dipole-bound state is therefore bound even at the equilibrium geometry of the neutral molecule, in contrast to the hydrogen halides shown in Figures 3.15 and 3.22. The valence state of the anion has, relative to the neutral molecule, longer N–O bonds and it is pyramidal around the N atom (Gutsev and Bartlett 1996). That means that the active path from the dipole-bound to the valence states is primarily along the symmetrical N–O stretch and the out-of-plane NO_2-wagging modes.

An elastic and two representative inelastic cross sections, measured at 135°, are shown in Figure 3.32 (author's unpublished). The elastic cross section is very large and rises rapidly at low energies, in agreement with the findings of Lunt et al. (2001). The visibility of the sharp structures is improved in the vertically expanded traces with a smooth background subtracted. The structure is dense and not all features can be uniquely assigned, but sharp dips are evidently observed slightly below the energies of vibrations, which include those identified as important for the path between the dipole-bound and the valence states. (Vibrational frequencies were taken from Deak et al. 1999.) Experimentally the dips are 5 meV below the thresholds, which agrees with the dipole-bound electron affinity of 12 meV in view of the fact that the present energy-scale calibration is reliable within ±20 meV. The structures become broader—boomerang oscillations—at energies around 0.5–0.7 eV. The cross section for exciting the NO_2 out-of-plane wagging vibration has sharp structures below 0.3 eV. The cross section for exciting the NO_2 symmetrical stretch vibration has sharp structures below 0.3 eV and broad boomerang oscillations in the 0.4–0.8 eV range. The sharp structures below 0.3 eV in all three cross sections are assigned to excited vibrational states of the

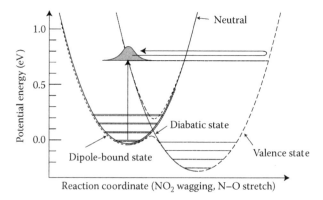

FIGURE 3.31   Schematic potential curves for nitromethane and its anion. (Based on Sommerfeld, T. 2002. *Phys. Chem. Chem. Phys. 4*, 2511–2516.)

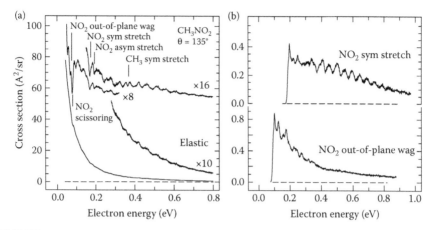

FIGURE 3.32 (a) Elastic DCS of nitromethane at 135°. The lower two curves are shown as recorded. Smoothly rising background was subtracted from the top two curves to enhance the visibility of the sharp structures. Energies of selected vibrations of neutral nitromethane are marked. (b) Cross sections for exciting the two vibrations indicated.

dipole-bound anion (i.e., to VFRs). Some of these states appear broadened, possibly by rapid interconversion into the valence state of the anion. The broad structures in the 0.4–0.8 eV range are assigned to a vertical attachment to the valence state, and boomerang oscillations on the (diabatic) potential of this state, as indicated by the horizontal returning arrow in Figure 3.31. Motion on a diabatic potential is appropriate in view of, and supportive of, the theoretical conclusion of Sommerfeld (2002) that the nitromethane anion is clearly in the weak-coupling regime, and any description of transitions between the dipole-bound and the valence state should be based on the diabatic picture.

It would be desirable to measure a high resolution study of DEA which would reveal whether the VFR seen in the elastic and the VE cross sections also appear in the DEA spectrum.

3.11.2 Pyrrole

Pyrrole may serve as a simple prototype for the nucleobases and porphyrin, containing a hydrogen atom linked to a nitrogen atom in planar configuration in an unsaturated ring.

The cross section for excitation of the C=C and C–N stretch vibrations, shown in Figure 3.33, reveals the two π^* shape resonances due to the temporary occupation of orbitals consisting essentially of the in-phase and out-of-phase combinations of the C=C π^* bond orbitals. The expected selectivity of VE indicates that the two bands at 2.52 and 3.6 eV in the C=C and C–N stretch excitation are π^* resonances, and the 2.5 eV band in the N–H stretch excitation could be an N–H σ^* resonance predicted at about this energy by the scaled Koopmans model. The latter band could also be due to the π_1^* resonance—it has the same energy—although the hydrogen atom is in the nodal plane of the π_1^* orbital and would not be expected to be strongly effected by its occupation. Interesting is the cusp in the N–H stretch excitation cross

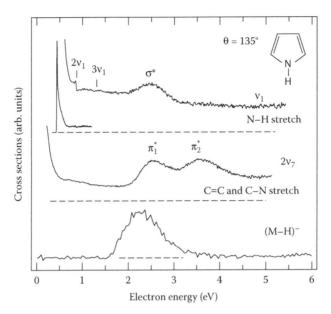

FIGURE 3.33 Cross sections for the excitation of the N–H stretch and ring breathing (C=C and C–N stretch) vibrations, and DEA (loss of H-atom).

section, at the energy of the $2v_1$ vibration, suggesting threshold phenomena related to those of the hydrogen halides. Note that this cusp is related to the N–H bond; it is absent in the C=C and C–N stretch cross section. DEA signal sets in at the threshold at 1.61 eV (derived from the gas phase acidity) and shows no structure in the present low-resolution spectrum (Skalický and Allan 2004).

3.12 SUMMARY AND CONCLUSIONS

This chapter describes instrumental progress in the measurement of absolute cross sections, primarily for elastic scattering, vibrational excitation, and dissociative electron attachment. The progress permitted new insight, particularly into threshold structures which require that the instrument performs well at low energies, that it is sensitive because the threshold structures often appear only in the excitation of high vibrational levels, and that it has a resolution of around 10 meV.

Several compounds are then presented which may, in some respects, serve as prototypes for various phenomena. N_2 and CO are the prototypes for "classical" shape resonances. They also serve as a validation of the experimental procedures—important is in particular the agreement of the "grand total" cross section obtained "the hard way," by integrating all differential cross sections and summing all partial cross sections, with the grand total cross section from the transmission experiment.

Hydrogen halides are important because they exhibit a number of phenomena which are challenging to describe theoretically—they require nonlocal resonance methods. Measurements of the cross sections provide data for testing these theories. Hydrogen halides also serve as prototypes for the threshold phenomena because they exhibit many of them with great clarity and simplicity inherent to diatomic molecules: Threshold peaks, vibrational Feshbach resonances, cusps at vibrational thresholds, boomerang structures converging to the DEA threshold, outer-well resonances, downward steps in DEA cross sections due to interchannel coupling, dipole and polarizability binding of an electron, and nonadiabatic effects. These concepts are applicable also to much larger molecules, in particular biomolecules, as has been shown both experimentally and theoretically (Scheer et al. 2004, 2005, 2007, Abouaf and Dunet 2005, Burrow et al. 2006, Gallup et al. 2009a, Fabrikant 2009, 2010, Gallup and Fabrikant 2011, Sommerfeld 2007).

Another prototype case is chlorobenzene, where the experiment shows that vertical attachment to a σ^* orbital does not lead to DEA, but that the π^* resonances, with their narrow autodetachment width, act as doorway states, whereby a distortion of the geometry bypasses the energetic barrier. Although threshold peaks are found in the VE cross sections and indicate that dipole-binding similar to the hydrogen halides case may occur, it does not appear to be important for DEA.

A π^* resonance is the doorway state also in acetylene, and quantitative calculations of the nuclear dynamics on *ab initio* resonant potential energy surface yielded results in good agreement with experiment, both in terms of the absolute value of the cross section and of the isotope effect. A major progress of these calculations is that they were carried out in several dimensions, and this is essential when symmetry is lowered during dissociation. The theory provided valuable insight into the time-dependent motion of the nuclear wave packet.

In systems with both a narrow π^* resonance and long-range binding the mutual interplay of the two mechanisms is not *a priori* clear and an extension of the nonlocal methods into several dimensions would be desirable in this respect. In pyrrole, as an example, a sharp cusp at the energy of the N–H stretch excitation cross section indicates phenomena related to those in hydrogen halides, but DEA could also occur from the π^* resonance, enabled by geometry distortion similarly to the chlorobenzene and acetylene cases. Formic acid is another molecule with both cusps in VE cross sections and a π^* resonance with boomerang structure, where the DEA mechanism is not *a priori* clear.

The long-range binding of the electron in nitromethane is stronger than in the above cases. The existence of both a dipole bound and valence states of the anion were established based on experiments on Rydberg electron transfer and anion photoelectron spectroscopy, and the ground vibrational levels of both states are not resonances but are bound. Higher vibrational levels of both states appear as narrow resonances and as boomerang structure in the elastic and VE cross sections. Their role in DEA is not clear and high-resolution DEA spectra would be interesting.

An important prototype is also CO_2, showing that many of the threshold phenomena are possible even without a permanent dipole moment.

N_2O could also be called a prototype because it explicitly shows VFRs acting as doorway states for DEA. This phenomenon cannot occur in the hydrogen halides because (in absence of energy barriers) in the diatomic molecules the VFRs converge to the DEA threshold, that is, they are energetically below it. (VFRs can dissociate if they are behind a barrier and thus above the dissociation limit, such as in methyl iodide (Schramm et al. 1999). In N_2O, the VFRs are sustained by the bending motion, up to energies which are above the dissociation threshold, since the dissociation path starts in an orthogonal direction, along the N–O stretch. An interesting problem to be solved by theory is to clarify the nature of the coupling between the initial bending mode involved in the VFR and the final N–O stretch leading to dissociation.

3.13   APPENDIX: ABBREVIATIONS

CS	Cross section
DCS	Differential cross section
DEA	Dissociative electron attachment
ETS	Electron transmission spectroscopy
eV	Electron volt
HOMO	Highest (fully) occupied molecular orbital
ICS	Integral cross section
IR	Infrared
LUMO	Lowest (fully) occupied molecular orbital
MAC	Magnetic angle changer
PES	Potential energy surface
SCF	Self-consistent field
σ_T	Grand total cross section
TOF	Time-of-flight
VE	Vibrational excitation
VFR	Vibrational Feshbach Resonance

ACKNOWLEDGMENT

This research was supported by project No. 200020-131962/1 of the Swiss National Science Foundation, by project SBF No. C07.0018 of the State Secretariat for Education and Research and by COST Action CM0601.

REFERENCES

Abouaf, R. 2008. Low energy electron impact in gas phase glycine, alanine and propanoic acid: Electronic, vibrational excitations and negative ions. *Chem. Phys. Lett.* 451(1–3), 25–30.

Abouaf, R., L. Andrić, R. Azria, and M. Tronc. 1981. Dissociative attachment in C_2N_2 and C_2H_2. In S. Datz (Ed.), *Proc. 12th Int. Conf. on the Physics of Electronic and Atomic Collisions*, North-Holland: Amsterdam.

Abouaf, R. and H. Dunet. 2005. Structures in dissociative electron attachment cross-sections in thymine, uracil and halouracils. *Eur. Phys. J. D 35*, 405–410.

Abouaf, R. and D. Teillet-Billy. 1977. Fine structure in dissociative-attachment cross sections for hydrogen chloride and deuterium chloride. *J. Phys. B: At. Mol. Phys. 10*, 2261–2268.

Abouaf, R. and D. Teillet-Billy. 1980. Fine structure in the dissociative attachment cross sections for HBr and HF. *Chem. Phys. Lett.* 73(1), 106–109.

Adaniya, H., B. Rudek, T. Osipov, D. J. Haxton, T. Weber, T. N. Rescigno, C. W. McCurdy, and A. Belkacem. 2009. Imaging the molecular dynamics of dissociative electron attachment to water. *Phys. Rev. Lett. 103*(23), 233201.

Aflatooni, K., A. M. Scheer, and P. D. Burrow. 2006. Total dissociative electron attachment cross sections for molecular constituents of DNA. *J. Chem. Phys.* *125*, 054301.

Allan, M. 1982. Forward electron scattering in benzene; forbidden transitions and excitation functions. *Helv. Chim. Acta 65*, 2008–2023.

Allan, M. 1985. Experimental observation of structures in the energy dependence of vibrational excitation in H_2 by electron impact in the $^2\Sigma_u^+$ resonance region. *J. Phys. B 18*(13), L451.

Allan, M. 1992. Measurement of differential cross sections for excitation of helium by electron impact within the first 4 eV above threshold. *J. Phys. B: At. Mol. Opt. Phys. 25*, 1559.

Allan, M. 2000. Excitation of the 2^3S state of helium by electron impact from threshold to 24 eV: Measurements with the magnetic angle changer. *J. Phys. B: At. Mol. Opt. Phys. 33*, L215.

Allan, M. 2001. Selectivity in the excitation of Fermi-coupled vibrations in CO_2 by impact of slow electrons. *Phys. Rev. Lett. 87*, 033201/1–033201/4.

Allan, M. 2002. Vibrational structures in electron–CO_2 scattering below the $^2\Pi_u$ shape resonance. *J. Phys. B: At. Mol. Opt. Phys. 35*, L387.

Allan, M. 2004. Threshold phenomena in electron–molecule scattering. *Phys. Scripta T110*(3), 161–165.

Allan, M. 2005. Measurement of the elastic and $v = 0 \rightarrow 1$ differential electron–N_2 cross sections over a wide angular range. *J. Phys. B: At. Mol. Opt. Phys. 38*, 3655–3672.

Allan, M. 2006. Study of resonances in formic acid by means of vibrational excitation by slow electrons. *J. Phys. B: At. Mol. Opt. Phys. 39*, 2939.

Allan, M. 2007a. Absolute angle-differential elastic and vibrational excitation cross sections for electron collisions with tetrahydrofuran. *J. Phys. B: At. Mol. Opt. Phys. 40*, 3531–3544.

Allan, M. 2007b. Electron collisions with formic acid monomer and dimer. *Phys. Rev. Lett. 98*, 123201.

Allan, M. 2010. Electron collisions with CO: Elastic and vibrational excitation cross sections. *Phys. Rev. A 81*, 042706/1–042706/9.

Allan, M. and L. Andrić. 1996. σ*-resonances in electron impact-induced vibrational excitation of *n*-propane, cyclopropane, ethylene oxide, cyclopentane, and cyclohexane. *J. Chem. Phys. 105*, 3559–3568.

Allan, M., M. Čížek, J. Horáček, and W. Domcke. 2000. Electron scattering in cooled HCl: Boomerang structures and outer-well resonances in elastic and vibrational excitation cross sections. *J. Phys. B: At. Mol. Opt. Phys. 33*, L209.

Allan, M. and I. I. Fabrikant. 2002. Threshold peaks and structures in vibrational excitation of CH_3I by electron impact. *J. Phys. B: At. Mol. Opt. Phys. 35*, 1025–1034.

Allan, M. and T. Skalický. 2003. Structures in elastic, vibrational, and dissociative electron attachment cross sections in N_2O near threshold. *J. Phys. B 36*, 3397.

Andrić, L. and R. I. Hall. 1988. Resonance phenomena observed in electron scattering from acetylene. *J. Phys. B 21*, 355–366.

Asmis, K. R. and M. Allan. 1997a. Excess energy partitioning between electrons departing at 0° and 180° in the ionization of helium near threshold. *J. Phys. B: At. Mol. Opt. Phys. 30*, L167–L173.

Asmis, K. R. and M. Allan. 1997b. Measurement of absolute differential cross sections for the excitation of the $n = 2$ states of helium at 0° and 180°. *J. Phys. B: At. Mol. Opt. Phys. 30*(8), 1961–1974.

Azria, R. and F. Fiquet-Fayard. 1972. Attachement électronique dissociatif sur C_2H_2 et C_2D_2. *J. Phys. 33*, 663–667.

Azria, R., L. Roussier, P. Paineau, and M. Tronc. 1974. Attachment electronique dissociatif sur HCl et DCl. *Rev. Phys. Apl. 9*, 469.

Azria, R., S. F. Wong, and G. J. Schulz. 1975. Vibrational excitation in N_2O *via* the 2.3-eV shape resonance. *Phys. Rev. A 11*(4), 1309–1313.

Bachorz, R. A., M. Haranczyk, I. Dabkowska, J. Rak, and M. Gutowski. 2005. Anion of the formic acid dimer as a model for intermolecular proton transfer induced by a π^* excess electron. *J. Chem. Phys. 122*, 204304.

Barsotti, S., E. Leber, M.-W. Ruf, and H. Hotop. 2002. High resolution study of cluster anion formation in low-energy electron collisions with molecular clusters of CO_2, CS_2 and O_2. *Int. J. Mass Spectrom. Ion Proc. 220*, 313–330.

Bazin, M., S. Ptasinska, A. D. Bass, and L. Sanche. 2009. Electron induced dissociation in condensed-phase nitromethane I: Desorption of ionic fragments. *Phys. Chem. Chem. Phys. 11*, 1610–1618.

Bazin, M., S. Ptasinska, A. D. Bass, L. Sanche, E. Burean, and P. Swiderek. 2010. Electron induced dissociation in the condensed-phase nitromethane: II. Desorption of neutral fragments. *J. Phys. Cond. Mat. 22*(8), 084003.

Bettega, M. H. F., C. Winstead, and V. McKoy. 2006. Low-energy electron scattering by N_2O. *Phys. Rev. A 74*(2), 022711.

Birtwistle, D. T. and A. Herzenberg. 1971. Vibrational excitation of N_2 by resonance scattering of electrons. *J. Phys. B 4*, 53–70.

Boness, M. J. W. and G. J. Schulz. 1974. Vibrational excitation in CO_2 *via* the 3.8-eV resonance. *Phys. Rev. A 9*(5), 1969–1979.

Brunger, M. J., P. J. O. Teubner, A. M. Weigold, and S. J. Buckman. 1989. Vibrational excitation of nitrogen in the $^2\Pi_g$ resonance region. *J. Phys. B: At. Mol. Opt. Phys. 22*, 1443–1453.

Buckman, S. J. and B. Lohmann. 1986. Electron scattering from CO in the $^2\Pi$ resonance region. *Phys. Rev. A 34*, 1561–1563.

Bulliard, C., M. Allan, and E. Haselbach. 1994. Intramolecular competition of phenylic and benzylic CX bond breaking in dissociative electron attachment to dihalotoluenes. *J. Phys. Chem. 98*, 11040–5.

Burrow, P. D., G. A. Gallup, A. M. Scheer, S. Denifl, S. Ptasińska, T. Märk, and P. Scheier. 2006. Vibrational feshbach resonances in uracil and thymine. *J. Chem. Phys. 124*, 124310/1–124310/7.

Campbell, L. and M. J. Brunger. 2008. Electron cooling by carbon monoxide in the atmospheres of Mars and Venus. *PMC Phys. B 1*:3.

Campbell, L. and M. J. Brunger. 2009a. Electron impact excitation of carbon monoxide in comet Hale-Bopp. *Geophys. Res. Lett. 36*, LO3101.

Campbell, L. and M. J. Brunger. 2009b. On the role of electron-driven processes in planetary atmospheres and comets. *Phys. Scripta 80*, 058101.

Chaney, E. L. and L. G. Christophorou. 1969. Electron attachment to N_2O. *J. Chem. Phys. 51*, 883–893.

Chantry, P. J. 1969. Temperature dependence of dissociative attachment in N_2O. *J. Phys. Chem. 51*, 3369–3380.

Cho, H., Y. S. Park, H. Tanaka, and S. J. Buckman. 2004. Measurements of elastic electron scattering by water vapour extended to backward angles. *J. Phys. B 37*, 625.

Chourou, S. T. and A. E. Orel. 2008. Dissociative attachment to acetylene. *Phys. Rev. A 77*(4), 042709.

Chourou, S. T. and A. E. Orel. 2009a. Dissociative electron attachment to HCCH, HCN and HCCCN. *J. Phys. Conf. Ser. 194*(5), 052032.

Chourou, S. T. and A. E. Orel. 2009b. Dissociative electron attachment to HCN and HNC. *Phys. Rev. A 80*, 032709.

Chourou, S. T. and A. E. Orel. 2009c. Improved calculation on the isotope effect in dissociative electron attachment to acetylene. *Phys. Rev. A 80*(3), 034701.

Čížek, M., J. Horáček, M. Allan, and W. Domcke. 2002. Resonances and threshold phenomena in low-energy electron collisions with hydrogen halides: New experimental and theoretical results. *Czechosl. J. Phys. 52*, 1057–1070.

Čížek, M., J. Horáček, M. Allan, I. I. Fabrikant, and W. Domcke. 2003. Vibrational excitation of hydrogen fluoride by low-energy electrons: Theory and experiment. *J. Phys. B 36*, 2837–2849.

Čížek, M., J. Horáček, M. Allan, A.-C. Sergenton, D. Popović, W. Domcke, T. Leininger, and F. X. Gadea. 2001. Inelastic low-energy electron collisions with the HBr molecule: experiment and theory. *Phys. Rev. A 63*, 062710.

Compton, R. N., H. S. J. Carman, C. Desfrancois, H. Abdoul-Carmine, J. P. Schermann, J. H. Hendricks, S. A. Lyapustina, and K. H. Bowen. 1996. On the binding of electrons to nitromethane: Dipole and valence bound anions. *J. Chem. Phys. 105*(9), 3472–3478.

Cvejanović, S. 1993. In T. Andersen et al. (Eds.), *The Physics of Electronic and Atomic Collisions*, pp. 390. *Proceedings of the XVIII ICPEAC*, Aarhus: American Institute of Physics, New York.

Cvejanovic, S. and J. Jureta. 1989. In *3rd Eur. Conf. on Atomic and Molecular Physics (Bordeaux)*, unpublished, pp. 638.

Deak, J. C., L. K. Iwaki, and D. D. Dlott. 1999. Vibrational energy redistribution in polyatomic liquids: Ultrafast IR-raman spectroscopy of nitromethane. *J. Phys. Chem. A 103*(8), 971–979.

Denifl, S., V. Vizcaino, T. D. Märk, E. Illenberger, and P. Scheier. 2010. High resolution electron attachment to CO_2 clusters. *Phys. Chem. Chem. Phys. 12*, 5219–5224.

Dessent, C. E. H., J. Kim, and M. A. Johnson. 2000. Spectroscopic observation of vibrational Feshbach resonances in near-threshold photoexcitation of $X^-{\cdot}CH_3NO_2$ ($X^- = I^-$ and Br^-). *Faraday Discuss. 115*, 395–406.

Domcke, W. 1989. In A. Herzenberg (Ed.), *Aspects of Electron-Molecule Scattering and Photoionization*, Number 204 in AIP Conference Proceedings, pp. 169–180. American Institute of Physics.

Domcke, W. 1991. Theory of resonance and threshold effects in electron-molecule collisions: The projection-operator approach. *Phys. Rep. 208*, 97.

Dorman, F. H. 1966. Negative fragment ions from resonance capture processes. *J. Chem. Phys. 44*, 3856–3863.

Dressler, R. and M. Allan. 1987. A dissociative electron attachment, electron transmission, and electron energy-loss study of the temporary negative ion of acetylene. *J. Chem. Phys. 87*, 4510–4518.

Dubé, L. and A. Herzenberg. 1975. Resonant electron-molecule scattering: The impulse approximation in N_2O. *Phys. Rev. A 11*(4), 1314–1325.

Dubé, L. and A. Herzenberg. 1979. Absolute cross sections from the "boomerang model" for resonant electron-molecule scattering. *Phys. Rev. A 20*(1), 194–213.

Estrada, H., L. S. Cederbaum, and W. Domcke. 1986. Vibronic coupling of short-lived electronic states. *J. Chem. Phys. 84*, 152.

Estrada, H. and W. Domcke. 1985. On the virtual state effect in low-energy electron-CO_2 scattering. *J. Phys. B: At. Mol. Phys. 18*, 4469–4479.

Fabrikant, I. I. 1990. Resonance processes in e-HCl collisions; comparison of the R-matrix and the nonlocal-complex potential methods. *Comm. At. Mol. Phys. 24*, 37.

Fabrikant, I. I. 2009. Dissociative electron attachment: Threshold phenomena and multimode effects. *J. Phys. Conf. Ser. 192*(1), 012002.

Fabrikant, I. I. 2010. Recent progress in the theory of dissociative attachment: From diatomics to biomolecules. *J. Phys. Conf. Ser. 204*, 012004.

Fedor, J., O. May, and M. Allan. 2008. Absolute cross sections for dissociative electron attachment to HCl, HBr and their deuterated analogues. *Phys. Rev. A 78*, 032701.

Fedor, J., C. Winstead, V. McKoy, M. Čížek, K. Houfek, P. Kolorenč, and J. Horáček. 2010. Electron scattering in HCl: An improved nonlocal resonance model. *Phys. Rev. A 81*, 042702.

Feng, H., W. Sun, and M. A. Morrison. 2003. Importance of nonresonant scattering in low-energy dissociative electron attachment to molecular hydrogen. *Phys. Rev. A 68*, 062709.

Gallup, G. A., P. D. Burrow, and I. I. Fabrikant. 2009a. Electron-induced bond breaking at low energies in HCOOH and glycine: The role of very short-lived σ* anion states. *Phys. Rev. A 79*(4), 042701.

Gallup, G. A., P. D. Burrow, and I. I. Fabrikant. 2009b. Reply to "Comment on electron-induced bond breaking at low energies in HCOOH and glycine: The role of very short-lived σ* anion states." *Phys. Rev. A 80*(4), 046702.

Gallup, G. A. and I. I. Fabrikant. 2007. Resonances and threshold effects in low-energy electron collisions with methyl halides. *Phys. Rev. A 75*(3), 032719.

Gallup, G. A. and I. I. Fabrikant. 2011. Vibrational Feshbach resonances in dissociative electron attachment to uracil. *Phys. Rev. A 83*(1), 012706.

Gauyacq, J. P. 1987. *Dynamics of Negative Ions.* Singapore, New Jersey, Hong Kong: World Scientific Lecture Notes in Physics.

Gianturco, F. A. and R. R. Lucchese. 2004. Nanoscopic models for radiobiological damage: Metastable precursors of dissociative electron attachment to formic acid. *New J. Phys. 6*, 66.

Gianturco, F. A., R. R. Lucchese, J. Langer, I. Martin, M. Stano, G. Karwasz, and E. Illenberger. 2005. Modelling electron-induced processes in condensed formic acid: Resonant states of $(HCOOH)_2$ at low energies. *Eur. Phys. J. D 35*, 417.

Gibson, J. C., L. A. Morgan, R. J. Gulley, M. J. Brunger, C. T. Bundschu, and S. J. Buckman. 1996. Low energy electron scattering from CO: Absolute cross section measurements and R-matrix calculations. *J. Phys. B: At. Mol. Opt. Phys. 29*, 3197–3214.

Gilbert, S. J., C. Kurz, R. G. Greaves, and C. M. Surko. 1997. Creation of a monoenergetic pulsed positron beam. *Appl. Phys. Lett. 70*, 1944–1946.

Gopalan, A., J. Bömmels, S. Götte, A. Landwehr, K. Franz, M. W. Ruf, H. Hotop, and K. Bartschat. 2003. A novel electron scattering apparatus combining a laser photoelectron source and a triply differentially pumped supersonic beam target: Characterization and results for the He^- ($1s2s^2$) resonance. *Eur. Phys. J. D 22*, 17.

Gribakin, G. F., J. A. Young, and C. M. Surko. 2010. Positron-molecule interactions: Resonant attachment, annihilation, and bound states. *Rev. Mod. Phys. 82*(3), 2557–2607.

Gutsev, G. L. and R. J. Bartlett. 1996. A theoretical study of the valence- and dipole-bound states of the nitromethane anion. *J. Chem. Phys. 105*(19), 8785–8792.

Hake, Jr., R. D. and A. V. Phelps. 1967. Momentum-transfer and inelastic-collision cross sections for electrons in O_2, CO, and CO_2. *Phys. Rev. 158*, 70–84.

Horáček, J. 2000. Inelastic low-energy electron collisions with hydrogen halides. In Y. Itikawa et al. (Eds.), *The Physics of Electronic and Atomic Collisions*, Volume 500, pp. 329. *Proceedings of the XXIICPEAC*, Sendai, 1999. American Institute of Physics, New York.

Horáček, J., W. Domcke, and H. Nakamura. 1997. Electron attachment and vibrational excitation in hydrogen iodide: Calculations based on the nonlocal resonance model. *Z. Phys. D 42*, 181–185.

Hotop, H., M.-W. Ruf, M. Allan, and I. I. Fabrikant. 2003. Resonance and threshold phenomena in low-energy electron collisions with molecules and clusters. *Adv. At. Mol. Opt. Phys. 49*, 85.

Ibanescu, B. C. and M. Allan. 2009. Selectivity in bond cleavage in amines and thiols by dissociative electron attachment. *J. Phys.: Conf. Ser. 194*, 012030.

Ibănescu, B. C., O. May, A. Monney, and M. Allan. 2007. Electron-induced chemistry of alcohols. *Phys. Chem. Chem. Phys. 9*, 3163–3173.

Itikawa, Y. 2002. Cross sections for electron collisions with carbon dioxide. *J. Phys. Chem. Ref. Data 31*, 749–767.

Jung, K., T. Antoni, R. Müller, K.-H. Kochem, and H. Ehrhardt. 1982. Rotational excitation of N_2, CO and H_2O by low-energy electron collisions. *J. Phys. B: At. Mol. Phys. 15*, 3535–3555.

Kanik, I., S. Trajmar, and J. C. Nickel. 1993. Total electron scattering and electronic state excitations cross sections for O_2, CO, and CH_4. *J. Geophys. Res. 98*, 7447.

Knoth, G., M. Gote, M. Rädle, K. Jung, and H. Ehrhardt. 1989. Nuclear-excited Feshbach resonances in the electron scattering from hydrogen halides. *Phys. Rev. Lett. 62*(15), 1735–1737.

Knudsen, M. 1910. Eine Revision der Gleichgewichtsbedingung der Gase. Thermische Molekularströmung. *Ann. Phys. 31*, 205–229.

Ko, Y. J., H. Wang, D. Radisic, S. T. Stokes, S. N. Eustis, K. H. Bowen, K. Mazurkiewicz et al. 2010. Barrier-free proton transfer induced by electron attachment to the complexes between 1-methylcytosine and formic acid. *Mol. Phys. 108*, 2621–2631.

Krishnakumar, E. and S. K. Srivastava. 1990. Dissociative attachment of electrons to N_2O. *Phys. Rev. A 41*(5), 2445–2452.

Kwan, C. K., Y. F. Hsieh, W. E. Kauppila, S. J. Smith, T. S. Stein, M. N. Uddin, and M. S. Dababneh. 1983. $e^{\pm}$–CO and $e^{\pm}$–CO_2 total cross-section measurements. *Phys. Rev. A 27*(3), 1328–1336.

Lecomte, F., S. Carles, C. Desfrancois, and M. A. Johnson. 2000. Dipole bound and valence state coupling in argon-solvated nitromethane anions. *J. Chem. Phys. 113*(24), 10973.

Lunt, S. L., D. Field, J.-P. Ziesel, N. C. Jones, and R. J. Gulley. 2001. Very low energy electron scattering in nitromethane, nitroethane, and nitrobenzene. *Int. J. Mass Spectr. 205*(1–3), 197–208.

Martin, I., T. Skalický, J. Langer, H. Abdoul-Carime, G. Karwasz, E. Illenberger, M. Stano, and S. Matejcik. 2005. Low energy electron driven reactions in single formic acid molecules (HCOOH) and their homogeneous clusters. *Phys. Chem. Chem. Phys. 7*, 2212.

May, O., J. Fedor, and M. Allan. 2009. Isotope effect in dissociative electron attachment to acetylene. *Phys. Rev. A 80*, 012706.

May, O., J. Fedor, B. C. Ibănescu, and M. Allan. 2008. Absolute cross sections for dissociative electron attachment to acetylene and diacetylene. *Phys. Rev. A 77*, 040701(R).

May, O., D. Kubala, and M. Allan. 2010. Absolute cross sections for dissociative electron attachment to HCN and DCN. *Phys. Rev. A 82*(1), 010701.

Michaud, M., E. M. Hébert, P. Cloutier, and L. Sanche. 1997. Charge trapping and the desorption of anionic and metastable fragments by dissociative electron attachment to condensed N_2O. *J. Phys. B: At. Mol. Opt. Phys. 30*, 3527–3541.

Michaud, M., E. M. Hébert, P. Cloutier, and L. Sanche. 2007. Electron photoemission from charged films: Absolute cross section for trapping 0–5 eV electrons in condensed CO_2. *J. Chem. Phys. 126*(2), 024701.

Modelli, A. 2005. Empty level structure and dissociative electron attachment cross section in (bromoalkyl)benzenes. *J. Phys. Chem. A 109*(28), 6193–6199.

Moore, J. H., P. Swiderek, S. Matejcik, and M. Allan. 2010. Fundamentals of interactions of electrons with molecules. In P. Russell, I. Utke, and S. Moshkalev (Eds.), *Nanofabrication Using Focused Ion and Electron Beams: Principles and Applications*, pp. in press. Oxford University Press.

Morgan, L. A. 1991. Low-energy electron scattering by CO. *J. Phys. B: At. Mol. Opt. Phys. 24*(21), 4649–4660.

Morgan, L. A. 1998. Virtual states and resonances in electron scattering by CO_2. *Phys. Rev. Lett. 80*, 1873–1875.

Motlagh, S. and J. H. Moore. 1998. Cross sections for radicals from electron impact on methane and fluoroalkanes. *J. Chem. Phys. 109*, 432.

Mündel, C., M. Berman, and W. Domcke. 1985. Nuclear dynamics in resonant electron-molecule scattering beyond the local approximation: Vibrational excitation and dissociative attachment in H_2 and D_2. *Phys. Rev. A 32*(1), 181–193.

Murphy, T. J. and C. M. Surko. 1992. Positron trapping in an electrostatic well by inelastic collisions with nitrogen molecules. *Phys. Rev. A 46*, 5696–5705.

Nandi, D., V. S. Prabhudesai, B. M. Nestmann, and E. Krishnakumar. 2011. Dissociative electron attachment to NO probed by velocity map imaging. *Phys. Chem. Chem. Phys. 13*, 1542–1551.

Nesbet, R. K. 1979. Variational calculations of accurate e^-–He cross sections below 19 eV. *Phys. Rev. A 20*, 58–70.

Nickel, J. C., P. W. Zetner, G. Shen, and S. Trajmar. 1989. Principles and procedures for determining absolute differential electron-molecule (atom) scattering cross sections. *J. Phys. E: Sci. Instrum. 22*, 730.

Orient, O. J. and S. K. Srivastava. 1983. Production of O^- from CO_2 by dissociative electron attachment. *Chem. Phys. Lett. 96*, 681–684.

Pavlovic, Z., M. J. W. Boness, A. Herzenberg, and G. J. Schulz. 1972. Vibrational excitation in molecular nitrogen by electron impact in the 15–35 eV region. *Phys. Rev. A 6*, 676–685.

Pelc, A., W. Sailer, P. Scheier, and T. D. Märk. 2005. Generation of (M-H)$^-$ ions by dissociative electron attachment to simple organic acids M. *Vacuum 78*, 631–634.

Pelc, A., W. Sailer, P. Scheier, N. J. Mason, E. Illenberger, and T. D. Märk. 2003. Electron attachment to simple organic acids. *Vacuum 70*, 429.

Pelc, A., W. Sailer, P. Scheier, N. Mason, and T. Märk. 2002. Low energy electron attachment to formic acid. *Eur. Phys. J. D 20*, 441–444.

Pelc, A., W. Sailer, P. Scheier, M. Probst, N. J. Mason, E. Illenberger, and T. D. Märk. 2002. Dissociative electron attachment to formic acid (HCOOH). *Chem. Phys. Lett. 361*(3–4), 277–284.

Pichou, F., A. Huetz, G. Joyez, and M. Landau. 1978. Near threshold ionization of helium by electron impact. *J. Phys. B 11*, 3683–3692.

Poparić, G. B., D. S. Belić, and M. D. Vićić. 2006. Resonant vibrational excitation of CO by low-energy electrons. *Phys. Rev. A 73*, 062713/1–062713/6.

Poparić, G. B., D. S. Belić, and M. M. Ristić. 2010. Resonant vibrational excitation of H_2 by electron impact: Full-range differential cross sections. *Phys. Rev. A 82*(1), 012706/1–012706/5.

Poparić, G. B., S. M. D. Galijaš, and D. S. Belić. 2004. Forward-to-backward differential-cross-section ratio in electron-impact vibrational excitation *via* the $^2\Pi$ resonance of CO. *Phys. Rev. A 70*(2), 024701.

Prabhudesaia, V. S., D. Nandia, A. H. Kelkara, R. Parajulib, and E. Krishnakumar. 2005. Dissociative electron attachment to formic acid. *Chem. Phys. Lett. 405*, 172–176.

Ptasinska, S., S. Denifl, P. Scheier, E. Illenberger, and T. D. Märk. 2005. Bond- and site-selective loss of H atoms from nucleobases by very-low-energy electrons (<3 eV). *Angew. Chem. Int. Ed. 44*, 6941–6943.

Ramsauer, C. and R. Kollath. 1929. Über den Wirkungsquerschnitt der Edelgasmoleküle gegenüber Elektronen unterhalb 1 Volt. *Ann. Physik (Leipzig) 3*, 536.

Ramsauer, C. and R. Kollath. 1930. Über den Wirkungsquerschnitt der Nichtedelgasmoleküle gegenüber Elektronen unterhalb 1 Volt. *Ann. Physik (Leipzig) 4*, 91.

Rapp, D. and D. D. Briglia. 1965. Total cross sections for ionization and attachment in gases by electron impact. II. Negative-ion formation. *J. Chem. Phys. 43*, 1480–1490.

Read, F. H. 1975. Doppler and other broadening effects in electron scattering experiments. *J. Phys. B 8*, 1034–1040.

Read, F. H. and N. J. Bowring. 2005. The CPO-3D program. www.electronoptics.com.

Read, F. H. and J. M. Channing. 1996. Production and optical properties of an unscreened but localized magnetic field. *Rev. Sci. Instrum. 67*, 2373.

Rescigno, T. N., C. W. McCurdy, D. J. Haxton, C. S. Trevisan, and A. E. Orel. 2007. Nuclear dynamics in resonant electron collisions with small polyatomic molecules. *J. Phys. Conf. Ser. 88*, 012027–012035.

Rescigno, T. N., C. S. Trevisan, and A. E. Orel. 2006. Dynamics of low-energy electron attachment to formic acid. *Phys. Rev. Lett. 96*, 213201.

Rescigno, T. N., C. S. Trevisan, and A. E. Orel. 2009. Comment on electron-induced bond breaking at low energies in HCOOH and glycine: The role of very short-lived σ^* anion states. *Phys. Rev. A 80*(4), 046701.

Rohr, K. 1978. Interaction mechanisms and cross sections for the scattering of low-energy electrons from HBr. *J. Phys. B 11*(10), 1849.

Rohr, K. and F. Linder. 1975. Vibrational excitation in e-HCl collisions at low energies. *J. Phys. B: At. Mol. Phys. 8*, L200.

Sailer, W., A. Pelc, S. Matejcik, E. Illenberger, P. Scheier, and T. D. Märk. 2002. Dissociative electron attachment study to nitromethane. *J. Chem. Phys. 117*, 7989–7994.

Schafer, O. and M. Allan. 1991. Measurement of near-threshold vibrational excitation of HCl by electron impact. *J. Phys. B: Atom. Mol. and Optic. Phys. 24*(13), 3069.

Scheer, A. M., K. Aflatooni, G. A. Gallup, and P. D. Burrow. 2004. Bond breaking and temporary anion states in uracil and halouracils: Implications for the DNA bases. *Phys. Rev. Lett. 92*, 068102.

Scheer, A. M., P. Możejko, G. A. Gallup, and P. D. Burrow. 2007. Total dissociative electron attachment cross sections of selected amino acids. *J. Chem. Phys. 126*, 174301.

Scheer, A. M., C. Silvernail, J. A. Belot, K. Aflatooni, G. A. Gallup, and P. D. Burrow. 2005. Dissociative electron attachment to uracil deuterated at the N_1 and N_3 positions. *Chem. Phys. Lett. 411*(1–3), 46–50.

Schramm, A., I. I. Fabrikant, J. M. Weber, E. Leber, M.-W. Ruf, and H. Hotop. 1999. Vibrational resonance and threshold effects in inelastic electron collisions with methyl iodide molecules. *J. Phys. B: At. Mol. Opt. Phys. 32*, 2153–2171.

Schulz, G. J. 1973. Resonances in electron impact on diatomic molecules. *Rev. Mod. Phys. 45*, 423.

Sergenton, A.-C. and M. Allan. 2000. Excitation of vibrational levels of HI up to $v = 8$ by electron impact. *Chem. Phys. Lett. 319*(1–2), 179–183.

Sergenton, A.-C., L. Jungo, and M. Allan. 2000. Excitation of vibrational levels of HF up to $v = 4$ by electron impact. *Phys. Rev. A 61*(6), 062702.

Shi, X., T. M. Stephen, and P. D. Burrow. 1993. Differential cross sections for elastic scattering of electrons from molecular nitrogen at 0.55, 1.5, and 2.2 eV. *J. Phys. B: At. Mol. Opt. Phys. 26*, 121–128.

Skalický, T. and M. Allan. 2004. The assignment of dissociative electron attachment bands in compounds containing hydroxyl and amino groups. *J. Phys. B 37*, 4849.

Skalický, T., C. Chollet, N. Pasquier, and M. Allan. 2002. Properties of the π^* and σ^* states of chlorobenzene anion determined by electron impact spectroscopy. *Phys. Chem. Chem. Phys. 4*, 3583.

Sohn, W., K.-H. Kochem, K. Jung, H. Ehrhardt, and E. S. Chang. 1985. Electron scattering from CO below resonance energy. *J. Phys. B: At. Mol. Phys. 18*, 2049–2055.

Sommerfeld, T. 2002. Coupling between dipole-bound and valence states: The nitromethane anion. *Phys. Chem. Chem. Phys. 4*, 2511–2516.

Sommerfeld, T. 2003. A fresh look at the $^2A_1$ CO_2^- potential energy surface. *J. Phys. B: At. Mol. Opt. Phys. 36*, L127–L133.

Sommerfeld, T. 2007. Doorway mechanism for dissociative electron attachment to fructose. *J. Chem. Phys. 126*, 124301.

Sommerfeld, T. 2008. Dipole-bound states as doorways in (dissociative) electron attachment. *J. Chem. Phys. 4*, 245.

Sommerfeld, T., H.-D. Meyer, and L. S. Cederbaum. 2003. Potential energy surface of the CO_2^- anion. *Phys. Chem. Chem. Phys. 6*, 42–45.

Stamatovic, A. and G. J. Schulz. 1968. Trochoidal electron monochromator. *Rev. Sci. Instrum. 39*, 1752.

Stricklett, K. L., S. C. Chu, and P. D. Burrow. 1986. Dissociative attachment in vinyl and allyl chloride, chlorobenzene and benzyl chloride. *Chem. Phys. Lett. 131*, 279–284.

Sullivan, J. P., A. Jones, P. Caradonna, C. Makochekanwa, and S. J. Buckman. 2008. A positron trap and beam apparatus for atomic and molecular scattering experiments. *Rev. Sci. Instrum. 79*, 113105/1–113105/5.

Sun, W., M. A. Morrison, W. A. Isaacs, A. D. T. Trail, W K, R. J. Gulley, M. J. Brennan, and S. J. Buckman. 1995. Detailed theoretical and experimental analysis of low-energy electron-N_2 scattering. *Phys. Rev. A 52*, 1229.

Sweeney, C. J. and T. W. Shyn. 1997. Measurement of absolute differential cross sections for the vibrational excitation of molecular nitrogen by electron impact in the $^2\Pi_g$ shape resonance region. *Phys. Rev. A 56*, 1348.

Szmytkowski, C., K. Maciag, and G. Karwasz. 1996. Absolute electron-scattering total cross section measurements for noble gas atoms and diatomic molecules. *Phys. Scripta 54*(3), 271–280.

Tanaka, H., H. Toyda, and H. Sugai. 1998. Cross sections for electron-impact dissociation of alternative etching gas, C_3HF_7O. *Jpn. J. Appl. Phys. 37*, 5053.

Teillet-Billy, D. and J. P. Gauyacq. 1984. Dissociative attachment in e^--HCl, DCl collisions. *J. Phys. B: At. Mol. Phys. 17*, 4041–4058.

Vanroose, W., Z. Zhang, C. W. McCurdy, and T. N. Rescigno. 2004. Threshold vibrational excitation of CO_2 by slow electrons. *Phys. Rev. Lett. 92*, 053201.

Vizcaino, V., S. Denifl, T. D. Märk, E. Illenberger, and P. Scheier. 2010. Low energy (0–4 eV) electron impact to N_2O clusters: Dissociative electron attachment, ion-molecule reactions, and vibrational Feshbach resonances. *J. Chem. Phys. 133*, 154512/1–6.

Walker, I. C., A. Stamatovic, and S. F. Wong. 1978. Vibrational excitation of ethylene by electron impact: 1–11 eV. *J. Chem. Phys. 69*, 5532.

Weber, J. M., E. Leber, M.-W. Ruf, and H. Hotop. 1999. Nuclear-excited Feshbach resonances in electron attachment to molecular clusters. *Phys. Rev. Lett. 82*(3), 516–519.

Wilde, R. S., G. A. Gallup, and I. I. Fabrikant. 2000. Comparative studies of dissociative electron attachment to methyl halides. *J. Phys. B: At. Mol. Opt. Phys. 33*, 5479–5492.

Zubek, M., N. Gulley, G. C. King, and F. H. Read. 1996. Measurements of elastic electron scattering in the backward hemisphere. *J. Phys. B: At. Mol. Opt. Phys. 29*, L239.

Zubek, M., B. Mielewska, and G. C. King. 2000. Absolute differential cross sections for electron elastic scattering and vibrational excitation in nitrogen in the angular range from 120° to 180°. *J. Phys. B: At. Mol. Opt. Phys. 33*, L527.

4 Nonlocal Theory of Resonance Electron– Molecule Scattering

Martin Čížek and Karel Houfek

CONTENTS

4.1 INTRODUCTION

Theoretical description of the dynamics of the molecular motion usually comprises two steps. First, quantum–chemical calculations of potential energy surfaces are performed nowadays using well-established packages of codes. The second step involves calculation of the vibrational dynamics on these potential energy surfaces within the Born–Oppenheimer approximation.

Similarly, the calculation of the *dynamics of electron–molecule collisions* involving nuclear* dynamics proceeds in two steps. The appropriate generalization of the Born–Oppenheimer approximation is called the *nonlocal resonance theory*. The term is historic; the resonances actually do not have to be involved, although they (or virtual states) appear in most of the cases, at least for some molecular geometries. The calculation of the electron dynamics for the fixed molecular geometry has to provide not only the potential energy surfaces, but also the electron-scattering quantities. A set of all quantities needed to perform the calculation of the nuclear dynamics is called the *nonlocal resonance model*. For a specific molecule, we need the potential energy surfaces for both the neutral molecule and the anion and also the transition elements for releasing the electron from the molecular anion into the continuum (at fixed molecular geometry). The last quantities complicate the definition of the nonlocal resonance model, since the transition elements (called the *coupling amplitudes*) depend on both the molecular geometry and the energy of the released electron. The same amplitude also controls the electron capture into the anionic state.

The separation of the electron and the nuclear motion is a very important assumption. Each of the two steps is already quite difficult and the problem to treat both the electronic and nuclear motion simultaneously would be immense. The success of this divide-and-conquer strategy depends on the proper separation of the electronic problem to a discrete state and the continuum. If it is successful, the full nonadiabatic nature of the exchange of energy between electrons and nuclei is properly captured.

We would also like to stress that once the nonlocal resonance model is set up, the number of processes can be treated on the equal theoretical level. The process

$$e^- + AB(v_i) \rightarrow e^- + AB(v_f) \tag{4.1}$$

includes both *elastic scattering* (but with full account of inelastic interaction) and *vibrational excitation*. It also comprises the *collisional detachment*

$$e^- + AB(v_i) \rightarrow A + B + e^-, \tag{4.2}$$

which is nothing else but the vibrational excitation in the dissociation continuum state of the molecule. Here, A and B are two (monatomic or polyatomic) parts of the studied molecule. The proper treatment of the vibrational excitation must also include the competing process of *dissociative electron attachment*

$$e^- + AB(v_i) \rightarrow A + B^-. \tag{4.3}$$

The same calculation can, of course, provide also the full information about the reverse processes of *associative detachment* or *three-particle recombination* which is the inverse of Equation 4.2.

In this chapter, we will assume that the fixed-nuclei electron-scattering problem was solved (see Chapter 2 for a review of methods dealing with the fixed-nuclei electron scattering) and we focus on the separation of the electronic Hilbert space into the

---

* We use the term nuclear dynamics when we talk about the dynamics of motion of nuclei in the molecule here. When talking about the electron–molecule collisions, we try to avoid the term "vibrational dynamics," because some processes include the dissociation of the molecule.

discrete state and the continuum and on the derivation of the equations describing the nuclear (= vibrational or dissociation) dynamics of the resonance electron–molecule collisions. We will also discuss numerical methods for solving these equations. An application of the theory to particular systems and a discussion of various structures appearing in the cross sections due to nuclear dynamics will be presented in Chapter 5.

4.1.1 BRIEF OVERVIEW OF THE METHODS

Treatment of the vibrational excitation or dissociative attachment requires the inclusion of both electronic and nuclear dynamics of the system. The straightforward basis expansion of the vibrational (and rotational) degrees of freedom together with electronic wave functions leads to coupled channel expansions (see Morrison and Sun 1995 and references therein). This approach has been applied to several systems but it is not suitable for treatment of the resonant or dissociative dynamics, where the nuclear motion gets very far from the vibrations in the initial molecular state and the basis would be too large for the calculation to be feasible.

Closely related to the close coupling approach is the R-matrix method for electron scattering (discussed in more detail in Chapter 2, including the references) with extension to include the vibrational dynamics by Schneider et al. (1979). Although the introduction of the R-matrix sphere may lead to smaller basis, the method itself also falls in the class of methods using direct expansion in combined electronic and vibrational space and does not remove the necessity to use a very large basis in the case of strongly coupled electron and vibrational motions.

The simplest case of the separation of the electronic and vibrational degrees of freedom is straightforward application of the Born–Oppenheimer approximation (applicability of the adiabatic approximation for scattering processes was discussed, for example, by Chase (1956)). This is possible only in the case of elastic scattering and vibrational excitation in the absence of resonances. The fixed-nuclei electron-scattering problem is solved and sandwiched between bra-vector describing the final vibrational state and ket-vector for the initial vibrational state. The resulting T-matrix describes the vibrational excitation process in the adiabatic nuclei approximation. The electron–vibrational correlation is completely ignored in this approach and it is therefore also not suited for the description of the processes involving resonances.

Fortunately the strong correlation of electronic and vibrational motion can often be attributed to one or few electronic states. It is therefore feasible to separate the electronic Hilbert space in the part containing these states and the rest that can be treated within the Born–Oppenheimer approximation. The projection-operator method of Feshbach developed in nuclear physics (Feshbach 1958, 1962) is ideally suited to perform such separation. The approach is basically equivalent to Fano's theory of discrete states embedded in the continuum (Fano 1961). The projection-operator approach was introduced to describe molecular processes involving the electron continuum by Chen (1966), O'Malley (1966), Bardsley (1968), and Nakamura (1971) in the late 1960s and early 1970s. As pointed out earlier, the projection-operator approach only divides the electronic states into two subspaces for each molecular geometry. This division is useful only if it restores the conditions of validity of the Born–Oppenheimer approximation as discussed by O'Malley (1971). The new basis in the electronic Hilbert space is thus called *diabatic basis*, in contrast to the conventional adiabatic basis that diagonalizes the electronic Hamiltonian for each molecular geometry. Recently, the performance of such generalized Born–Oppenheimer approximation was tested on a numerically solvable model containing one electronic and one nuclear coordinates (Houfek et al. 2006). It was shown (Houfek et al. 2008) that the nonlocal resonance theory can give correct results contrary to some other approximations.

The equations governing the nuclear motion in the projection-operator-based theories include the nonlocal, complex, and energy-dependent effective potential which is difficult to deal with. Therefore, the early treatments of collisions employed the *local complex potential approximation*. The nonlocality of the full theory is closely related to energy dependence of the effective potential. The local complex potential approximation is not able to describe the threshold effects properly as discussed by Bardsley (1968).

The role of the threshold effects in the full nonlocal version of the theory was studied in detail by Domcke (1991). He showed that the nonlocal theory is capable of describing correctly the various types of singularities in the fixed-nuclei electron–molecule S-matrix (bound and virtual states, resonances) and to include their influence on the vibrational dynamics. Cederbaum and Domcke (1981) furthermore introduced an analytic model solvable in terms of continued fractions. The numerical treatment of the full nonlocal dynamics was furthermore developed by Horáček (1995) and applied to several diatomic molecules (see Chapter 5 for the details).

The projection-operator formulation leads to a parametrization of the Hamiltonian which is then used for the calculation of the dynamics. The similar approach was developed by Fabrikant (1985) based on the single-pole approximation within the R-matrix theory. Although the definition of the discrete state is somewhat special in this case (and it is not clear that it gives the proper diabatic state), the resulting equations for the vibrational dynamics have been shown to be equivalent to the projection-operator formulation of the nonlocal resonance model (see Fabrikant 1990 or Hotop et al. 2003 and references therein). Similar approach is also the zero-range (or effective-range) approximation theory (Gauyacq and Herzenberg 1982), which parameterizes the electron dynamics based on the low-energy expansion formulas, with parameters depending on the molecular geometry. By tuning the parameters, the vibrational dynamics can also give similar results to full nonlocal calculations, but the zero-range expansions do not provide the accurate fixed-nuclei electron–molecule scattering data.

The validity of the projection-operator approach leading to the nonlocal resonance model is not limited to electron–molecule scattering processes. Employing the time-reversal symmetry, the description of the dissociative attachment is easily extended to reverse processes of associative detachment, that is, to ion–molecule collisions. The generalization to positive ions is also possible and the approach has also been applied to dissociative recombination and Penning and associative ionization (see, e.g., Bieniek 1978).

4.2 NONLOCAL RESONANCE THEORY

The basic assumption usually made to derive the nonlocal resonance theory of electron–molecule collisions is the formation of a metastable molecular anion M^- during the collision. The wave function of the whole system is then projected on this state and the resulting dynamics in the nonlocal, energy-dependent, and complex potential is solved. The intermediate state M⁻ is usually the resonance (or virtual state), hence the term the nonlocal resonance theory.

Although this picture is essentially correct, there are some subtleties that occasionally led to misunderstanding. The typical situation is depicted in Figure 4.1. Here the case of a diatomic molecule is considered for simplicity. The picture is a schematic representation of the spectrum of the electronic Hamiltonian as it depends on the internuclear distance R in the molecule. The solid red curve is the potential energy curve $V_0(R)$ for the neutral molecule in the ground electronic state. The area above this curve is shaded to indicate that we have one additional electron in the electron–molecule collisions that can add arbitrary positive amount of energy to this potential energy. (The potential energy curve $V_0(R)$ is thus called the continuum threshold in this context.) In addition to this curve, we have the resonance state with energy marked by dotted blue curve inside the continuum. This is a metastable state and the electron can leave the system. The usual complication, present in most of the systems, is that this resonance state disappears above some critical internuclear distance and becomes the (electronically) bound anion state (solid blue line in Figure 4.1). The power of the nonlocal resonance theory is that it can correctly describe this transition. During this transition, as the internuclear distance R is varied, the metastable state can change its character from the resonance (directly or via a virtual state) to a bound state. We therefore prefer to call it the discrete state instead of the resonance.

To understand the situation a little bit deeper, we have to speak about the determination of the discrete state |d⟩ with the wavefunction $\varphi_d(\mathbf{r}, \mathbf{q}; R)$ where $\mathbf{r}$ are the coordinates of the incoming electron and $\mathbf{q}$ and R denote collectively the coordinates of the target electrons and nuclei, respectively. In our case represented by Figure 4.1, it is rather straightforward to get φ_d for larger R, where it represents the bound state of one additional electron captured on the neutral molecule. For smaller distances, it is more complicated since there is no square integrable function associated with

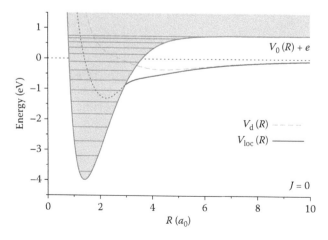

FIGURE 4.1 **(See color insert.)** Typical potential–energy curves for the electron collisions with diatomic molecules.

the resonance. The resonances (or virtual states) can still be defined as generalized eigenstates of the electronic Hamiltonian, but resulting states are not square integrable and they thus do not belong to the electronic Hilbert space. Practically, it is sufficient that the function φ_d is a square integrable function with a significant overlap with the true metastable state. This creates a large degree of arbitrariness in the theory, and we have infinitely many choices for the discrete state. This does not have to worry us, since the possible changes in the choice of the discrete state are compensated by the respective changes in the couplings to the continuum. The second criterion for the validity of the theory is the smoothness of the dependence of φ_d on the internuclear distances $\mathbf{R}$ since the diabatic approximation is used for the description of the nuclear dynamics as we will discuss later in this chapter. (See also the results of Houfek et al. (2008) on probing the nonlocal resonance model by choosing different discrete states.)

Once the discrete state φ_d is specified, the nonlocal resonance theory is developed employing the Feshbach projection-operator formalism (Feshbach 1958, 1962) with the operators Q and P chosen to project on the discrete-state and continuum parts of the electronic Hilbert space, respectively. The essence of the nonlocal resonance theory is in solving the whole dynamics in the simpler Q-part of the space. In the following, we will develop the theory in several steps. First, we define the projection operators Q and P and an appropriate basis in the electronic Hilbert space to expand these operators and the electronic Hamiltonian. The matrix elements of the electronic Hamiltonian play a central role in the construction of the nonlocal resonance models. Then, we will decompose the scattering T-matrix into the resonant and background (nonresonant) terms and give explicit formulas for them. Next, we derive the equations describing the full dynamics of the system within the Q-part of the space, which we obtain by the projection of the full wavefunction of the system on the discrete state. This quantity contains all information needed for calculation of resonance contributions to the vibrational excitation and dissociative attachment cross sections within the diabatic approximation.

The dynamics of the collision between the electron and the molecule with the fixed molecular geometry is governed by the Hamiltonian operator

$$H_{el} = T_{el} + V_{int}(\mathbf{R},\mathbf{r}), \qquad (4.4)$$

where $T_{el} = -(1/2)\Delta_{\mathbf{r}}$ is the kinetic energy of the incoming electron and the second term is the effective one-electron interaction with the molecule, which is in general an energy-dependent, nonlocal in $\mathbf{r}$, and (above the inelastic threshold) non-Hermitian operator (Domcke 1991). By introducing the optical potential V_{int}, the reference to internal electrons in the molecule (through their coordinates $\mathbf{q}$) disappears from the description, although they are fully included through the many-body theory. It is possible to repeat the whole following derivation with explicit reference to internal molecular electrons, but in our case, below the threshold for ionization, we find it more convenient to suppress the explicit reference to other electrons. The molecular geometry is specified by a set of coordinates of all nuclei denoted here collectively by $\mathbf{R}$. The target electrons are projected out and are implicitly included in the optical potential V_{int}. The complete dynamics of the electron–molecule collision, including the motion of the nuclei, is then described using the Hamiltonian operator

$$H = T_N + V_0(\mathbf{R}) + H_{el} \equiv H_0 + V_{int}, \tag{4.5}$$

where T_N is the kinetic energy of nuclei and $V_0(\mathbf{R})$ is the potential energy surface of the ground electronic state of the molecule of the interest. The separation of H to

$$H_0 = T_N + V_0(\mathbf{R}) + T_{el} \tag{4.6}$$

and the interaction V_{int} is a natural choice since H_0 is the channel Hamiltonian for the electron–molecule scattering (see, e.g., Taylor (1972) for use of the channel Hamiltonian in the multichannel scattering theory).

4.2.1 Projection Operators Q and P

We start by defining the projection operators in the fixed-nuclei approximation.[*] All states and operators are thus understood as operating on the electronic Hilbert space H_{el}.

The operator Q projecting on the resonant part Q of H_{el} is given by

$$Q = |\,d\rangle\langle d\,|, \tag{4.7}$$

where $|d\rangle$ approximately describes the electron after being captured by the molecule. While the vector $|d\rangle$ represents the state in the electronic Hilbert space H_{el}, it can still parametrically depend on the molecular geometry, specified by the set of coordinates of all nuclei denoted here collectively as $\mathbf{R}$. We will sometimes omit the explicit reference to this dependence, but it should be kept in mind that $|d\rangle$ can depend on $\mathbf{R}$. The vector $|d\rangle$ is always given by a normalized, square-integrable, and in general complex wave function $\varphi_d(\mathbf{r}, \mathbf{R}) = \langle\mathbf{r}|d\rangle$ although (as discussed above) the state may represent a resonance (for some $\mathbf{R}$). The correct resonance wave function is not normalizable, but large overlap of this function with $|d\rangle$ is sufficient for good performance of the theory.

To simplify our derivation, we assume that there is a single isolated electronic resonance state becoming a bound state for large internuclear distances, which is true for most of the diatomic molecules. The generalization to more discrete states (or more continua) is rather straightforward.

The projector P on the complementary (nonresonant or background) part, P, of H_{el} is simply

$$P = 1 - Q. \tag{4.8}$$

It is now convenient to introduce the energy-normalized electronic states $|\mathbf{k}^{(+)}\rangle$ (*background scattering states*) that diagonalize the electronic Hamiltonian restricted on the P space, that is,

$$PH_{el}P\,|\,\mathbf{k}^{(+)}\rangle = \frac{1}{2}k^2\,|\,\mathbf{k}^{(+)}\rangle, \tag{4.9}$$

---

[*] This approximation is used for the definition of bases and projection operators but not in the final dynamical calculation where motion of nuclei is allowed (see the end of this section).

where a boundary condition for $\phi_k^{(+)}(\mathbf{r},\mathbf{R}) \equiv \langle \mathbf{r} | \mathbf{k}^{(+)} \rangle$ is determined by the incoming plane wave

$$\langle \mathbf{r} | \mathbf{k} \rangle = (2\pi)^{-\frac{3}{2}} e^{i\mathbf{k}\cdot\mathbf{r}}. \tag{4.10}$$

The projector P can then be expressed in the form

$$P = \int | \mathbf{k}^{(+)} \rangle\langle \mathbf{k}^{(+)} | \, d^3\mathbf{k}. \tag{4.11}$$

The construction of the background scattering states is straightforward once the fixed-nuclei electron-scattering problem is solved. The details are discussed in Domcke (1991) and references therein. In a similar way, we define electronic states $|\mathbf{k}^{(-)}\rangle$ but with the plane wave (4.10) defining the outgoing boundary condition. The P operator expressed in these states has the same form as in Equation 4.11. It should be noted here that even though $\varphi_d(\mathbf{r};\mathbf{R})$ can be chosen to be independent of $\mathbf{R}$ the background states $\phi_k^{(\pm)}(\mathbf{r};\mathbf{R})$ are always parametrically dependent on $\mathbf{R}$ because of explicit $\mathbf{R}$-dependence of the operator H_{el}.

Once the fixed-nuclei electron-scattering problem is solved, all the information about the electronic dynamics can be coded in the matrix elements of H_{el}. To be more specific, we introduce the notation

$$V_d(R) = V_0(R) + \langle d | H_{el} | d \rangle, \tag{4.12}$$

$$V_{dk}^{(+)}(R) = \langle d | H_{el} | \mathbf{k}^{(+)} \rangle, \tag{4.13}$$

$$V_{kk'} \equiv \left\{ V_0(R) + \frac{1}{2}k^2 \right\} \delta\left(\frac{1}{2}k^2 - \frac{1}{2}k'^2 \right) = \langle \mathbf{k}^{(+)} | \{V_0(\mathbf{R}) + H_{el}\} | \mathbf{k}'^{(+)} \rangle, \tag{4.14}$$

where the last equality follows from Equation 4.9 and the energy-normalization of the continuum states. It is important to realize that $V_{kk'}$ is diagonal only because of the appropriate choice of $|\mathbf{k}^{(+)}\rangle$. As we will see later, this choice greatly simplifies the nuclear dynamics leading to a single equation describing the nuclear motion in the discrete state governed by the effective nonlocal potential. In addition to the matrix element $V_{dk}^{(+)}(\mathbf{R})$, we will need a matrix element

$$V_{dk}^{(-)}(\mathbf{R}) = \langle d | H_{el} | \mathbf{k}^{(-)} \rangle \tag{4.15}$$

to define the background and resonant T-matrix for vibrational excitation.

The main reason to perform the decomposition of the electronic space to the discrete state and continuum is twofold. First, the separation to the discrete state and continuum allows specifying the proper boundary condition for the dissociative attachment process. We should mention at this point that although the choice of $\mathbf{R}$-dependence of the discrete state $|d\rangle$ is to a large degree arbitrary, $|d\rangle$ should go to the proper electronic bound state describing $A + B^-$ fragments (see Equation 4.3) when $\mathbf{R}$ attains the corresponding values ($R \to \infty$ in the case of diatomics). The second reason is to restore the Born–Oppenheimer approximation. All the electron–molecule-scattering systems, where a large dissociative attachment signal is observed, feature the electronic bound states (anions) disappearing in the continuum for some molecular geometries. The Born–Oppenheimer approximation is inherently wrong in such situations. But with a proper choice of the discrete state, we can still assume

$$\frac{\partial \phi_d(\mathbf{r}; \mathbf{R})}{\partial \mathbf{R}} \simeq 0, \quad \frac{\partial \phi_k^{(+)}(\mathbf{r}; \mathbf{R})}{\partial \mathbf{R}} \simeq 0, \tag{4.16}$$

which allows us to use the relations

$$[T_N, P] = [T_N, Q] = 0, \tag{4.17}$$

$$T_N = PT_N P + QT_N Q, \tag{4.18}$$

$$T_N \mid d\rangle \mid \psi_{nucl}\rangle = \mid d\rangle T_N \mid \psi_{nucl}\rangle, \tag{4.19}$$

$$T_N \mid \mathbf{k}^{(+)}\rangle \mid \psi_{nucl}\rangle = \mid \mathbf{k}^{(+)}\rangle T_N \mid \psi_{nucl}\rangle. \tag{4.20}$$

These relations will be used in the subsequent sections. The expansion of the electronic Hilbert space and relations (4.16) define the *diabatic approximation*, which is the proper generalization of the Born–Oppenheimer approximation for our case.

Let us note that by introducing T_N, we started to work in the full Hilbert space of the problem and vectors like $\mid d\rangle \mid \psi_{nucl}\rangle$ are a direct product of the electronic and nuclear part. The fixed-nuclei projection operators $P(\mathbf{R})$ and $Q(\mathbf{R})$ are diagonal in the nuclear Hilbert space, that is, in combined space of electrons and nuclei, they become

$$Q = \int d\mathbf{R} \mid \mathbf{R}\rangle Q(\mathbf{R})\langle \mathbf{R} \mid, \tag{4.21}$$

$$P = \int d\mathbf{R} \mid \mathbf{R}\rangle P(\mathbf{R})\langle \mathbf{R} \mid. \tag{4.22}$$

4.2.2 Background and Resonant T-Matrices for Vibrational Excitation

We begin by writing the full Hamiltonian given by Equation 4.5 in the form

$$H = H_0 + V_1 + V_2, \tag{4.23}$$

where H_0 is given by Equation 4.6 and

$$\begin{aligned} V_1 &= PH_{el}P + QH_{el}Q - T_{el} \\ &= PH_{el}P + QH_{el}Q - H_{el} + V_{int}, \end{aligned} \tag{4.24}$$

$$V_2 = PH_{el}Q + QH_{el}P. \tag{4.25}$$

The motivation for the choice of V_1 is to express the scattering T-matrix (more precisely only its *resonant* part) in terms of $\mid v\rangle \mid \mathbf{k}^{(+)}\rangle$ instead of the unperturbed initial or final state $\mid v\rangle \mid \mathbf{k}\rangle$, where $\mid v\rangle$ is the Born–Oppenheimer vibrational wave function $\chi_v(\mathbf{R})$ of the molecule (i.e., an eigenstate of $T_N + V_0$). We can easily see that the state $\mid v\rangle \mid \mathbf{k}^{(+)}\rangle$ is an eigenfunction of

$$H_1 \equiv H_0 + V_1 = T_N + V_0 + PH_{el}P + QH_{el}Q$$
$$= Q[T_N + V_d(\mathbf{R})]Q + P[T_N + V_0(\mathbf{R}) + H_{el}]P = QHQ + PHP \qquad (4.26)$$

within the diabatic approximation defined by Equation 4.16.

By employing the two-potential formula for the scattering T-matrix (see Goldberger and Watson (1964), p. 202), we get

$$T^{VE}_{v_i \to v_f} = \langle v_f \,|\, \langle \mathbf{k}_f^{(-)} \,|\, V_1 \,|\, v_i \rangle \,|\, \mathbf{k}_i \rangle + \langle v_f \,|\, \langle \mathbf{k}_f^{(-)} \,|\, V_2 \,|\, \Psi^{(+)} \rangle, \qquad (4.27)$$

where $|\Psi^{(+)}\rangle$ is the full (including both electronic and vibrational degrees of freedom) scattering wave function of the system

$$|\Psi^{(+)}\rangle = |v_i\rangle \,|\, \mathbf{k}_i^{(+)}\rangle + \frac{1}{E - H + i\varepsilon} V_2 \,|\, v_i\rangle \,|\, \mathbf{k}_i^{(+)}\rangle. \qquad (4.28)$$

The last equation shows that we can take the product $|v_i\rangle |\mathbf{k}_i^{(+)}\rangle$ as the initial state of the system to determine the wave function $|\Psi^{(+)}\rangle$ which we will use later to derive the effective equation for the nuclear dynamics.

The second term of Equation 4.27 corresponds to the resonant part of the T-matrix as defined in Domcke (1991) and is fully determined by the resonant part $Q|\Psi^{(+)}\rangle$ of the full wave function. Using Equations 4.8, 4.13, 4.25, and the orthogonality $\langle \mathbf{k}^{(-)}|d\rangle = 0$, we obtain

$$T^{res}_{v_i \to v_f} = \langle v_f \,|\, \langle \mathbf{k}_f^{(-)} \,|\, PH_{el}Q + QH_{el}P \,|\, \Psi^{(+)}\rangle$$
$$= \langle v_f \,|\, \langle \mathbf{k}_f^{(-)} \,|\, PH_{el}Q \,|\, \Psi^{(+)}\rangle. \qquad (4.29)$$

This expression can further be simplified if we define $\Psi_d(\mathbf{R}) = \langle d|\Psi^{(+)}\rangle_r$ where $\langle \cdots \rangle_r$ means an integration over the electronic coordinate r only. The resulting function is thus a state belonging to the nuclear Hilbert space. In terms of the resonant nuclear wavefunction Ψ_d, for which we will derive the effective Schrödinger equation in the following subsection, the resonant part of the T-matrix can be written as

$$T^{res}_{v_i \to v_f} = \langle v_f \,|\, V_{dk_f}^{(-)*} \,|\, \Psi_d \rangle. \qquad (4.30)$$

Note that this expression differs slightly from the result of Domcke (1991). Namely, in Equation 4.14 of this work the matrix V_{dk} has no superscript, which corresponds to the matrix element $V_{dk}^{(+)}$ defined by Equation 4.13. This small difference becomes important only when the background terms defined below are added to the resonant T-matrix (which was not usually the case in previous studies of resonant electron–molecule collisions), since the coupling matrix elements $V_{dk}^{(\pm)}$ are in general complex even when the discrete state is real. The reason why we cannot use $V_{dk}^{(+)}$ instead of $V_{dk}^{(-)}$ is that in general, in spite of the fact that $\phi_k^{(-)}$ belongs to P space,

$$\langle \mathbf{k}_f^{(-)} \,|\, \mathbf{k}^{(+)}\rangle \neq \delta\!\left(\frac{1}{2}k_f^2 - \frac{1}{2}k^2\right)\delta(\Omega_{k_f} - \Omega_k) \qquad (4.31)$$

(the quantity in fact equals to the background scattering matrix) and therefore

$$\langle \mathbf{k}_f^{(-)} \,|\, PH_{el}Q \,|\, d\rangle \neq \langle \mathbf{k}_f^{(+)} \,|\, H_{el} \,|\, d\rangle. \qquad (4.32)$$

Instead, if we consider a special case of the *real* discrete state and one dimensional scattering with radial coordinate r

$$\phi_k^{(-)}(r) = (\phi_k^{(+)}(r))^*, \tag{4.33}$$

we can simplify the matrix element between electronic wave functions in Equation 4.29 as

$$\langle \mathbf{k}_f^{(-)} \mid PH_{el}Q \mid d \rangle = \langle \mathbf{k}_f^{(-)} \mid H_{el} \mid d \rangle = \langle d \mid H_{el} \mid \mathbf{k}_f^{(+)} \rangle = V_{d\mathbf{k}_f}^{(+)} \tag{4.34}$$

where we assumed that H_{el} is a Hermitian operator. Note that in this special case we can use the matrix element $V_{dk}^{(+)}$ but *without complex conjugation*. In the three-dimensional case, Equation 4.33 must be modified to

$$\phi_{\mathbf{k}}^{(-)} = (\phi_{-\mathbf{k}}^{(+)})^* \tag{4.35}$$

and thus V_{dk}^* in Equation 4.14 of Domcke (1991) should be replaced by $V_{d,-k}$ under the assumption that φ_d is real, otherwise $V_{dk}^{(-)*}$ must be used.

We now return to Equation 4.27. The first term is generally called the *background* scattering T-matrix and reads

$$T_{v_i \to v_f}^{bg} = \langle v_f \mid \langle \mathbf{k}_f^{(-)} \mid V_1 \mid \mathbf{k}_i \rangle \mid v_i \rangle = \langle v_f \mid T_{\mathbf{k}_i \to \mathbf{k}_f}^{el}(\mathbf{R}) \mid v_i \rangle, \tag{4.36}$$

where

$$T_{\mathbf{k}_i \to \mathbf{k}_f}^{el}(\mathbf{R}) = \langle \mathbf{k}_f^{(-)} \mid V_1 \mid \mathbf{k}_i \rangle = \langle \mathbf{k}_f \mid V_1 \mid \mathbf{k}_i^{(+)} \rangle \tag{4.37}$$

is the fixed-nuclei background scattering T-matrix. The resulting expression is by definition the adiabatic nuclei approximation. We obtained this expression as a consequence of the fact that we used the diabatic approximation for the projection–operator partitioned basis. Once we solve the background scattering elastic problem for each fixed $\mathbf{R}$, it is easy to calculate this quantity. All effects beyond the Born–Oppenheimer approximation are contained within the resonant contribution.

The background terms are nonzero even for inelastic vibrational excitation but generally small when compared with the resonant part of the T-matrix. For an example where these terms are not negligible, see the results for the F_2-like model in Houfek et al. (2008).

The wave function (4.28) contains also the information about the dissociative electron attachment. The T-matrix for this process reads

$$T_{v_i,\mathbf{k}_i}^{DA} = \langle \mathbf{K}_f^{(-)} \mid \langle d \mid V_2 \mid \Psi^{(+)} \rangle$$
$$= \langle \mathbf{K}_f^{(-)} \mid \langle d \mid QHP \mid \Psi^{(+)} \rangle, \tag{4.38}$$

where $\mid \mathbf{K}_f^{(-)} \rangle \mid d \rangle$ is the eigenstate of H_1 representing the final state for the disso-ciative attachment, with $\mid \mathbf{K}_f^{(-)} \rangle$ standing for the dissociative eigenstate of $T_N + V_d$ with the outgoing boundary condition and $\mathbf{K}_f$ giving the wave-vector of the dis-sociation products. The above expression does not contain the background contribution

$$T_{v_i,\mathbf{k}_i}^{\text{bg,DA}} = \langle \mathbf{K}_f^{(-)} | \langle d | V_1 | v_i \rangle | \mathbf{k}_i \rangle$$

$$= \langle \mathbf{K}_f^{(-)} | \langle d | \{H_1 - H_0\} | v_i \rangle | \mathbf{k}_i \rangle$$

$$= \langle \mathbf{K}_f^{(-)} | v_i \rangle \langle d | \mathbf{k}_i \rangle \left\{ \frac{1}{2\mu} K_f^2 - E_{v_i} - \frac{1}{2} k_i^2 \right\},$$

which equals zero due to the energy conservation. This is consistent with the well-known fact that it is not possible to describe the dissociative attachment process within the Born–Oppenheimer approximation.

4.2.3 INTEGRAL EQUATION FOR THE NUCLEAR WAVE FUNCTION

To get the dynamical equation for the evaluation of the wave function (4.28), we write the Lippmann–Schwinger equation corresponding to separation

$$H = H_1 + V_2$$

which reads

$$|\Psi^{(+)}\rangle = |\Phi_i^{(+)}\rangle + [E - H_P + H_Q + i\varepsilon]^{-1}(H_{PQ} + H_{QP})|\Psi^{(+)}\rangle, \qquad (4.39)$$

where ε is the usual positive infinitesimal. We also introduced the short hand notation $H_P = PHP$, $H_Q = QHQ$, $H_{PQ} = PHQ$, $H_{QP} = QHP$. The vector $|\Phi_i^{(+)}\rangle$ characterizes the initial state and must be an eigenstate of H_P for electron–molecule scattering and an eigenstate of H_Q for ion–atom scattering channel. Multiplying this equation with Q and P, we obtain the system of equations

$$Q|\Psi^{(+)}\rangle = Q|\Phi_i^{(+)}\rangle + G_Q H_{QP} P|\Psi^{(+)}\rangle, \qquad (4.40)$$

$$P|\Psi^{(+)}\rangle = P|\Phi_i^{(+)}\rangle + G_P H_{PQ} Q|\Psi^{(+)}\rangle, \qquad (4.41)$$

where we have introduced Green's functions $G_Q = Q(E - H_Q + i\varepsilon)^{-1}Q$ and $G_P = P(E - H_P + i\varepsilon)^{-1}P$. Substitution of Equation 4.41 into Equation 4.40 gives

$$Q|\Psi^{(+)}\rangle = Q|\Phi_i^{(+)}\rangle + G_Q H_{QP} P|\Phi_i^{(+)}\rangle + G_Q H_{QP} G_P H_{PQ} Q|\Psi^{(+)}\rangle. \qquad (4.42)$$

We thus reduced the scattering problem in the complete Hilbert space to a scattering problem in its small subspace Q. In the next section, we will show that the Q-space part of the wave function $Q|\Psi^{(+)}\rangle$ obeys the inhomogeneous Schrödinger equation with the effective Hamiltonian (O'Malley 1966)

$$H_{\text{eff}} = H_Q + H_{QP} G_P H_{PQ}. \qquad (4.43)$$

Employing the explicit expansion of the projectors in terms of the basis $|d\rangle$, $|\mathbf{k}^{(+)}\rangle$, we find

$$H_{\text{eff}} = T_N + V_d(R) + F(E), \qquad (4.44)$$

where $F(E)$ is the operator in the nuclear space which in coordinate representation reads

$$\langle \mathbf{R} \mid F(E) \mid \mathbf{R}' \rangle = \int V_{\mathrm{dk}}(\mathbf{R}) \left(E - T_{\mathrm{N}} - V_0 - \frac{1}{2}k^2 + i\varepsilon \right)^{-1} V_{\mathrm{dk}}^*(\mathbf{R}') \, d\Omega_k k \, dk. \quad (4.45)$$

The integration is performed over the direction Ω_k and the length k of the electron wavevector $\mathbf{k}$. We do not indicate the type of the coupling amplitude $V_{\mathrm{dk}}^{(\pm)}$, because both signs $(\pm)$ lead to the same result since they represent the same operator $H_{QP} G_P H_{PQ}$ in the two different bases $\mid\mathbf{k}^{(\pm)}\rangle$. The operator $F(E)$ is obviously nonlocal in the coordinate representation and it can be split in the Hermitian and the anti-Hermitian parts

$$F(E, \mathbf{R}, \mathbf{R}') = \Delta(E - T_{\mathrm{N}} - V_0, \mathbf{R}, \mathbf{R}') - \frac{i}{2}\Gamma(E - T_{\mathrm{N}} - V_0, \mathbf{R}, \mathbf{R}'), \quad (4.46)$$

where

$$\Gamma(e, \mathbf{R}, \mathbf{R}') = 2\pi \int d\Omega_k k \, dk \, V_{\mathrm{dk}}(\mathbf{R}) \delta\left(e - \frac{1}{2}k^2\right) V_{\mathrm{dk}}^*(\mathbf{R}'), \quad (4.47)$$

$$\Delta(e, \mathbf{R}, \mathbf{R}') = \frac{1}{2\pi} \wp \int de' \Gamma(e', \mathbf{R}, \mathbf{R}')/(e - e'). \quad (4.48)$$

We have separated the Hermitian and anti-Hermitian part of $F(E)$ using the relation $(x + i\varepsilon)^{-1} = \wp\frac{1}{x} - i\pi\delta(x)$. For $\mathbf{R} = \mathbf{R}'$ the functions (4.47) and (4.48) are called the width $\Gamma(e, \mathbf{R})$ and the level shift $\Delta(e, \mathbf{R})$ of the discrete state $\mid d\rangle$ due to interaction with the continuum. We would also like to point out the importance to distinguish between the functions $\Gamma(e, \mathbf{R})$ and $\Delta(e, \mathbf{R})$ and the Hermitian and anti-Hermitian components of the nonlocal potential (4.46). Whereas $\Gamma(e, \mathbf{R})$ and $\Delta(e, \mathbf{R})$ are just functions of e and $\mathbf{R}$, the components of $F(E)$ are operators acting on the nuclear Hilbert space. The important difference is that the operator $E - T_{\mathrm{N}} - V_0$ is substituted for e in Equations 4.47 and 4.48, which means, for example, that the ordering of terms inside integral (4.47) is important.

As we saw in Equation 4.29, the resonant part of the T-matrix for the process of vibrational excitation can be written in terms of $Q\mid\psi^{(+)}\rangle$, that is, in terms of $\mid\psi^{(+)}\rangle = \langle d\mid\psi^{(+)}\rangle_r$, where the integration is performed only over electronic degrees of freedom as indicated by a subscript r, that is, the result is still a state in the nuclear part of the full Hilbert space. Equation 4.42 together with the definition of the nonlocal potential $F(E)$ implies

$$\mid \psi^{(+)} \rangle = \mid \phi^{(+)} \rangle + G_Q^{(+)}(E) F(E) \mid \psi^{(+)} \rangle, \quad (4.49)$$

where $G_Q^{(+)}(E) = (E - V_{\mathrm{d}} - T_{\mathrm{N}} + i\varepsilon)^{-1}$ and the function

$$\mid \phi^{(+)} \rangle = \langle d \mid \Phi_i^{(+)} \rangle_r + \langle d \mid G_Q H_{QP} P \mid \Phi_i^{(+)} \rangle_r, \quad (4.50)$$

is determined by boundary conditions for the processes of interest. Formula (4.49) is the integral version of the fundamental equation describing uniquely dynamics of the nuclear degrees of freedom. All electronic degrees of freedom are projected out and the formula contains only operators and states in the nuclear Hilbert space of the molecule. The state $|\psi^{(+)}\rangle$ is the unique solution of this equation, but the vector $|\psi^{(+)}\rangle$ depends on the channel of interest. There are two important channels for low energies. The electron–molecule collisions $e^- + AB$ (channel I) and the ion–atom* collisions $A + B^-$ (channel II). For these two channels, we have

$$|\Phi_I^{(+)}\rangle = |v_i\rangle|k_i^{(+)}\rangle \Rightarrow |\phi_I^{(+)}\rangle = G_Q^{(+)}(E)V_{dk_i}^{(+)}|v_i\rangle, \tag{4.51}$$

$$|\Phi_{II}^{(+)}\rangle = |d\rangle|K_i^{(+)}\rangle \Rightarrow |\phi_{II}^{(+)}\rangle = |K_i^{(+)}\rangle, \tag{4.52}$$

where $|v_i\rangle$ is the wave function of the initial vibrational state $\chi_{v_i}(R)$ of the molecule AB, k_i is the momentum of the incoming electron and K_i is the center of mass relative momentum for the ion–atom collision. Solutions $|\psi_I^{(+)}\rangle$, $|\psi_{II}^{(+)}\rangle$ of the fundamental Equation 4.49 with incoming boundary conditions $|\phi_I^{(+)}\rangle$ and $|\phi_{II}^{(+)}\rangle$, respectively, give the full nuclear dynamics of our problem for the electron–molecule and ion–atom collisions. Similarly, we can define the solutions $|\psi_I^{(-)}\rangle$, $|\psi_{II}^{(-)}\rangle$, which have $e^- + AB$ and $A + B^-$ channels as outgoing asymptotes.

Finally, we write explicit formulas for the T-matrix for each of the processes (4.1) and (4.3) in terms of $|\psi^{(\pm)}\rangle$. The resonant contribution to the T-matrices for VE, DA, and AD are

$$T_{VE} \equiv \langle\Phi_I^{(-)}|PH_{PQ}Q|\Psi^{(+)}\rangle = \langle v_f|V_{dk_f}^{(-)*}|\psi_I^{(+)}\rangle, \tag{4.53}$$

$$T_{DA} \equiv \langle\Psi^{(-)}|QH_{QP}P|\Phi_I^{(+)}\rangle = \langle\psi_{II}^{(-)}|V_{dk_i}^{(+)}|v_i\rangle, \tag{4.54}$$

$$T_{AD} \equiv \langle\Phi_I^{(-)}|PH_{PQ}Q|\Psi^{(+)}\rangle = \langle v_f|V_{dk_f}^{(-)*}|\psi_{II}^{(+)}\rangle, \tag{4.55}$$

where $\langle\psi_{II}^{(-)}|$ is the unique solution of

$$\langle\psi_{II}^{(-)}| = \langle K_f^{(-)}| + \langle\psi_{II}^{(-)}|F(E)G_Q^{(+)}(E). \tag{4.56}$$

The expression for T_{DA} apparently differs from Equation 4.38. In fact, here we used the principle of microreversibility and calculated the dissociative attachment from the final state wave function. The expression (4.38) can also be written in terms of Q-projection as follows:

$$T_{DA} = \langle\Phi_I^{(-)}|QH_{QP}P|\Psi^{(+)}\rangle = \langle K^{(-)}|V_{dk_i}^{(+)}|v_i\rangle + \langle K^{(-)}|F(E)|\psi_I^{(+)}\rangle, \tag{4.57}$$

where we used formula (4.41) to express $P|\Psi^{(+)}\rangle$. Both expressions for T_{DA} should give identical results which can be used for cross-checking of numerical results. All the expressions for T-matrices are valid on energy shell, that is, for example, for dissociative attachment the total energy E is identical to initial and final energies

$$E = \frac{1}{2\mu}K_f^2 = E_{v_i} + \frac{1}{2}k_i^2. \tag{4.58}$$

---

* This general formulation also applies for A being polyatomic molecular fragment.

4.2.4 DIFFERENTIAL EQUATION FOR THE NUCLEAR WAVE FUNCTION

In this section, we briefly show the derivation of the differential form of the basic equation for the nonlocal dynamics. We start with Equation 4.28 written in the form applicable for both electron–molecule and ion–atom collisions

$$| \Psi^{(+)} \rangle = | \Phi^{(+)} \rangle + \frac{1}{E - H + i\varepsilon} V_2 | \Phi^{(+)} \rangle, \quad (4.59)$$

where $| \Phi^{(+)} \rangle = | \Phi_I^{(+)} \rangle$ or $| \Phi^{(+)} \rangle = | \Phi_{II}^{(+)} \rangle$ is the appropriate solution of $H_1 = H_P + H_Q$, see Equations 4.51 and 4.52. The boundary conditions for function (4.59) are a little bit complicated. It is easier to work with

$$| \Psi_{sc}^{(+)} \rangle = | \Psi^{(+)} \rangle - | \Phi^{(+)} \rangle, \quad (4.60)$$

which is purely the outgoing wave in infinity. This type of boundary conditions can then be enforced using some variant of complex absorption potential or complex scaling methods. Equation 4.59 implies

$$(E - H) | \Psi_{sc}^{(+)} \rangle = V_2 | \Phi^{(+)} \rangle = [H_{PQ} + H_{QP}] | \Phi^{(+)} \rangle. \quad (4.61)$$

We then project this equation using both projection operators P and Q

$$(E - H_Q) Q | \Psi_{sc}^{(+)} \rangle - H_{QP} P | \Psi_{sc}^{(+)} \rangle = H_{QP} P | \Phi^{(+)} \rangle, \quad (4.62)$$

$$(E - H_P) P | \Psi_{sc}^{(+)} \rangle - H_{PQ} Q | \Psi_{sc}^{(+)} \rangle = H_{PQ} Q | \Phi^{(+)} \rangle. \quad (4.63)$$

Finally, we express the P-part of $| \Psi_{sc}^{(+)} \rangle$ as

$$P | \Psi_{sc}^{(+)} \rangle = G_P(E) H_{PQ} [Q | \Psi_{sc}^{(+)} \rangle + Q | \Phi^{(+)} \rangle] \quad (4.64)$$

and substitute this in the first of the two-projected equations

$$[E - H_Q - F] Q | \Psi_{sc}^{(+)} \rangle = H_{QP} P | \Phi^{(+)} \rangle + FQ | \Phi^{(+)} \rangle. \quad (4.65)$$

To be more specific, we write this final equation in the coordinate representation for $\Psi_d^{(+)}(\mathbf{R}) = \langle d | \Psi_{sc}^{(+)} \rangle_r$ and for electron–molecule collisions, that is, for $| \Phi^{(+)} \rangle = | \Phi_I^{(+)} \rangle$,

$$(E - T_R - V_d(\mathbf{R})) \Psi_d^{(+)}(\mathbf{R}) - \int d\mathbf{R}'\, F(E, \mathbf{R}, \mathbf{R}') \Psi_d^{(+)}(\mathbf{R}') = V_{dk_i}^{(+)}(\mathbf{R}) \chi_{v_i}(\mathbf{R}) \quad (4.66)$$

with the nonlocal, complex, and energy-dependent potential $F(E) = H_{QP} G_P(E) H_{PQ}$ given by formula (4.45).

Once the solution $\Psi_d^{(+)}$ of Equation 4.66 is obtained, it can be used to evaluate the *resonant* part of the T-matrix with Equation 4.30 and the corresponding resonant contribution to the vibrational excitation cross section. The dissociative attachment can also be calculated using Equation 4.57 or from the asymptotic behavior of $\Psi_d^{(+)}$. (Note that normalization of the wave function is fixed by the right-hand side of the Schrödinger Equation 4.66.) For example, the s-wave contribution reads

$$\sigma_{v_i}^{DA}(E) = \frac{2\pi^2}{k_i^2} \frac{K_{DA}}{\mu} \lim_{R \to \infty} | \Psi_d^+(R) |^2. \quad (4.67)$$

4.3 DETAILED TREATMENT OF ANGULAR MOMENTUM FOR DIATOMIC MOLECULES

The treatment up to now was completely general regarding the molecular vibrational degrees of freedom. The variable $\mathbf{R}$ could represent coordinates for any number of atomic nuclei in the studied molecule. The nuclear motion is then taking place in the many-dimensional space. However, for more than 1–2 degrees of freedom, the numerical effort needed to solve the corresponding integral equations would be enormous. Up to now, there are only few attempts to go beyond one degree of freedom. In this chapter, we will solve completely the dynamics for the case of diatomic molecules in terms of partial wave expansions for both the electronic and nuclear parts of the wave function. As a result, the dynamics reduces to solution of a set of coupled one-dimensional equations. The numerical methods suitable to treat such problems are then discussed in the following sections.

The partial wave expansion follows to some extent the work of Bieniek (1978), originally designed for Penning ionization. We briefly give the equivalent derivation for AD and also the formulas for DA and VE.

We start with the partial wave expansion of the discrete–continuum coupling $V_{dk}(\mathbf{R})$ in the coordinate system fixed in the molecule and with the z-axis along the molecular symmetry axis. We will assume that the electronic state of the molecule AB has the Σ symmetry and the same symmetry is also assumed for the discrete state $|d\rangle$. This assumption considerably simplifies our derivation and most of the systems discussed in the next chapter are of this form. Generalization for different symmetry would be straightforward only with little longer expressions. When the Σ symmetry is assumed, the scattering problem exhibits the symmetry with respect to rotations along the internuclear axis so that the partial expansion of $|\mathbf{k}^{(+)}\rangle$ is of the form

$$\langle \mathbf{r} \mid \mathbf{k}^{(+)} \rangle = \sum_{l'lm} \phi_{l'lm}(k,r) Y_{l'm}(\hat{r}) Y_{lm}^*(\hat{k}). \tag{4.68}$$

With $\hat{k}$, we denote the unit vector in the direction of $\mathbf{k}$. Insertion of this expansion into the definition (4.13) of $V_{dk}(\mathbf{R})$ yields

$$V_{dk} = \sum_l Y_{l0}^*(\hat{k}) V_{dkl} = \sum_l \sqrt{\frac{2l+1}{4\pi}} P_l(\cos\theta) V_{dkl}, \tag{4.69}$$

where

$$V_{dkl} = \sum_{l'} \int dr dr' \phi_d^*(\mathbf{r}) H_{el} \phi_{ll'm}(k,r) Y_{l'm}(\hat{r}) \bigg|_{m=0} . \tag{4.70}$$

The terms with $m \neq 0$ vanish due to Σ symmetry. Expansion (4.69) was derived in the coordinate system fixed in the molecule and θ is the angle between $\hat{k}$ and $\hat{R}$. To rewrite the expansion into the center off mass system fixed in space, we employ the relation

$$\frac{2l+1}{4\pi} P_l(\hat{a}\cdot\hat{b}) = \sum_{m=-l}^{l} Y_{lm}(\hat{a}) Y_{lm}^*(\hat{b}) \tag{4.71}$$

to obtain

$$V_{\mathrm{dk}}(\mathbf{R}) = \sum_{lm} \sqrt{\frac{4\pi}{2l+1}} Y_{lm}(\hat{R}) Y_{lm}^*(\hat{k}) V_{\mathrm{dkl}}(R). \tag{4.72}$$

This formula, unlike the previous ones, employs the spherical harmonics $Y_{lm}(\hat{n})$ in arbitrary coordinate system, that is, spherical coordinates are no longer measured from the internuclear axis $\hat{R}$.

In the second step, we expand the nonlocal part F of the effective Hamiltonian H_{eff}. This has been done in detail elsewhere (Bieniek 1978) and we will only outline the derivation briefly. The potential $V_0(R)$ is spherically symmetric, which means that the Green's function in the definition (4.45) is diagonal in the angular momentum representation

$$\left\langle \mathbf{R} \left| (E - T_{\mathrm{N}} - V_0 + i\varepsilon)^{-1} \right| \mathbf{R}' \right\rangle = \sum_{JM} Y_{JM}(\hat{R}) \frac{1}{R} G_{0J}(E, R, R') \frac{1}{R'} Y_{JM}^*(\hat{R}') \tag{4.73}$$

Employing this formula and expansion (4.72) in the definition (4.45) yields

$$\int d\hat{R}\, d\hat{R}'\, Y_{JM}^*(\hat{R}) F(E, \mathbf{R}, \mathbf{R}') Y_{J'M'}^*(\hat{R}') = \int dk\, d\hat{k}\, d\hat{R}\, d\hat{R}' \sum_{l_1 m_1} \sum_{J_0 M_0} \sum_{l_2 m_2}$$

$$Y_{JM}^*(\hat{R}) \sqrt{\frac{4\pi}{2l_1+1}} Y_{l_1 m_1}(\hat{R}) Y_{l_1 m_1}^*(\hat{k}) V_{\mathrm{dkl}_1}(R)$$

$$\times Y_{J_0 M_0}(\hat{R}) \frac{1}{R} G_{0J_0}(E, R, R') \frac{1}{R'} Y_{J_0 M_0}^*(\hat{R}')$$

$$\times \sqrt{\frac{4\pi}{2l_2+1}} V_{\mathrm{dkl}_2}(R') Y_{l_2 m_2}(\hat{k}) Y_{l_2 m_2}^*(\hat{R}') Y_{J'M'}(\hat{R}') \tag{4.74}$$

Since the spherical harmonics are orthogonal, we can perform integration over $d\hat{k}$ and then by employing the relation for complex conjugation of spherical harmonics

$$Y_{lm}^*(\hat{n}) = (-1)^m Y_{l-m}(\hat{n}), \tag{4.75}$$

and

$$\int d\hat{n}\, Y_{l_1 m_1}(\hat{n}) Y_{l_2 m_2}(\hat{n}) Y_{l_3 m_3}(\hat{n})$$

$$= \frac{(2l_1+1)(2l_2+1)(2l_3+1)}{4\pi} \begin{pmatrix} l_1 & l_2 & l_3 \\ 0 & 0 & 0 \end{pmatrix} \begin{pmatrix} l_1 & l_2 & l_3 \\ m_1 & m_2 & m_3 \end{pmatrix} \tag{4.76}$$

for integrals of products of three harmonics the integrations over $d\hat{R}$ and $d\hat{R}'$ can be performed as well. Finally, using orthogonality of Wigner 3j symbols

$$\sum_{m_1 m_2} \begin{pmatrix} l_1 & l_2 & J \\ m_1 & m_2 & M \end{pmatrix} \begin{pmatrix} l_1 & l_2 & J' \\ m_1 & m_2 & M' \end{pmatrix} = \frac{1}{2J+1} \delta_{JJ'} \delta_{MM'} \delta(l_1, l_2, J), \tag{4.77}$$

we obtain

$$\int d\hat{R}\, d\hat{R}'\, Y_{JM}^*(\hat{R}) F(E, \mathbf{R}, \mathbf{R}') Y_{J'M'}^*(\hat{R}') = \frac{1}{RR'} \delta_{JJ'} \delta_{MM'} f_J(E, R, R'), \qquad (4.78)$$

where

$$f_J(E, R, R') = \sum_{lJ'} (2J' + 1) \begin{pmatrix} l & J' & J \\ 0 & 0 & 0 \end{pmatrix}^2$$
$$\int dk V_{dkl}(R) G_{0J'}(E - k^2/2, R, R') V_{dkl}^*(R') \qquad (4.79)$$

or

$$F(E, \mathbf{R}, \mathbf{R}') = \sum_{JM} Y_{JM}(\hat{R}) \frac{1}{R} f_J(E, R, R') \frac{1}{R'} Y_{JM}^*(\hat{R}'). \qquad (4.80)$$

This form of expansion of the nonlocal operator F is a consequence of its spherical symmetry. It is an important property since equations for partial waves with different J are not coupled through the nonlocal potential although they are coupled with several partial waves of the molecular channel (sum over J' in Equation 4.79) accounting thus for angular momentum of the electron released in the detachment or captured in the attachment process.

A next step is to define the partial wave expansion of the wave functions. It is simpler to start with the ion–atom wave function $|\psi_{\mathrm{II}}\rangle$. We define partial waves ψ_J and φ_J with

$$\langle \mathbf{R} | \mathbf{K}^{(+)} \rangle = N \sum_{JM} i^J \frac{1}{KR} \varphi_J^{\mathrm{II}}(R) Y_{JM}^*(\hat{E}) Y_{JM}(\hat{R}) \qquad (4.81)$$

$$\langle \mathbf{R} | \psi_{\mathrm{II}}^{(+)} \rangle = N \sum_{JM} i^J \frac{1}{KR} \psi_J^{\mathrm{II}}(R) Y_{JM}^*(\hat{E}) Y_{JM}(\hat{R}) \qquad (4.82)$$

and expand the Green's function

$$\langle \mathbf{R} | G_Q^{(+)}(E) | \mathbf{R}' \rangle = \sum_{JM} Y_{JM}(\hat{R}) \frac{1}{R} G_{QJ}(E, R, R') \frac{1}{R'} Y_{JM}^*(\mathbf{R}'). \qquad (4.83)$$

The Lippmann–Schwinger equation 4.49 yields

$$\psi_J^{\mathrm{II}}(R) = \varphi_J^{\mathrm{II}}(R) + \int dR' dR'' G_{QJ}(E, R, R') f_J(E, R', R'') \psi_J^{\mathrm{II}}(R''). \qquad (4.84)$$

We introduced the wave number $K = \sqrt{2\mu E}$ in Equations 4.81 and 4.82 and the normalization coefficient $N = 4\pi\sqrt{\mu K}(2\pi)^{-3/2}$ ensures the energy normalization

$$\langle \mathbf{K}^{(+)} | \mathbf{K}'^{(+)} \rangle = \delta\left(\frac{K^2}{2\mu} - \frac{K'^2}{2\mu} \right). \qquad (4.85)$$

By employing the partial wave expansions in formula (4.55) for the associative detachment T-matrix and writing the wave functions $|v\rangle$ for bound states of AB in the usual separated form

$$\langle \mathbf{R} \mid v \rangle = \frac{1}{R} \chi_{n_f J_f}(R) Y_{J_f M_f}(\hat{R}), \tag{4.86}$$

we obtain

$$
\begin{aligned}
T_{AD}(\mathbf{k}_f, E) &= \langle v_f | V_{dk_f}^{(-)*} | \psi_{II}^{(+)} \rangle \\
&= \int R^2 dR d\hat{R} \sum_{lm} \sum_{JM} \frac{1}{R} \chi_{n_f J_f}(R) Y_{J_f M_f}^*(\hat{R}) \\
&\quad \times \sqrt{\frac{4\pi}{2l+1}} Y_{lm}^*(\hat{R}) Y_{lm}(\hat{k}) V_{dk_f l}^{(-)*}(R) \\
&\quad \times N i^J \frac{1}{KR} \psi_J^{II}(R) Y_{JM}^*(\hat{E}) Y_{JM}(\hat{R}) \\
&= \frac{N}{K} \sum_{lmJM} t_{Jl}^{AD}(v_f, J_f, E) Y_{lm}(\hat{k}) Y_{JM}^*(\hat{E}) \\
&\quad \times i^J \sqrt{(2J_f+1)(2J+1)} \begin{pmatrix} J_f & l & J \\ M_f & m & M \end{pmatrix} \begin{pmatrix} J_f & l & J \\ 0 & 0 & 0 \end{pmatrix}, \tag{4.87}
\end{aligned}
$$

where

$$t_{Jl}^{AD}(v_f, J_f, E) = \int dR \chi_{n_f J_f}(R) V_{dk_f l}^{(-)*}(R) \psi_J^{II}(R). \tag{4.88}$$

The integration formula (4.76) has been used again. The differential cross section for the associative detachment reads

$$\frac{d\sigma_{AD}}{d\Omega}(\mathbf{k}_f, E) = (2\pi)^4 \frac{g_i}{K^2} |T_{AD}(\mathbf{k}_f, E)|^2, \tag{4.89}$$

where g_i is the statistical factor for the discrete state potential. This factor applies only in the case that more potentials are connected to the same asymptote. The number g_i then gives the statistical probability for particles to move along the potential curve V_d. This is, for example, the case of $H + H^-$, where there are two potentials connected to the asymptote. The attractive $^2\Sigma_u$ state and the repulsive $^2\Sigma_g$ state which contributes only little to the AD cross section. The statistical factor in this example is thus $g_i = 1/2$. Some caution is also needed in the case of the homonuclear systems like that described in Bieniek and Dalgarno (1979) for H_2, where J-dependent g-factor is present as a result of different nuclear spin. Formula (4.89) can further be simplified using the orthogonality relations (4.77) if we want to know only the integral cross section (integrated over $\hat{k}_f$ and averaged over $\hat{E}$)

$$\sigma_{AD}(k_f, E) = \frac{2\pi g_i}{KE} \sum_{lJ} (2J+1) \begin{pmatrix} J_f & l & J \\ 0 & 0 & 0 \end{pmatrix}^2 |t_{Jl}^{AD}(v_f, J_f, E)|^2. \tag{4.90}$$

The total AD cross section is obtained via summation over all accessible final states $(E_{vfJf} < E)$ or using the formula

$$\sigma_{tot}(E) = \frac{2\mu g_i}{KE} \sum_J (2J+1)$$
$$+ \int dR\, dR'\, \psi_J^{II}(R)^* [f_J(E,R,R') - f_J^*(E,R',R)]\psi_J^{II}(R'), \quad (4.91)$$

which can be derived from the total flux lost from the discrete state channel due to the non-Hermitian part of the nonlocal potential F. Comparison of the two expressions for σ_{tot} was used as a test for numerical solution of the resonant scattering problem by Bieniek (1980) and proved failure of the local complex potential approximation for associative detachment in H^- + H collisions. Using the expansion of $G_{0J}(E)$ in the sum over the bound states $\chi_{vJ}(R)$ in the definition (4.79), it is possible to show that the two expressions for σ_{tot} are equivalent if no further approximation is imposed on $F(E)$.

The T-matrix and the cross section for dissociative attachment can be obtained from the principle of detailed balance since the processes AD and DA are mutually reverse. It is

$$T_{DA}(\hat{E}, \hat{k}) = T_{AD}^*(\hat{k}, \hat{E}) \quad (4.92)$$

and the integral cross section is

$$\sigma_{DA}(E, k_i) = (2\pi)^4 \frac{g_i}{2k_i} |T_{DA}(\hat{E}, \hat{k}_i)|^2. \quad (4.93)$$

Let us now investigate the vibrational excitation process. Employing the relation

$$G^{(+)}(E) = G_Q^{(+)}(E) + G_Q^{(+)}(E)F(E)G^{(+)}(E) \quad (4.94)$$

for the complete Green's function $G^{(+)}(E) \equiv (E - H_{eff} + i\varepsilon)^{-1}$, Equations 4.49 and 4.51 yield

$$|\psi_i^{(+)}\rangle = G^{(+)}(E)V_{dk_i}^{(+)} |v_i\rangle \quad (4.95)$$

and (4.53) implies

$$T_{VE} = \langle v_f | V_{dk_f}^{(-)*} G^{(+)}(E)V_{dk_i}^{(+)} | v_i \rangle. \quad (4.96)$$

Since F is spherically symmetric, it is

$$\langle \mathbf{R} | G^{(+)}(E) | \mathbf{R}' \rangle = \sum_{JM} Y_{JM}(\hat{R}) \frac{1}{R} G_J(E,R,R') \frac{1}{R'} Y_{JM}^*(\mathbf{R}'). \quad (4.97)$$

We can thus use a similar procedure as for T_{AD} to obtain (for the resonance contribution)

$$T_{VE} = \sum_{l_1 m_1} \sum_{l_2 m_2} \sum_{JM} \int dR dR' \chi_{n_f J_f}(R) V_{dk_f l_1}^{(-)*}(R) G_J(E,R,R') V_{dk_i l_2}^{(+)}(R') \chi_{n_i J_i}(R')$$

$$\times (-1)^{M_f + m_1 + M} Y_{l_1 m_1}^*(\hat{k}_f) Y_{l_2 m_2}^*(\hat{k}_i)(2J+1)\sqrt{(2J_f+1)(2J_i+1)}$$

$$\times \begin{pmatrix} J_f & l_1 & J \\ 0 & 0 & 0 \end{pmatrix}\begin{pmatrix} J_f & l_1 & J \\ -M_f & -m_1 & M \end{pmatrix}\begin{pmatrix} J & l_2 & J_i \\ -M & m_2 & M_i \end{pmatrix}\begin{pmatrix} J & l_2 & J_i \\ 0 & 0 & 0 \end{pmatrix}. \tag{4.98}$$

The formula for the vibrational excitation process is more complicated than Equation 4.87 for associative detachment since the angular momentum is transferred twice between the electron and the molecule—once during the resonant capture of an electron into the discrete state and for the second time in the autoionization. Even if we start with the molecule AB in a state v_i with a defined angular momentum, several partial waves are populated in $|\psi_i^{(+)}\rangle$ if $\Gamma_l \neq 0$ for $l > 0$. The state $|\psi_i^{(+)}\rangle$ can subsequently decay in many states v_f (with many different J_f). This complication does not apply if we introduce the following approximation.

4.3.1 LOW-ENERGY S-WAVE APPROXIMATION

Since the electron is a light particle, it feels strongly the repulsive barrier $\{l(l+1)\}/2r^2$ ($m_e = 1$ in atomic units). For large r, we can neglect the other forces and assume that the potential is spherically symmetric,* that is, expansion (4.68) reduces to

$$\langle \mathbf{r}|\mathbf{k}^{(+)}\rangle = \sum_{lm} \phi_{lm}(k,r) Y_{lm}(\hat{r}) Y_{lm}^*(\hat{k}). \tag{4.99}$$

If the electron energy $(1/2)k^2$ is small (near threshold energies), it is reasonable to assume that the electron cannot penetrate the repulsive barrier and $\phi_{lm}(k,r) = 0$ for r small and $l > 0$, that is,

$$V_{dkl} = 0 \quad \text{for } l > 0 \tag{4.100}$$

and

$$V_{dk} = \sqrt{4\pi} Y_{00}(\hat{R}) Y_{00}(\hat{k}) V_{dk}(R), \tag{4.101}$$

where with $V_{dk}(R)$ we denoted $V_{dkl=0}(R)$. As a consequence, the sums over l can be dropped out in all previous formulas for near threshold processes, which simplifies the problem significantly. Thus, using relation

$$\begin{pmatrix} 0 & J' & J \\ 0 & 0 & 0 \end{pmatrix} = (-1)^J \delta_{JJ'}/\sqrt{2J+1}, \tag{4.102}$$

formula (4.79) yields

$$f_J(E,R,R') = \int dk V_{dk}(R) G_{0J}(E - k^2/2, R, R') V_{dk}^*(R'). \tag{4.103}$$

---

* This is not true in the case the molecule AB possesses an electric dipole. Influence of approximation (4.100) to the cross sections then needs further investigation.

Similarly employing (4.102) and

$$
\begin{pmatrix} J & J' & 0 \\ M & -M' & 0 \end{pmatrix} = (-1)^{J-M} \delta_{JJ'} \delta_{MM'} / \sqrt{2J+1}
\tag{4.104}
$$

the associative detachment T-matrix is found to be

$$
T_{AD}(\mathbf{k}_f, E) = \frac{N}{K} \frac{i^{J_f}}{\sqrt{4\pi}} Y^*_{J_f M_f}(\hat{E}) \int dR \chi_{n_f J_f}(R) V^{(-)*}_{dk_f}(R) \psi^{II}_{J_f}(R)
\tag{4.105}
$$

and the cross section

$$
\sigma_{AD}(k_f, E) = \frac{2\pi g_i}{KE} \left| \int dR \chi_{n_f J_f}(R) V^{(-)*}_{dk_f}(R) \psi^{II}_{J_f}(R) \right|^2 .
\tag{4.106}
$$

The description of vibrational excitation also considerably simplifies if we neglect the angular momentum of the electron. The T_{VE} thus reduces to

$$
T_{VE}(\mathbf{k}_f, \mathbf{k}_i) = \frac{1}{4\pi} \delta_{J_f J_i} \delta_{M_f M_i} t^{VE}_J(n_f, k_f, n_i, k_i),
\tag{4.107}
$$

where

$$
t^{VE}_J(n_f, k_f, n_i, k_i) = \int dR \, dR' \chi_{n_f J_f}(R) V^{(-)*}_{dk_f}(R) J(E, R, R') V^{(+)}_{dk_i}(R') \chi_{n_i J_i}(R').
\tag{4.108}
$$

The presence of Kronecker delta $\delta_{J_f J_i}$ and $\delta_{M_f M_i}$ is the consequence of the angular momentum conservation for the molecule AB, when angular momentum of the electron has been neglected. We thus do not have the possibility to describe rotational excitation unless the assumption (4.100) is rejected. The formula for the partial contribution to T_{VE} can easily be transformed into the form

$$
t^{VE}_J(n_f, k_f, n_i, k_i) = \int dR \chi_{n_f J}(R) V^{(-)*}_{dk_f}(R) \psi^J_J(R),
\tag{4.109}
$$

where $\psi^J_J(R)$ is the unique solution of

$$
\psi^J_J(R) = \varphi^J_J(R) + \int dR' dR'' G_{QJ}(E, R, R') f_J(E, R', R'') \psi^J_J(R'')
\tag{4.110}
$$

and we have introduced the following abbreviation

$$
\varphi^J_J(R) = \int dR' G_{QJ}(E, R, R') V^{(+)}_{dk_i}(R') \chi_{n_i J}(R').
\tag{4.111}
$$

Equation 4.110 is the partial wave expansion of Equation 4.49 under the assumption (4.100). Finally, the angle-integrated VE cross section reads

$$
\sigma_{VE}(n_f, k_f, n_i, k_i) = \frac{(2\pi)^3}{4k_i} |t^{VE}_J(n_f, k_f, n_i, k_i)|^2,
\tag{4.112}
$$

with J being the angular momentum of the initial state $|v_i\rangle$.

4.3.2 Parametrization of the Nonlocal Resonance Model

It is clear on the basis of the previous discussion of the theory of resonant collisions that the nuclear dynamics and the cross sections are fully determined if the functions

$$V_0(R), V_d(R) \quad \text{and} \quad V_{dkl}(R)$$

are known. As was shown in the last section $V_{dkl}(R)$ with $l > 0$ does not contribute significantly for small energies ε and only the function $V_{dk}(R)$ is important for the dynamics.[*] Further simplification can be achieved thanks to the extraction of all fast dependencies on ε and R from the background scattering due to the extraction of the resonance part of the wave function. It follows that the discrete-state-continuum coupling $V_{dk}(R)$ is a slowly varying function of ε and R. The coupling $V_{dk}(R)$ is a complex quantity, but the overall phase is not important, since $V_{dk}(R)$ is present together with $V_{dk}^*(R')$ in the nonlocal potential and only the absolute value of its matrix elements is needed for the cross sections. What is important is the relative phase between $V_{dk}(R)$ and $V_{dk}(R')$ in two different points R and R' but since the nonlocal resonant dynamics is restricted to a small region near the equilibrium distance R_0 of the molecule AB and since the coupling is slowly varying with R it is reasonable to assume that this relative phase can be neglected and $V_{dk}(R)$ can be regarded as being a real number.[†] This assumption is unnecessary simplification if the quantities $V_{dk}(R)$ are constructed directly from electron–molecule scattering calculation, but it is very helpful, when $V_{dk}(R)$ is constructed from incomplete information about the electron dynamics.

Expansion of the discrete-state-continuum coupling into a sum of separable terms in energy ε and internuclear distance R is often performed

$$V_{dk}(R) = \sum_i f_i(\varepsilon) g_i(R). \tag{4.113}$$

As will be discussed later, the separation of energy and distance is crucial for time demands of actual numerical calculation. It has been shown that even one term in the expansion often gives qualitatively correct results for the description of the resonant dynamics.

The nonlocal resonant dynamics described by the four functions $V_0(R)$, $V_d(R)$, $f_i(\varepsilon)$, and $g_i(R)$ will be further referred to as the nonlocal resonance model of the dynamics in this work. The calculation of the function $V_0(R)$ for the particular molecule AB is a routine task of quantum chemistry and a large variety of the methods have been developed in the last few decades to treat this problem. The direct determination of $V_d(R)$ and $V_{dk}(R)$ is still far from routine. The calculations for the range of both ε and R are available only for the most fundamental systems like $H_2 + e^-$ (Berman et al. 1985) and $HeH^+ + e^-$ (Movre and Meyer 1997). It is therefore useful to use some analytic knowledge about the behavior of the model functions like asymptotic expansions at low energy.

---

[*] There are some exceptional cases like H_2, where s-wave is forbidden by symmetry and $l = 1$ is the dominant contribution.

[†] The phase factor in $V_{dk}(R)$ equals to the square root of the background scattering S-matrix $e^{i\delta_{bg}}$.

The threshold behavior of the discrete-state-continuum coupling $V_{dk}(R)$ for the small energies ε is determined by the long-range electron–molecule forces in accordance with the threshold law of Wigner (1948). If the molecule AB does not possess an electric dipole, the long-range electron–molecule interaction vanishes faster than r^{-2} and the threshold behavior of the coupling is (Domcke 1991)

$$\Gamma_l(\varepsilon) \equiv 2\pi \, |V_{dkl}(R)|^2 \sim \varepsilon^{(2l+1)/2}. \tag{4.114}$$

As discussed before, only the lowest l allowed by symmetry is really important, which usually is $l = 0$. In the case of polar molecules, the threshold behavior depends on the dipole moment D of the molecule. For $D < D_{crit} = 1.625$ Debye, it is (Domcke 1991)

$$\Gamma(\varepsilon) \sim E^\alpha, \tag{4.115}$$

where

$$\alpha = \sqrt{d + \frac{1}{4}} \tag{4.116}$$

and d is a dimensionless-reduced dipole moment given by the lowest eigenvalue of an infinite-dimensional tridiagonal matrix (Crawford 1967). If the molecule possesses a supercritical dipole $D > D_{crit}$ bound states exist in electron–molecule channel and the threshold behavior is more complicated. Similar situation is faced in the case of Penning ionization or dissociative recombination, where the electron is subjected to Coulomb force in the $e^- + AB^+$ channel. Nonlocal dynamics for these cases will not be investigated in this work.

The threshold law (4.114) or (4.115) fixes the form of $f_i(\varepsilon)$ near the origin $\varepsilon = 0$ and both $f_i(\varepsilon)$ and $g_i(R)$ must vanish at infinity. The value of $V_d(R)$ at infinity is fixed since $E_d = V_d - V_0$ evaluated in $R = \infty$ equals to the electron affinity of the atom A (or B).

Methods for construction of $V_d(R)$, $V_{de}(R)$ from *ab initio* fixed-R electron–molecule scattering data and other *ab initio* information will be given in the following chapter.

4.4 NUMERICAL TREATMENT OF THE NUCLEAR DYNAMICS EQUATIONS

The problem of resonant electron–molecule or ion–atom scattering has been transformed into solution of the one-dimensional Lippmann–Schwinger equation in the previous sections. The equation contains a rather complicated quantity—the nonlocal potential, resulting from a coupling of the discrete resonant state with the continuum, which has to be calculated first. As soon as the solution of the equation is known, the cross sections are easily found by evaluating the matrix elements of the discrete-state-continuum coupling V_{dk}. The general scheme for the solution of the problem can be summarized as follows:

1. To find the scattering solution $\varphi_J(R)$ and the Green's function $G_{QJ}(E)$ for the local potential V_d.
2. To calculate the nonlocal potential $f_J(E)$. As shown later, this is closely related to calculation of bound states $|v\rangle$ in the potential V_0.

3. To solve the Lippmann–Schwinger equation for the wave functions $\psi_J^I(R)$ or $\psi_J^{II}(R)$, respectively.
4. To calculate the T-matrix elements and cross sections (or other quantities of interest) from $\psi_{IJ}(R)$ or $\psi_J^I(R)$.

Step 4 is a trivial application of the formulas derived earlier. Computational procedures to perform the other three steps are outlined below.

4.4.1   SOLUTION TO THE LOCAL PROBLEM

There have been developed large variety of methods to solve the Lippmann–Schwinger equation for a local potential in the past few decades and it is nowadays a routine task. First of all the Lippmann–Schwinger equation

$$\varphi_J(R) = u_J(R) - \frac{2\mu}{K} \int dR u_J(R_<) v_J(R_>) V_d(R') \varphi_J(R'), \qquad (4.117)$$

where $R_> = \max(R, R')$, $R_< = \min(R, R')$ and the functions u_J and v_J are given in terms of the spherical Bessel functions j_J and h_J

$$u_J(R) = KRj_J(KR), \qquad (4.118)$$

$$v_J(R) = iKRh_J(KR) \qquad (4.119)$$

are equivalent to the Schrödinger equation

$$\left(-\frac{1}{2\mu} \frac{d^2}{dR^2} + \frac{J(J+1)}{2\mu R^2} + V_d(R) \right) \varphi_J(R) = E\varphi_J(R) \qquad (4.120)$$

with corresponding boundary conditions. Standard finite-difference methods like algorithms of Numerov or Runge–Kutta can be used to solve this equation. Special methods for the case when $\varphi_J(R)$ is a fast oscillating function have also been developed (Gordon 1969, Gordon 1970). We prefer to solve the integral Equation 4.117 directly. The solution of Equation 4.117 gives the correct normalization automatically, whereas the normalization has to be found afterwards when solving the differential Schrödinger equation. As we will see later, by solving the Lippmann–Schwinger equation, we obtain also the T-matrix for the local problem as a byproduct. Another advantage of the integral equation formulation is the possibility of increasing the numerical precision by means of the Romberg quadrature (Horáček 1989). The long-range rotational energy term $J(J+1)/2\mu R^2$ is treated exactly in the Lippmann–Schwinger equation method. The price for this is the occurrence of the Bessel functions in the equation, which must be calculated at many points. Nevertheless, this does not slow the calculation considerably if the solution is required for a large number of J since a very efficient method for calculation of large number of the Bessel functions by means of stable recurrence relations has been developed (Gillman and Fiebig 1988).

The method presented here for solving Equation 4.117 is described in detail in the article of Horáček (1989). Here, we only outline the principles and generalize the method for finding the irregular solution and the Green's function.

4.4.2 REGULAR SOLUTION

We will omit the subscripts J and d in this and the following section. The Equation 4.117 can easily be rewritten in the form

$$\varphi(R) = Au(R) - \frac{2\mu}{K}\int_0^R dR'[v(R)u(R') - u(R)v(R')]V(R')\varphi(R'), \qquad (4.121)$$

where

$$A = 1 - \frac{2\mu}{K}\int_0^\infty v(R')V(R')\varphi(R')dR'. \qquad (4.122)$$

In other words, if we introduce the function $f(R) = \varphi(R)/A$, we obtain

$$f(R) = [1 - \beta(R)]u(R) + \alpha(R)v(R), \qquad (4.123)$$

where

$$\alpha(R) = -\frac{2\mu}{K}\int_0^R dR'u(R')V(R')f(R'), \qquad (4.124)$$

$$\beta(R) = -\frac{2\mu}{K}\int_0^R dR'v(R')V(R')f(R'). \qquad (4.125)$$

The constant A can be found from

$$A = (1 - \beta(\infty))^{-1} \qquad (4.126)$$

and the T-matrix for elastic scattering is given by the formula

$$t \equiv -\frac{2\mu}{K}\int_0^\infty dRu(R)V(R)\varphi(R) = \alpha(\infty)[1 - \beta(\infty)]^{-1}. \qquad (4.127)$$

Numerical solution of the integral Equation 4.117 on a grid $R = x_1,\ldots,x_n$ is then based on Equation 4.123. The trapezoidal rule is used for the integrals (4.124) and (4.125). We can observe that the terms with $f(R' = R)$ vanish in Equation 4.123 which can be used to calculate $f(x_{i+1})$ if $f(x_1),\ldots,f(x_i)$ are known. The use of the trapezoidal rule also facilitates the possibility to speed the convergence using the Romberg scheme.

4.4.3 IRREGULAR SOLUTION AND GREEN'S FUNCTION

The regular solution of the Schrödinger Equation 4.120 is the solution obeying the boundary condition $\varphi_{(r)}(R = 0) = 0$. Since $u(R = 0) = 0$, we can see from (4.121) that the solution of (4.117) is indeed the regular solution. The irregular solution on the other hand is the solution obeying the condition $\varphi_{(i)}(R \to \infty)\sim v(R)$. The Green's function $G_{QJ}(E, R, R')$ can be written in terms of these two solutions as

$$G_{QJ}(E,R,R') \equiv \left\langle R \left| \left(E + \frac{1}{2\mu}\frac{d^2}{dR^2} - \frac{J(J+1)}{2\mu R^2} - V(R) + i\varepsilon \right)^{-1} \right| R' \right\rangle$$

$$= -\frac{2\mu}{K}\varphi_{(r)}(R_<)\varphi_{(i)}(R_>). \tag{4.128}$$

It is easy to show that $\varphi_{(i)}(R)$ is the solution of

$$\varphi_{(i)}(R) = v(R) - \frac{2\mu}{K}\int_0^\infty dR u(R_<)v(R_>)V(R')\varphi_{(i)}(R'), \tag{4.129}$$

which can be rearranged into the form

$$\varphi_{(i)}(R) = Av(R) - \frac{2\mu}{K}\int_R^\infty dR'[v(R')u(R) - u(R')v(R)]V(R')\varphi_{(i)}(R'), \tag{4.130}$$

which is analogous to Equation 4.121 and it yields for $f(R) = \varphi_{(i)}(R)/A$

$$f(R) = [1 - \alpha(R)]v(R) + \beta(R)u(R) \tag{4.131}$$

with

$$\alpha(R) = -\frac{2\mu}{K}\int_R^\infty dR'u(R')V(R')f(R'), \tag{4.132}$$

$$\beta(R) = -\frac{2\mu}{K}\int_R^\infty dR'v(R')V(R')f(R'). \tag{4.133}$$

The constant A can be written in terms of α

$$A = 1 - \frac{2\mu}{K}\int_0^\infty dR u(R)V(R)\varphi_{(i)}(R) = 1 + A\alpha(0), \tag{4.134}$$

that is, $A = (1 - \alpha(0))^{-1}$. Equation 4.131 can be used for the numerical calculation of $f(R)$ on a grid in the same way as Equation 4.123 in the case of the regular solution. The only difference is that now the function $f(R)$ has to be calculated in the decreasing sequence of grid points starting from sufficiently large x_n, so that $f(x_n) = v(x_n)$ can be assumed. The irregular solution $\varphi_{(i)}(R)$ is then obtained by multiplication of $f(R)$ with A.

The procedure for the calculation of the regular and irregular solutions and the Green's function outlined earlier can be generalized for the coupled channel problem.

4.4.4 Evaluation of the Nonlocal Potential

The key element of evaluation of the nonlocal potential

$$F(E) = \int k dk \, d\hat{k}\, V_{dk} G_0(E - \varepsilon)V_{dk}^*$$

is the knowledge of the Green's function $G_0(E) \equiv (E - T_N - V_0 + i\varepsilon)^{-1}$ for the neutral molecule AB. The formula similar to Equation 4.128 is of little use here, because the regular and irregular solutions depend on energy and the integration over ε cannot easily be performed. A better idea is to expand the $G_0(E)$ in terms of bound states of molecule AB

$$\langle \mathbf{R} | v \rangle = \frac{1}{R} \chi_{vJ}(R) Y_{JM}(\hat{R}) \tag{4.135}$$

to obtain

$$G_0(E, \mathbf{R}, \mathbf{R}') = \sum_v \frac{\langle \mathbf{R} | v \rangle \langle v | \mathbf{R}' \rangle}{E - E_v + i\varepsilon}$$

$$= \sum_{JM} Y_{JM}(\hat{R}) \frac{1}{R} \sum_v \frac{\chi_{vJ}(R) \chi_{vJ}(R')}{E - E_v^J + i\varepsilon} \frac{1}{R'} Y_{JM}(\hat{R}'). \tag{4.136}$$

Comparison of this expansion with (4.73) yields

$$G_{0J}(E, R, R') \equiv \left(E + \frac{1}{2\mu} \frac{d^2}{dR^2} - \frac{J(J+1)}{2\mu R^2} - V_0(R) + i\varepsilon \right)^{-1}$$

$$= \sum_v \frac{\chi_{vJ}(R) \chi_{vJ}(R')}{E - E_v^J + i\varepsilon}. \tag{4.137}$$

This expression is convenient for the evaluation of $F(E)$ since the integration over de can be performed analytically (see (4.47) and (4.48))

$$F(E, \mathbf{R}, \mathbf{R}') = \sum_v \langle \mathbf{R} | v \rangle [\Delta(E - E_v, \mathbf{R}, \mathbf{R}') - \frac{i}{2} \Gamma(E - E_v, \mathbf{R}, \mathbf{R}')] \langle v | \mathbf{R}' \rangle \tag{4.138}$$

or after the partial wave expansion (see formula 4.79)

$$f_J(E, R, R') = \sum_{lJ'} (2J'+1) \begin{pmatrix} l & J' & J \\ 0 & 0 & 0 \end{pmatrix}^2$$

$$\times \sum_v \chi_{vJ'}(R) [\Delta_l(E - E_v^{J'}, R, R') - \frac{i}{2} \Gamma_l(E - E_v^{J'}, R, R')] \chi_{vJ'}(R), \tag{4.139}$$

where

$$\Gamma_l(e, R, R') = 2\pi V_{del}(R) V_{del}^*(R'), \tag{4.140}$$

$$\Delta_l(e, R, R') = \frac{1}{2\pi} \wp \int de' \Gamma_l(e', R, R') / (e - e') \tag{4.141}$$

If we neglect the angular momentum of the released/captured electron (see Section 4.3.1), we get

$$f_J(E, R, R') = \sum_v \chi_{vJ}(R) [\Delta_0(E - E_v^J, R, R') - \frac{i}{2} \Gamma_0(E - E_v^J, R, R')] \chi_{vJ}(R'). \tag{4.142}$$

Minor difficulty for application of this expansion to actual numerical calculation is the presence of continuum functions in the sum over ν, which becomes important near the dissociation limit for the neutral molecule AB. This problem has been overcome by an appropriate discretization of the continuum spectrum by Horáček et al. (1996) who suggested to use the Fourier DVR method for the discretization, which is equivalent to putting the molecule AB into a box, that is, imposing the boundary condition $\chi_{\nu J}(R_c) = 0$ at a certain cutoff radius R_c. The sum over ν is then cut at a certain $\nu = \nu_{max}$. One may wonder about the role of the box boundary condition on the calculation, since the correct boundary condition should be given by prescribing the outgoing wave at some large R_c. The correct behavior is however ensured by the presence of the infinitesimal part $i\varepsilon$ in the expansion (4.137).

4.4.5 LOCAL APPROXIMATIONS

A local approximation of the nonlocal dynamics is based on the replacement of the nonlocal potential $f_J(E, R, R')$ with a certain local quantity. The possibility of such a replacement is based on different timescales for electronic and nuclear dynamics. The nuclei are moving in the presence of an electron captured in the discrete resonant state $|d\rangle$. The resonant state $|d\rangle$ can decay at any point R and the electron can again be captured into $|d\rangle$ at a different point R'. If the electron released in the decay is fast, the difference between R and R' cannot be large, because otherwise the electron would be gone before nuclei succeed to travel the distance $|R-R'|$ and could not be captured again. The nonlocality of the interaction is therefore pronounced only if resonance energy $E_d = V_d - V_0$ is small, that is, if the potential curves $V_d(R)$ and $V_0(R)$ come close together or even cross each other. If they are well separated (as is often the case for the Penning ionization), it is possible to replace $V_d + f_J$ with the local approximation (Bieniek 1978, Domcke 1991, Morgner 1990)

$$V_{loc}(R) - \frac{i}{2}\Gamma_{loc}(R), \tag{4.143}$$

where

$$V_{loc}(R) = V_0(R) + E_{res}(R), \tag{4.144}$$

$$\Gamma_{loc}(R) = \Gamma(E_{res}(R), R, R), \tag{4.145}$$

and

$$E_{res}(R) = V_d(R) - V_0(R) + \Delta_J(E_{res}(R), R, R). \tag{4.146}$$

A transparent derivation of this approximation including the possibility of first-order corrections* has been done in the time-dependent formalism (Domcke 1991). It is also possible to show that if $E_{res}(R) < 0$ then $V_{loc}(R)$ is the bound state of H_{el} (see Equation 4.4).

---

* In the time-dependent picture, the effective potential (after projecting out the continuum) is not only nonlocal, but contains also the memory effects. The first-order corrections depend only on the first derivative of the wave function with respect to time and R.

The local approximation (4.144) cannot give good results for the systems where the curves $V_0(R)$ and $V_d(R)$ come close together. Nevertheless, local approximations are still useful to accelerate the convergence of iteration methods for solving the nonlocal Lippmann–Schwinger equation

$$| \psi \rangle = | \varphi \rangle + G_{QJ} f_J | \varphi \rangle, \tag{4.147}$$

which corresponds to the partitioning of the potential $V_{eff} = V_d + f_J$. We can use the different partitioning $V_{eff} = V_{loc} + f_J'$ instead, with $f_J' = f_J + V_d - V_{loc}$ being a "smaller" operator than the original f_J and having a chance for faster convergence of methods for solving the corresponding Lippmann–Schwinger equation (see the next section). This subtraction of the local approximation from the full nonlocal problem is a special case of the method of *preconditioning* known in the matrix iteration methods.

Use of the approximation (4.144) is limited also by the fact that the Equation 4.146 often does not have a unique solution. A different local approximation has been proposed by Horáček et al. (1996) to accelerate the convergence (in the way suggested above). This approximation is convenient in the cases where the wave function $\psi_J(R)$ is not strongly oscillating. The first term in the Taylor expansion

$$\int f_J(E,R,R')\psi_J(R')dR' = \int f_J(E,R,R')[\psi_J(R) + (R'-R)\psi_J'(R) + \cdots]dR' \tag{4.148}$$

yields the energy-dependent local approximation

$$f_{loc}(E,R) = \int f_J(E,R,R')dR'. \tag{4.149}$$

4.4.6 SOLUTION OF THE LIPPMANN–SCHWINGER EQUATION WITH NONLOCAL POTENTIAL

A very efficient method for computing the T-matrix elements in the presence of nonlocal interactions, the Schwinger–Lanczos algorithm, was proposed by Meyer et al. (1991). This method has been successfully used in several realistic applications (Horáček and Domcke 1996, Horáček et al. 1996, Gemperle and Horáček 1997) to calculate the cross sections for vibrational excitation, associative detachment, and dissociative attachment based on the use of the nonlocal resonance model. The essence of the method has been outlined already in Chapter 2 in connection with the solution of the fixed-nuclei electronic scattering problem. Originally, the method was designed for calculation of the diagonal T-matrix elements. In principle, it can also be used for calculation of the nondiagonal T-matrix elements resulting from multichannel collisions, but the calculation is rather cumbersome and the incoming and outgoing states are not treated symmetrically. It is the purpose of this section to present a generalization of the Schwinger–Lanczos algorithm for the multichannel case and to establish its relation to other methods.

4.4.6.1 Short Review of the Schwinger–Lanczos Approach

According to the Schwinger variational principle (Lippmann and Schwinger 1950), the T-matrix element

$$T_{\text{fi}} \equiv \langle \phi_{\text{f}} \mid T \mid \phi_{\text{i}} \rangle = \langle \phi_{\text{f}} \mid V(V - VG_0V)^{-1}V \mid \phi_{\text{i}} \rangle \qquad (4.150)$$

is given by the stationary value of the functional

$$T[\psi_-, \psi_+] = \langle \phi_{\text{f}} \mid V \mid \psi_+ \rangle + \langle \psi_- \mid V \mid \phi_{\text{i}} \rangle - \langle \psi_- \mid V - VG_0V \mid \psi_+ \rangle, \qquad (4.151)$$

where V is an interaction potential and G_0 the free particle Green's function. This stationary value is achieved for $|\psi_\pm\rangle$ being solutions of the corresponding Lippmann–Schwinger equations. Considering $|\psi_\pm\rangle$ in the form

$$\mid \psi_\pm \rangle = \sum_{k=1}^{N} c_k^{(\pm)} \mid g_k \rangle, \qquad (4.152)$$

with variational parameters $c_k^{(\pm)}$ we obtain an approximation to the T-matrix

$$T_{\text{fi}}^N = \sum_{k,l=1}^{N} \langle \phi_{\text{f}} \mid V \mid g_k \rangle (M^{-1})_{kl} \langle g_l \mid V \mid \phi_{\text{i}} \rangle, \qquad (4.153)$$

where the matrix M is given by $M_{kl} = \langle g_k|V - VG_0V|g_l\rangle$. The set of vectors $\{|g_k\rangle\}_{k=1}^{N}$ can be chosen arbitrarily (not necessarily orthogonal) provided that M_{kl} is a regular matrix. The Schwinger–Lanczos method (SLM) was proposed for calculation of the diagonal T-matrix elements (Meyer et al. 1991), that is, $|\phi_f\rangle = |\phi_i\rangle = |\phi\rangle$, needed for evaluation of elastic cross sections. In this method, T_{fi}^N is calculated according to Equation 4.153, with $|g_1\rangle = |\phi\rangle\langle\phi|V|\phi\rangle^{-1/2}$ and the set $\{|g_k\rangle\}_{k=1}^{N}$ is taken as V-orthogonal

$$\langle g_k \mid V \mid g_l \rangle = \delta_{kl} \qquad (4.154)$$

and such that the matrix VG_0V is tridiagonal

$$\langle g_{k-1} \mid VG_0V \mid g_k \rangle = \langle g_k \mid VG_0V \mid g_{k-1} \rangle = \beta_{k-1}, \qquad (4.155)$$

$$\langle g_k \mid VG_0V \mid g_k \rangle = \alpha_k, \qquad (4.156)$$

$$\langle g_k \mid VG_0V \mid g_l \rangle = 0 \quad \text{for} \quad |k - l| \geq 2. \qquad (4.157)$$

Let us note that the complex-symmetric scalar product (i.e., without complex conjugation) is used throughout this section and not the usual (Hermitian) one, since G_0 is a symmetric but non-Hermitian operator. Only the matrix element $(M^{-1})_{11}$ is needed in Equation 4.153 and this element is for a tridiagonal matrix M easily expressible in the form of a continued fraction. The T-matrix then reads

$$T^N = \langle \phi \mid V \mid g_1 \rangle (M^{-1})_{11} \langle g_1 \mid V \mid \phi \rangle = \frac{\langle \phi \mid V \mid \phi \rangle}{1 - \alpha_1 -} \frac{\beta_1^2}{1 - \alpha_2 -} \frac{\beta_2^2}{1 - \alpha_3 -} \cdots \frac{\beta_{N-1}^2}{1 - \alpha_N}. \qquad (4.158)$$

The numbers α_k, β_k and the vectors $|g_k\rangle$ with properties (4.154) through (4.157) are constructed according to the recurrence (see Meyer et al. 1991 and references therein for more details about the Lanczos algorithm)

$$\mid r_k \rangle = G_0V \mid g_k \rangle - \beta_{k-1} \mid g_{k-1} \rangle, \qquad (4.159)$$

$$\alpha_k = \langle g_k | V | r_k \rangle, \tag{4.160}$$

$$| s_k \rangle = | r_k \rangle - \alpha_k | g_k \rangle, \tag{4.161}$$

$$\beta_k = \langle g_k | V | g_k \rangle^{1/2}, \tag{4.162}$$

$$| g_{k+1} \rangle = \beta_k^{-1} | s_k \rangle, \tag{4.163}$$

with $|g_1\rangle = |\varphi\rangle \langle \varphi | V | \varphi \rangle^{-1/2}$ and $\beta_0 = 0$.

The off-diagonal matrix elements of the T-matrix are needed for the calculation of resonant electron–molecule and ion–atom collisions. The same Lanczos basis with the properties (4.154) through (4.157) generated by the algorithm (4.159) through (4.163) and with the starting vector $|g_1\rangle = |\varphi_i\rangle\langle\varphi_i|V|\varphi_i\rangle^{-1/2}$ can be used in such a case. The formula for the T-matrix element resulting from Equation 4.153 will be slightly more complicated than in the previous case

$$T_{fi}^N = \sum_{k=1}^{N} \langle \phi_f | V | g_k \rangle (M^{-1})_{k1} \langle g_1 | V | \phi_i \rangle. \tag{4.164}$$

Note that now the first column of M^{-1} is needed instead of the single element M_{11}^{-1}. Using elementary algebra, it turns out that

$$(M^{-1})_{k \cdot 1} = \frac{\beta_1}{f_1} \frac{\beta_2}{f_2} \cdots \frac{\beta_{k-1}}{f_{k-1}} \frac{1}{f_k}, \tag{4.165}$$

where the quantities

$$f_k = 1 - \alpha_k - \beta_k^2 / f_{k+1}, \quad f_N = 1 - \alpha_N \tag{4.166}$$

are involved also in the calculation of the continued fraction (4.158). As pointed out in Meyer et al. (1991), this approach does not treat in and out states in a symmetrical manner. For this reason, the so-called band Lanczos algorithm (Meyer and Pal 1989) was proposed by Meyer et al. (1991) which leads to a banded instead of a tridiagonal matrix. Another approach preserving tridiagonality of the matrix, but using different basis sets for in- and outgoing states was proposed by Čížek et al. (2000).

It is also shown by Meyer et al. (1991) that SLM is equivalent to the method of continuous fractions of Horáček and Sasakawa (1983), see also Horáček and Sasakawa (1984, 1985).

4.4.7 ANOTHER METHOD OF SOLUTION OF NONLOCAL DYNAMICS

We tried to present the details of all the steps needed to solve the nonlocal nuclear dynamics in the preceding sections. Before discussing the results of the calculations for several systems in the following chapter, we would also like to mention briefly some alternative methods of solution of the equations of the nonlocal resonance theory.

The first step of the abovementioned procedure was solution of the local problem. The alternative to the direct solution of the Lippmann–Schwinger Equation

4.117 in the coordinate representation is to use the R-matrix basis. The advantage of this approach is that the R-matrix is found in one step and it can be then used for all energies (Kolorenč et al. 2002), whereas the numerical solution on the grid must be repeated for each energy.

The solution of the Lippmann–Schwinger equation for the nonlocal problem with the potential written in terms of the separable expansion (4.142) can be found directly by expansion in the vibrational basis (extended with the discretized continuum). This approach converges slower than Schwinger–Lanczos method, but it was used, for example, by Hickman (1991) for vibrational excitation of H_2. Alternative separable approximation to the nonlocal resonance potential (the Bateman approximation) was also discussed by Houfek et al. (2002).

To complete the list of the methods, we must also briefly mention the time-dependent formulation of the nonlocal resonance theory (see Domcke 1991 and references therein), the semiclassical method of Kazansky and Kalin (1990), Kalin and Kazansky (1990), the discretization of the full dynamics in the P-space (Kazansky 1996) or direct solution of the Schrödinger Equation 4.66 using the exterior complex scaling methods (Houfek et al. 2008).

ACKNOWLEDGMENTS

Support from the Czech Science Foundation (GAČR) by Grant No. 208/10/1281 and by Záměr MSM0021620860 and by the Center of Theoretical Astrophysics LC06014 of the Ministry of Education, Youth and Sports of the Czech Republic is gratefully acknowledged.

REFERENCES

Bardsley, J. N. 1968. Configuration interaction in the continuum states of molecules. *J. Phys. B 1*(3), 349–364.

Berman, M., C. Mündel, and W. Domcke. 1985. Projection-operator calculations for molecular shape resonances: The $^2\Sigma_u^+$ resonance in electron-hydrogen scattering. *Phys. Rev. A 31*(2), 641–651.

Bieniek, R. J. 1978. Complex potential and electron spectrum in atomic collisions involving fast electronic transitions: Penning and associative ionization. *Phys. Rev. A 18*(2), 392–413.

Bieniek, R. J. 1980. A source of errors in cross sections of curve-crossing processes. *J. Phys. B 13*(22), 4405–4416.

Bieniek, R. J. and A. Dalgarno. 1979. Associative detachment in collisions of H and H⁻. *Astrophys. J. 228*, 635–639.

Cederbaum, L. S. and W. Domcke. 1981. Local against non-local complex potential in resonant electron-molecule scattering. *J. Phys. B 14*(23), 4665–4689.

Chase, D. M. 1956. Adiabatic approximation for scattering processes. *Phys. Rev. 104*(3), 838–842.

Chen, J. C. Y. 1966. Dissociative attachment in rearrangement electron collision with molecules. *Phys. Rev. 148*(1), 66–73.

Čížek, M., J. Horáček, and H.-D. Meyer. 2000. Schwinger-Lanczos algorithm for calculation of off-shell T-matrix elements and Wynn's epsilon algorithm. *Comput. Phys. Comm. 131*(1–2), 41–51.

Crawford, O. H. 1967. Bound states of a charged particle in a dipole field. *Proc. Phys. Soc.* *91*(2), 279–284.

Domcke, W. 1991. Theory of resonance and threshold effects in electron-molecule collisions: The projection-operator approach. *Phys. Rep.* *208*(2), 97–188.

Fabrikant, I. I. 1985. *R*-matrix theory of vibrational excitation of the HCl molecule by slow electrons in the adiabatic approximation. *J. Phys. B* *18*(9), 1873–1879.

Fabrikant, I. I. 1990. Resonance processes in e-HCl collisions: Comparison of the *R*-matrix and the nonlocal-complex-potential methods. *Comments At. Mol. Phys.* *24*(1), 37–52.

Fano, U. 1961. Effects of configuration interaction on intensities and phase shifts. *Phys. Rev.* *124*(6), 1866–1878.

Feshbach, H. 1958. Unified theory of nuclear reactions. *Ann. Phys.* *5*(4), 357–390.

Feshbach, H. 1962. A unified theory of nuclear reactions. II. *Ann. Phys.* *19*(2), 287–313.

Gauyacq, J. P. and A. Herzenberg. 1982. Nuclear-excited Feshbach resonances in e+HCl scattering. *Phys. Rev. A* *25*(6), 2959–2967.

Gemperle, F. and J. Horáček. 1997. Threshold peak structures in the vibrational excitation of HCl by low-energy electron impact. *Czech. J. Phys.* *47*(3), 305–315.

Gillman, E. and H. R. Fiebig. 1988. Accurate recursive generation of spherical Bessel and Neumann functions for a large range of indices. *Comput. Phys.* *2*(1), 62–74.

Goldberger, M. L. and K. M. Watson. 1964. *Collision Theory*. New York, NY: John Wiley & Sons.

Gordon, R. G. 1969. New method for constructing wavefunctions for bound states and scattering. *J. Chem. Phys.* *51*(1), 14–25.

Gordon, R. G. 1970. Constructing wavefunctions for nonlocal potentials. *J. Chem. Phys.* *52*(12), 6211–6217.

Hickman, A. P. 1991. Dissociative attachment of electrons to vibrationally excited H_2. *Phys. Rev. A* *43*(7), 3495–3502.

Horáček, J. 1989. Extrapolation technique for solution of scattering integral equations. *J. Phys. A* *22*(4), 355–363.

Horáček, J. 1995. Application of the Schwinger-Lanczos method to the calculation of dissociative attachment of electrons to diatomic molecules. *J. Phys. B* *28*(8), 1585–1591.

Horáček, J. and W. Domcke. 1996. Calculation of cross sections for vibrational excitation and dissociative attachment in electron collisions with HBr and DBr. *Phys. Rev. A* *53*(4), 2262–2271.

Horáček, J., F. Gemperle, and H.-D. Meyer. 1996. Calculation of dissociative attachment of electrons to diatomic molecules by the Schwinger-Lanczos approach. *J. Chem. Phys.* *104*(21), 8433–8441.

Horáček, J. and T. Sasakawa. 1983. Method of continued fractions with application to atomic physics. *Phys. Rev. A* *28*(4), 2151–2156.

Horáček, J. and T. Sasakawa. 1984. Method of continued fractions with application to atomic physics. II. *Phys. Rev. A* *30*(5), 2274–2277.

Horáček, J. and T. Sasakawa. 1985. Method of continued fractions for on- and off-shell *T* matrix of local and nonlocal potentials. *Phys. Rev. C* *32*(1), 70–75.

Hotop, H., M. W. Ruf, M. Allan, and I. I. Fabrikant. 2003. Resonance and threshold phenomena in low-energy electron collisions with molecules and clusters. *Adv. At. Mol. Phys.* *49*, 85–216.

Houfek, K., M. Čížek, and J. Horáček. 2002. Dissociative attachment of low-energy electrons to vibrationally excited hydrogen molecules. *Czech. J. Phys.* *52*(1), 29–40.

Houfek, K., T. N. Rescigno, and C. W. McCurdy. 2006. Numerically solvable model for resonant collisions of electrons with diatomic molecules. *Phys. Rev. A* *73*(3), 032721.

Houfek, K., T. N. Rescigno, and C. W. McCurdy. 2008. Probing the nonlocal approximation to resonant collisions of electrons with diatomic molecules. *Phys. Rev. A* *77*(1), 012710.

Kalin, S. A. and A. K. Kazansky. 1990. The semiclassical version of the non-local resonance theory of electron-molecule collisions. *J. Phys. B 23*, 4377–4400.

Kazansky, A. K. 1996. Discretization of the electron continuum in the non-local theory of inelastic electron-molecule collisions. *J. Phys. B 29*, 4709–4725.

Kazansky, A. K. and S. A. Kalin. 1990. Semiclassical approximation in the non-local theory of resonance processes in slow collisions of two atoms and an electron. *J. Phys. B 23*, 809–819.

Kolorenč, P., M. Čížek, J. Horáček, G. Mil'nikov, and H. Nakamura. 2002. Study of dissociative electron attachment to HI molecule by using *R*-matrix representation for Green's function. *Phys. Scr. 65*(4), 328–335.

Lippmann, B. A. and J. Schwinger. 1950. Variational principles for scattering processes. I. *Phys. Rev. 79*(3), 469–480.

Meyer, H.-D., J. Horáček, and L. S. Cederbaum. 1991. Schwinger and anomaly-free Kohn variational principles and a generalized Lanczos algorithm for nonsymmetric operators. *Phys. Rev. A 43*(7), 3587–3596.

Meyer, H.-D. and S. Pal. 1989. A band-Lanczos method for computing matrix elements of a resolvent. *J. Chem. Phys. 91*(10), 6195–6204.

Morgner, H. 1990. The validity of the local approximation in Penning ionization as studied by model calculations. *Chem. Phys. 145*(2), 239–252.

Morrison, M. A. and W. Sun. 1995. How to calculate rotational and vibrational cross sections for low-energy electron-molecule scattering from diatomic molecules using close-coupling techniques. In W. M. Huo and F. A. Gianturco (Eds.), *Computational Methods for Electron-Molecule Collisions*, pp. 131–190. New York, NY: Plenum Press.

Movre, M. and W. Meyer. 1997. Theoretical investigation of the autoionization process in molecular collision complexes: Computational methods and applications to He* (2^3S) + H(1^2S). *J. Chem. Phys. 106*(17), 7139–7160.

Nakamura, H. 1971. Associative ionization in slow collisions between atoms. I. *J. Phys. Soc. Jap. 31*(2), 574–583.

O'Malley, T. F. 1966. Theory of dissociative attachment. *Phys. Rev. 150*(1), 14–29.

O'Malley, T. F. 1971. Diabatic states of molecules—Quasistationary electronic states. *Adv. At. Mol. Phys. 7*, 223–249.

Schneider, B. I., M. Le Dourneuf, and P. G. Burke. 1979. Theory of vibrational excitation and dissociative attachment: An *R*-matrix approach. *J. Phys. B 12*(12), L365–L369.

Taylor, J. R. 1972. *Scattering Theory*. New York, NY: Wiley.

Wigner, E. P. 1948. On the behavior of cross sections near thresholds. *Phys. Rev. 73*(9), 1002–1009.

5 Applications of the Nonlocal Resonance Theory to Diatomic Molecules

Karel Houfek and Martin Čížek

CONTENTS

5.1 INTRODUCTION

In Chapter 4, we summarized the nonlocal theory of resonant low-energy electron–molecule collisions and provided its basic equations when applied to collisions with diatomic molecules. In this chapter, we focus on the specific applications of the theory and provide examples of results obtained using the nonlocal theory for processes of vibrational excitation (VE), dissociative attachment (DA), and associative detachment (AD), see Equations 4.1 and 4.3.

The basic idea behind the construction of the nonlocal resonance model (NRM) for various molecular species was given in Section 4.3.2. Let us summarize here that the nonlocal resonance model is given by the potential–energy curve of the neutral molecule, $V_0(R)$, the discrete-state potential of the molecular anion, $V_d(R)$, and the discrete-state-continuum coupling, $V_{de}(R)$, which serve as input into the effective equations for nuclear dynamics of electron–molecule collisions derived in Chapter 4. For most of the systems discussed in this chapter, the NRM was obtained by the least-squares fitting of the *ab initio* fixed-nuclei scattering data, and only as an exception, parameters of the model were determined directly (e.g., for Cl_2) or by fitting the experimental cross section (for HI). References to original papers are provided in the text and we summarize the formulas of nonlocal resonance models for specific diatomic molecules with numerical values of all parameters of these models in Appendix (Section 5.5).

5.2 OVERVIEW OF THE NONLOCAL RESONANCE MODELS FOR DIATOMIC MOLECULES

NRMs were constructed for several diatomic systems. Here we give a summary of the NRMs that are available in the literature for several homogeneous diatomic molecules (H_2, N_2, F_2, and Cl_2) and for hydrogen halides (HF, HCl, HBr, and HI) and show examples of the VE and DA cross sections.

5.2.1 HYDROGEN MOLECULE

The low-energy collisions of electrons with the hydrogen molecules are of importance in several fields of physics, especially in astrophysics, and therefore this system has been the most studied one, both experimentally and theoretically by many different approaches including the nonlocal theory. Although it is the simplest system (at least considering the number of electrons), an accurate calculation of the dynamics of the collision and interpretation of the structures found in the cross sections have proved to be quite difficult.

Here, we discuss the vibrational excitation and dissociative attachment for electron energies up to 6 eV where only the H_2^- $^2\Sigma_u^+$ shape resonance contributes significantly to the cross sections. For this system, two NRMs based on *ab initio* calculations were constructed. The first one by Mündel, Berman, and Domcke (1985), the MBD model, was based on the *ab initio* data for the resonance width $\Gamma(E, R)$ and the level-shift $\Delta(E, R)$ by Berman et al. (1985). The second NRM by Čížek, Horáček, and Domcke (1998), the CHD model, improved the MBD model by taking into account a new data for the H_2^- potential–energy curve calculated by Senekowitsch et al. (1984).

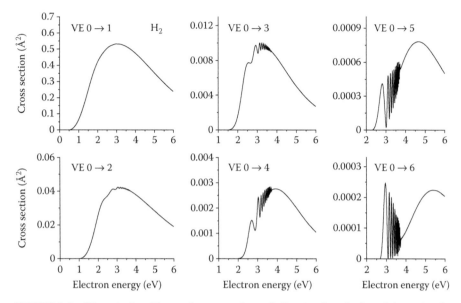

FIGURE 5.1   The calculated integral cross sections of vibrational excitation of the molecule H_2 in the ground rovibrational state by the electron impact. Calculations are based on the nonlocal resonance model of Čížek et al. (1998).

The oscillatory structure which appears in the VE cross sections for transitions to higher vibrational levels as we can see in Figure 5.1 was predicted for the $e^- + H_2$ system by the MBD model and later confirmed by the experiment of Allan (1985). Comparison of the oscillations in the experimental $0 \rightarrow 4$ VE cross section with the results of the CHD model is shown in Figure 5.2. The origin of these oscillations was

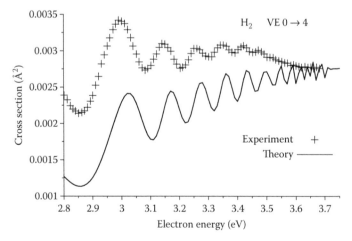

FIGURE 5.2   The detail of the oscillatory structure in the $0 \rightarrow 4$ VE cross sections of H_2. The calculated cross section of Čížek et al. (1998) is compared with the experimental one of Allan (1985).

a puzzle since the paper by Gertitschke and Domcke (1993) was published because the authors proposed that the oscillations in the VE cross sections of H_2 are of a different origin than the well-known *boomerang* oscillations observed in the $e^- + N_2$ system. Although alternative, but quite complicated explanations of the oscillatory structure were proposed in the literature, see for example, Narevicius and Moiseyev (1998, 2000), a simple interpretation based on the original *boomerang* model was finally provided by Horáček et al. (2006). We return to this topic in Section 5.3.2.

There has been a long discussion in the literature about the discrepancy between various experimental and theoretical values of the threshold dissociative attachment cross section. The comparison of the DA cross section for H_2 in its ground rovibrational state ($v = 0$, $J = 0$) obtained using the CHD model with several other theoretical results is shown in Figure 5.3.

The shape of all theoretical cross sections is very similar, perpendicular onset followed by exponential decrease. The magnitudes of the cross sections, however, differ significantly ranging from the smallest value of 1.6×10^{-5} Å^2 to 5×10^{-5} Å^2. The largest value is obtained by the CHD model calculation denoted by a solid line. Next to it is the DA cross section of the MBD model calculated by Gertitschke and Domcke (1993). In the CHD model, the long-range part of the H–H$^-$ potential at large and intermediate internuclear distances possesses the correct polarization asymptotics and thus it is more attractive than the potential used in the MBD model which leads to the increase of the DA cross section near the threshold. Calculation of Fabrikant et al. (2002) uses a similar treatment of the nonlocal dynamics but the parameters of the model are slightly different and the semiclassical approach is used for the solution of the dynamical equation. The cross section obtained by Gauyacq (1985) represents a completely different approach based on the zero-range model of the interaction and a resonance is not explicitly introduced. The theory of Wadehra and Bardsley (1978) and Bardsley and Wadehra (1979) uses the local complex potential approximation and the parameters fitted to the experimental data. Finally, the calculation of Hickman (1991) treats the nonlocal problem in the approximation of

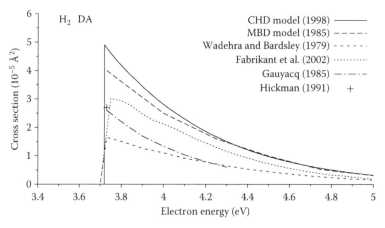

FIGURE 5.3 Selection of the theoretical DA cross sections for the ground rovibrational state of H_2.

open channels. This list of calculations is not complete, see for example, Launay et al. (1991), Mazevet et al. (1999), and so on.

First measurement of the DA cross section in the 4 eV region was carried out by Schulz and Asundi (1965). The H^- formation was observed with a very sharp onset at the electron energy 3.73 ± 0.07 eV. The measured cross section was found to be steeply rising, peaked very close to its onset but small, of the order of 10^{-5} Å^2. The magnitude of the 4 eV peak was measured relative to the height of the 14 eV peak. Schulz and Asundi found that the ratio of the magnitudes of both peaks is approximately 8%. Two absolute measurements of the 14 eV peak exist in the literature: The value of 3.5×10^{-4} Å^2 was measured by Schulz (1959) and the value 2.1×10^{-4} Å^2 was measured by Rapp et al. (1965). According to this uncertainty, two values of the peak cross sections, namely 2.8×10^{-5} Å^2 and 1.6×10^{-5} Å^2, can be found in the literature. The measurement of Schulz and Asundi was done at the temperature of about 300 K with an energy resolution of about 450 meV. If the CHD model cross section is evaluated at this temperature and convoluted with the assumed experimental energy resolution, one obtains a value of 2.8×10^{-5} Å^2 in full agreement with the measurement of Schulz and Asundi. For more details and comparison with other measurements, see Horáček et al. (2004).

Although the absolute value of the DA cross section is still an open problem and other independent experiments and calculations are needed, a recent measurement of the inverse process of associative detachment (Kreckel et al. 2010) supports the results obtained using the most recent nonlocal resonance model of Čížek et al. (1998) which was also used for calculations of the AD cross sections.

As another success of the nonlocal theory, we consider the prediction of the long-lived states of the molecular anions H_2^- and D_2^- (Golser et al. 2005) due to the nuclear motion in the potential well outside the autodetachment region for high rotational quantum numbers ($J > 20$). These states appear in the VE and DA cross sections as very narrow resonances as we can see in Figure 5.4, where the DA cross sections for the initial rovibrational states $v = 0, J = 23$, and $J = 24$ are plotted. These structures are usually too narrow (<1 meV) to be observable by experiment. See Section 5.3.3 for a more detailed discussion on the outer-well resonances and long-lived resonance states.

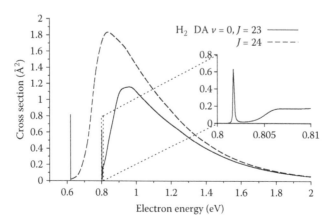

FIGURE 5.4 Narrow resonances in the DA cross sections for high rotational states of H_2.

5.2.2 HYDROGEN HALIDES

The importance of electron collisions with hydrogen halides and their deuterated analogs lies in the richness of structures that were observed in the VE and DA cross sections (threshold peaks and Wigner cusps, oscillatory structures, narrow resonances, see a detailed discussion in Section 5.3). These phenomena demanded for explanation and thus motivated the theoretical work on the resonant electron–molecule collisions. One of the first papers on the full nonlocal theory of electron–molecule collisions by Domcke et al. (1979) was devoted to hydrogen halides. In this paper, the authors constructed the first, rather qualitative nonlocal resonance models for HF, HBr, and HCl and showed that the nonlocal theory is general enough to explain different types of threshold phenomena.

To get an idea of how the nonlocal resonance models describe hydrogen halides and how parameters of the models change from system to system, we plotted the potential–energy curves of these systems in Figure 5.5. The neutral molecule HX (X = F, Cl, Br, I) potential–energy curve $V_0(R)$ together with vibrational states is shown (red curves) and the shaded area corresponds to the electronic continuum.

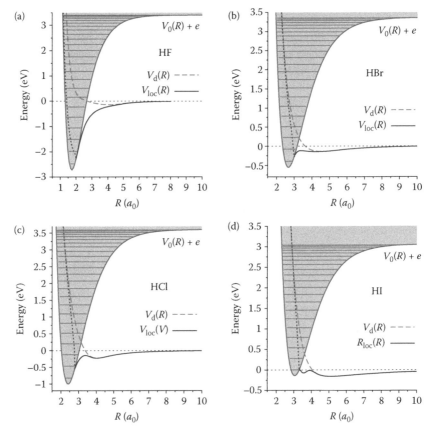

FIGURE 5.5 (**See color insert.**) Potential–energy curves of the nonlocal resonance models for the following hydrogen halides: (a) HF molecule, (b) HBr molecule, (c) HCl molecule, and (d) HI molecule.

The discrete-state potential $V_d(R)$ of the NRM (dashed curves) tends asymptotically as $R \to \infty$ to the negative molecular ion HX^- potential which is shown as $V_{loc}(R)$ (blue curves). The outer wells in the potentials of HCl, HBr, and HI can be easily identified (see the Section 5.3.3 for a discussion on outer-well resonances).

The most studied hydrogen-halide system has been $e^- + HCl$ for which several nonlocal resonance models based on the *ab initio* fixed-nuclei scattering data and quantum-chemical calculations of lowest potential–energy curves of HCl and HCl^- have been constructed. The VE cross section for transition $0 \to 1$ calculated using different models are compared with the experimental data of Schafer and Allan (1991) and of Allan et al. (2000) in Figure 5.6 where improvements of models can be noticed. The first full NRM for HCl was constructed by Domcke and Mündel (1985), the DM model. The model was based on the scattering data of Padial and Norcross (1984) and its results were in quantitative agreement with the available experiments. The next two models improved the DM model in two respects. First, a variable threshold exponent $\alpha(R)$ (see Equation 4.115 and model functions in the Appendix) was added ($DM\alpha(R)$ model) to take into account the internuclear dependence of the molecular dipole moment (Horáček et al. 1998) which turned out to be important to get a proper threshold behavior of the VE cross section where a sharp onset peak appears (the top right panel of Figure 5.6). Later, the DM model was modified by changing the long-range part of the discrete-state potential energy $V_d(R)$ (Čížek et al. 1999) to take into account a new HCl^- potential–energy function that was determined by the accurate *ab initio* calculations of Åstrand and Karlström (1990). The new HCl^- potential is more attractive than the potential used in the DM model resulting in more pronounced oscillatory structure as one can see in the bottom left panel of Figure 5.6 where the results of this new model, denoted as Domcke, Mündel, Horáček, and Čížek (DMHC) model, are shown. Moreover, due to the presence of the outer well at internuclear distances around $4\ a_0$ (see Figure 5.5), two narrow resonances at energies about 0.63 and 0.7 eV appear in the VE cross section. The most recent NRM (Fedor model) for HCl was built on new fixed-nuclei scattering data (Fedor et al. 2010). The motivation for this new model was disagreement of the experimental DA cross section measured recently by Fedor et al. (2008) with the calculated one by Čížek et al. (1999). The DA cross section from the ground vibrational state of HCl obtained by the Fedor model was about twice less than in the previous calculations in agreement with the experiment. Relative magnitudes of the oscillatory structure right below the DA threshold in the VE cross section, which are closely related to the DA cross sections at the threshold (see a detailed discussion in Section 5.3.2), also decreased as we can notice in the last panel of Figure 5.6.

Another well-studied system is $e^- + HBr$. There are also several NRMs for this system. The first NRM based on the *ab initio* scattering data of Fandreyer et al. (1993) was the model of Horáček and Domcke (1996) who published the cross sections for transitions between the lowest vibrational states and also the DA cross section from $v_i = 0, 1, 2$. They also discussed the isotope effect by calculating the VE and DA cross sections for DBr using the same model. Their results were in good agreement with available experimental data of Abouaf and Teillet-Billy (1980) for dissociative electron attachment but the experimental data of Rohr (1978) for the VE cross section were significantly larger. The NRM for HBr of Horáček and Domcke (1996) was

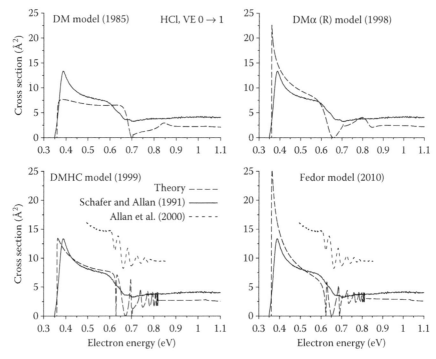

FIGURE 5.6 The calculated integral cross section of vibrational excitation $0 \to 1$ of the molecule HCl for different nonlocal resonance models (see text for references). The experimental cross sections of Schafer and Allan (1991), scaled to the threshold peak of the DMHC model, and of Allan et al. (2000), scaled and shifted, are also shown.

later modified using a variable threshold exponent (Horáček et al. 1998) resulting in the more pronounced threshold peaks in the VE cross sections and the threshold onset of the DA cross section and finally by including a newly calculated bound part of the HBr⁻ potential–energy function (Čížek et al. 2001). Results of this last model for both HBr and DBr were found in very good agreement with the measurements of Sergenton, Popović, and Allan published in the same paper. Comparison of the theoretical and experimental DA cross sections for HBr and DBr is shown in Figure 3.17 and of the $0 \to 1$ VE cross section for HBr in Figure 3.18, see Chapter 3. As we can see, agreement of the theory with experiment is almost perfect. In Figure 5.7, a representative sample of the cross sections for HBr is shown together with a description of the main features observed in the resonant electron–molecule collisions. (We will discuss these structures later.) The temperature-averaged cross sections of dissociative electron attachment to HBr were discussed by Fedor et al. (2007). Again they found very good agreement of predictions of the nonlocal theory and measurements.

The system $e^- + HF$ is an example of a system where the molecule possesses a supercritical dipole moment. As a consequence, the threshold exponent α in Equation 4.115 becomes complex and the energy dependence of the resonance width $\Gamma(E)$ must be modified resulting in rapid oscillations as $E \to 0$, see Fabrikant (1977) and Gallup et al. (1998) for details. To avoid the problems of determining the resonance

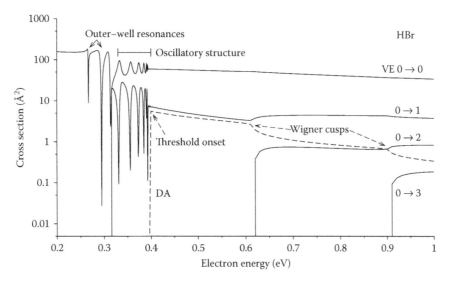

FIGURE 5.7   The calculated integral cross sections of vibrational excitation and dissociative attachment for the molecule HBr up to 1 eV. The results of the nonlocal resonance model of Čížek et al. (2001) are shown.

shift $\Delta(E, R)$ as the Hilbert transform of $\Gamma(E, R)$, the NRM of Čížek et al. (2003) for HF fitted both $\Gamma(E, R)$ and $\Delta(E, R)$ independently using the *ab initio* calculations of these quantities by Gallup et al. (1998) and neglected the oscillations of $\Gamma(E, R)$ (see the Appendix for details on model functions). As an example, both theoretical and experimental VE cross sections of transitions $0 \rightarrow 3$ are shown in Figure 5.8, see Čížek et al. (2003) for more cross sections.

The molecule HI is the only hydrogen halide for which parameters of the NRM were fitted to the experimental cross section because *ab initio* electron scattering calculations had not been available. Horáček et al. (1997) built the NRM for HI by fitting the DA cross section of Klar et al. (1994) in the energy range 0–200 meV and used this model to calculate the DA cross section for higher energies and the VE cross sections up to 4.5 eV for various transitions. Predictions of the NRM for vibrational excitation to higher levels were later found in rather good agreement with the experiment as it is shown for the $0 \rightarrow 2$ transition in Figure 3.19a, see Chapter 3. The dependence of the DA cross section on the initial rotational state of the molecule HI was discussed in some more extent by Kolorenč et al. (2002). Although the DA and VE cross sections for DI were also presented by Horáček et al. (1997), a rather unexpectedly large resonant contribution to elastic electron scattering of DI at energies below the DA threshold was calculated and published later by Horáček et al. (2007). A series of sharp peaks of very large amplitudes of the order of hundreds and even thousands of Å^2 appears in the elastic cross section for the molecule in its ground rovibrational state. These structures can be interpreted as long-lived resonant states in the outer well of the negative molecular ion potential (see Figure 5.9a). If a finite temperature and finite resolution of the experiment are

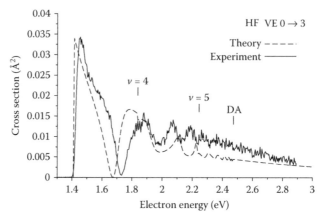

FIGURE 5.8 Calculated (dashed) and measured (solid), scaled to the threshold peak, VE $0 \rightarrow 3$ cross sections for HF. (Adapted from Čížek, M. et al. 2003. *J. Phys. B 36*(13), 2837–2849.)

taken into account, these sharp peaks are broadened (as in Figure 5.9b) but still these structures should be observable experimentally with the beam energy resolution in the range of few meV.

Let us conclude this section by giving references to other works discussing related topics for hydrogen halides. The associative detachment cross sections and electron spectra were published by Čížek et al. (2001) and Živanov et al. (2002) and comparison with electron-energy-loss experiments for HCl and HBr was reviewed in Čížek et al. (2002). Isotopic effects for DCl and DBr were discussed by Horáček et al. (2005).

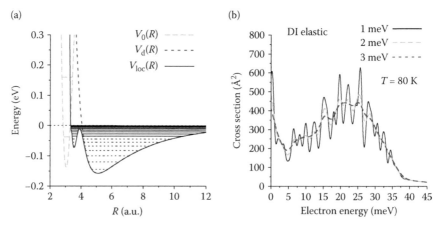

FIGURE 5.9 (a) Potential–energy curves of the e^- + DI system with the long-lived resonant states (solid lines) and the bound states (short dashed lines) of DI$^-$ in the outer well of the potential $V_{loc}(R)$. (b) Calculated resonance contribution to the elastic electron scattering cross section for nonzero temperature of DI, $T = 80$ K, and electron beams with FWHM 1, 2, and 3 meV. Data were obtained using the NRM of Horáček et al. (1997).

5.2.3 OTHER DIATOMIC MOLECULES

There are several other diatomic systems, except for H_2 and hydrogen halides and their deuterated analogs, for which the NRMs were constructed to calculate the VE and/or DA cross sections. As a benchmark system, one can consider the electron collisions with the molecule N_2. A pronounced oscillatory structure in the VE cross sections in the vicinity of the resonance energy 2.3 eV, first observed by Schulz (1962), motivated an introduction of the local complex potential approximation by Herzenberg (1968), which provided a qualitative explanation of the observed cross sections, and inspired a series of theoretical works employing different approaches to the description of nuclear dynamics, see, for example, Chandra and Temkin (1976) and Schneider et al. (1979). The nonlocal resonance model constructed by Berman et al. (1983), based on the data from *ab initio* calculations of the $^2\Pi_g$ shape resonance of N_2^- by Hazi et al. (1981), was probably the first complete quantum-mechanical calculation of nuclear dynamics beyond the local complex potential (LCP) approximation for a real system.

Other theoretically and experimentally interesting processes are electron collisions with halogen molecules. The dissociative electron attachment to these molecules has attracted a lot of attention for there has been long-time disagreement between the theoretical behavior of the DA cross sections at very low energies and experimental measurements. As the consequence of the symmetry of the lowest electronic $^1\Sigma_g^+$ states of halogen molecules and $^2\Sigma_u^+$ resonance state of halogen-molecule anions, the theory predicted that the DA cross section threshold behavior should be decreasing as $E^{1/2}$ for $E \to 0$ (Wigner 1948) because the zero angular momentum of the incoming electron does not contribute. On the other hand, the experiment found the DA cross sections increasing at energies close to zero for both F_2 (Chutjian and Alajajian 1987) and Cl_2 (Tam and Wong 1978, Azria et al. 1982). This discrepancy remained unexplained until the *p*-wave threshold behavior of the DA cross sections was recently confirmed by new high-resolution measurements first for Cl_2 by Barsotti et al. (2002) which were found in very good agreement with calculations based on the semiempirical *R*-matrix theory (Fabrikant et al. 2000, Ruf et al. 2004) and later also for F_2 by Braun et al. (2007). The molecules F_2 and Cl_2 were the first systems for which the discrete-state energy $V_d(R)$ and discrete-state-continuum coupling $V_{d\varepsilon}(R)$ of the NRM were directly extracted from *ab initio* fixed-nuclei *R*-matrix calculations by the Fano–Feshbach-*R*-matrix method (Nestmann 1998, Kolorenč et al. 2005). Brems et al. (2002) and Kolorenč and Horáček (2006) applied this approach to determine the DA and VE cross sections for various initial and final vibrational states of F_2 and Cl_2, respectively.

5.3 STRUCTURES IN THE CROSS SECTIONS AND THEIR INTERPRETATION

Resonant electron collisions with molecules attracted molecular physicists primarily because of richness of structures appearing in the VE and DA cross sections. Demand for explanation of these diverse features stimulated research for several decades. Nowadays, it seems that finally both qualitative and quantitative agreement between theory and experiment was achieved for several systems and the origin of these phenomena can be considered as known. In the following paragraphs, we provide plausible explanations of some of these structures based on the nonlocal theory.

To illustrate the richness of the structures in the VE and DA cross sections, we show in Figure 5.10 a collection of the cross sections for the molecule HCl in the initial ground state as calculated using the latest NRM of Fedor et al. (2010). Pronounced threshold peaks and a sharp onset at the threshold often appear in the VE and DA cross sections, respectively, for polar molecules such as HCl. Wigner cusps appear in the DA cross sections at the openings of higher VE channels and oscillatory structures and narrow, outer-well resonances appearing below the DA threshold in the VE cross sections result from a shape of the long-range part of the negative molecular ion potential–energy curve which governs, together with the complicated, short-range nonlocal potential, the nuclear dynamics of the system.

5.3.1 Threshold Effects and Wigner Cusps

The first experimental evidence of the pronounced threshold peaks in the electron-impact vibrational excitation cross sections for hydrogen halides and H_2O was provided by Rohr and Linder (1976) and Seng and Linder (1976) which was later confirmed by several subsequent experiments (see, e.g., Čížek et al. 2002 and references therein). The manifestation of the threshold effects in the dissociative attachment cross sections for electron collisions with diatomic molecules is twofold. First there is a sharp onset of the cross sections at the DA threshold, second there are abrupt decreases at the openings of the VE channels which are called Wigner cusps (see, e.g., Abouaf and Teillet-Billy 1977, 1980).

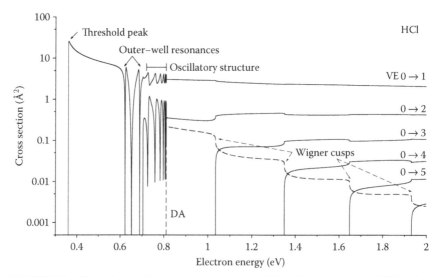

FIGURE 5.10 The calculated integral cross sections of vibrational excitation (full lines) and dissociative attachment (dashed line) for the molecule HCl up to 2 eV. The results of the nonlocal resonance model of Fedor et al. (2010) are shown.

Several possible mechanisms of the enhancement of the cross sections at the threshold were initially proposed which included the effects of virtual states (Dubé and Herzenberg 1977, Kazanskii 1982), s-wave-bound states (Teillet-Billy and Gauyacq 1984, Gauyacq and Herzenberg 1982), or resonances (Domcke et al. 1979, Domcke and Cederbaum 1980). But it was soon recognized that the difference between these mechanisms is vague since all of them are included in a unified way within the nonlocal theory, which provides a particularly suitable theoretical framework to describe and explain all these threshold phenomena and to distinguish among them is neither necessary nor sometimes possible. A detailed discussion of analysis of potential–energy curves involved in electron–molecule collisions as singularities of the fixed-nuclei electron–molecule scattering matrix was given by Domcke (1991) and it is out of the scope of this book. Let us only mention that the basic ingredient of a success of the nonlocal theory is a correct energy-dependence of the discrete-state-continuum coupling $V_{de}(R)$ (or resonance width $\Gamma(E, R)$) as discussed briefly in Section 4.3.2.

5.3.2 Oscillatory Structure in the Vibrational Excitation Cross Sections

The oscillatory structure that appears in the vibrational excitation cross sections of the resonant electron–molecule collisions was for the first time observed experimentally in the system $e^- + N_2$ at energies around 2.3 eV (Schulz 1962). This notable feature was qualitatively explained by the so-called *boomerang model* (Herzenberg and Mandl 1962, Herzenberg 1968) as interference between a direct electron detachment and a time-delayed electron detachment due to the boomerang-like motion of the nuclei to larger internuclear distances. Subsequent calculations based on both the LCP approximation with an empirical complex potential–energy curve for N_2^- (Birtwistle and Herzenberg 1971, Dubé and Herzenberg 1979) and the nonlocal resonance model (Berman et al. 1983) constructed from the *ab initio* calculations of the fixed-nuclei resonance parameters (Berman and Domcke 1984) confirmed the validity of this picture.

A similar oscillatory structure was also observed in several other diatomic systems, for example, in CO (Schulz 1964), NO (Schulz 1964), and O_2 (Schulz and Dowell 1962). As the resolution of experiments was increasing, it was finally possible to measure even fine oscillatory structures in the cross sections of vibrational excitation of H_2 (to higher final vibrational states) at energies around 3 eV (Allan 1985) and of HCl and other hydrogen halides (Allan et al. 2000, Čížek et al. 2001, 2003), which were first predicted by the theory based on the nonlocal resonance model for H_2 (Mündel et al. 1985) and HCl (Domcke and Mündel 1985).

Although the detailed properties of the oscillatory structures appearing in the VE cross sections can be fully quantitatively determined using the nonlocal theory, most of its basic characteristics are satisfactorily explained within the boomerang model. According to this simple model, the oscillations can be understood as a consequence of the interference of two processes. The first one is the direct decay of the intermediate anion (usually a shape resonance) in the autodetachment region with little change in the molecular geometry. This process determines the overall envelope of the cross section and its contribution to the VE cross section is sometimes called the adiabatic approximation (Chase 1956). The second process is the motion of the nuclei away from the autodetachment region (extension of the internuclear distance R beyond the crossing point of the potential–energy curves of the neutral molecule and the molecular anion) until the attractive part of the potential $V_d(R)$ inverts their motion, sending them back into autodetachment region, where the molecular anion again decays by releasing an electron. If the width of the resonance electronic state is relatively small, that is, the decay time of the molecular anion is comparable with the period of its vibrational motion, then a part of the wave packet moves again to larger internuclear distances and the process repeats, for some system even several times, till majority of the wave packet decays. As the wave packet has to be reflected for interference to appear, the second process is allowed only below the DA threshold which explains an abrupt end of the oscillations for example in the cross section of vibrational excitation of HCl (Figure 5.10) or H_2 (Figure 5.11).

The magnitude of the oscillations is given by the relative amplitudes of the two interfering processes and if correct quantitative results are to be obtained one has to take full account of nonlocal effects. In the system like $e^- + H_2$, we find that the absolute magnitude of the oscillations near the DA threshold is almost the same in the VE cross sections from a given initial vibrational state v_i to different final states v_f, but is rapidly growing when the vibrational or rotational quantum number v_i of the initial state is increasing. The former behavior is illustrated in Figure 5.11, where the VE cross sections $v_i = 0 \rightarrow v_f = 0-6$ for energies near the DA threshold are shown. The latter effect can be observed in Figure 5.12. We can explain both behaviors if we realize that the magnitude of the oscillations is proportional to the amplitude of the reflected wave and that this amplitude depends strongly on the initial vibrational

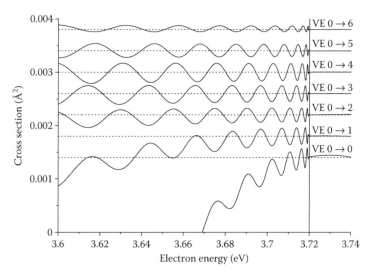

FIGURE 5.11 The cross sections of vibrational excitation of the molecule H_2 in the ground rovibrational state by the electron impact shown in the vicinity of the DA threshold (vertical line). A smooth background (given by linear approximation of the energy dependence of the cross section in the range $E = 3.72-3.74$ eV) was subtracted from each curve, and the results are vertically displaced for better visibility of the oscillations.

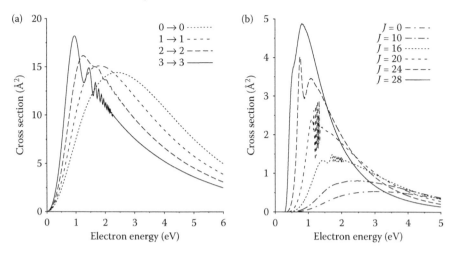

FIGURE 5.12   The oscillatory structure in the VE cross section for H_2 is more pronounced as the vibrational (a) and rotational (b) initial state of the molecule is increasing.

state, but not at all on the final one. The kinetic energy of the nuclear motion of the molecule in a higher initial vibrational state is larger and consequently it is more probable for the molecular anion to leave the autodetachment region before it releases an electron. The amplitude of the reflected wave will therefore be larger relative to the amplitude of the directly decaying wave.

It is noteworthy to mention here the connection between the magnitude of the oscillations in the VE cross section and the magnitude of the DA cross section near the threshold. The amplitude of the reflected nuclear wave functions right below the DA threshold is almost energy independent (see Figure 5.11) and is comparable to the amplitude of the outgoing waves in the DA channel near the threshold. We can thus conclude that the magnitude of the oscillations will be proportional to the DA cross section near the threshold for the same initial state. This is actually observed in H_2 when increasing the rotational or vibrational quantum number of the initial state. (Both the magnitude of the oscillations and the DA cross section are rapidly increasing.)

Other important features of the oscillations are their narrowing with increasing electron energy toward the DA threshold and the amplitude inversion that was observed early in the system $e^- + N_2$ by Herzenberg (1968) and which means that if the VE cross section in a given vibrational channel, $v_i \rightarrow v_f$, has a maximum at energy E, then the vibrational excitation cross section to the next higher channel, $v_i \rightarrow v_f + 1$, has a dip at almost the same energy (see again Figure 5.11). The narrowing of the oscillations is related to the relative phases of the reflected waves for different energies. Qualitatively speaking, the wave with energy approaching the DA threshold has to travel farther in the long-range part of the potential $V_d(R)$. The energy difference of two waves which will be reflected with the same relative phase is decreasing toward the DA threshold, similarly as the energy distance of two bound states in the long-range potential. The amplitude inversion appears as a consequence of the overlap of the reflected wave function (the same for all final states) with the different final molecular vibrational wave functions. Qualitatively, if the vibrational state v_f is in phase with the reflected wave at a given energy (maximum in the VE cross section), then the vibrational state $v_f + 1$ is almost out of phase with that reflected wave.

5.3.3 OUTER-WELL RESONANCES AND LONG-LIVED STATES OF THE MOLECULAR ANIONS

Outer-well resonances are narrow structures in the VE cross sections that appear usually as irregularities in the otherwise rather regular oscillatory structure discussed above. We have seen examples of such resonances in the VE cross section for HCl in Figure 5.10 and their effect on the large increase of the elastic cross section for DI in Figure 5.9. For higher rotational molecular states such resonances can also lie above the DA threshold and they can thus be observed in DA cross section as demonstrated in Figure 5.4.

The origin of these resonances is connected, as their name indicates, to the shape of the negative molecular ion potential–energy curve at larger internuclear distances. (The outer well lies outside the so-called autodetachment region where the probability of electron escaping the molecule is nonzero.) The outer well can be a direct result of electronic interaction as in hydrogen halides (see Figures 5.5 and 5.9). It can also appear as a dynamical effect when the molecule rotates. For example, there is no outer well on the H_2^- potential curve (Figure 5.13a), but such an outer well clearly appears if the angular momentum $J = 20,\dots,27$ (Figure 5.13b). For nonzero angular momentum, there is also a potential–energy barrier at large internuclear distances, which prevents the molecular anion to dissociate. It should be clear that if the outer well is not separated from the autodetachment region by the sufficiently high potential barrier the resonance states in this well are shortlived and the difference between the oscillations as described in the previous section and the outer-well resonance in the VE cross section becomes blurred. A more detailed discussion of

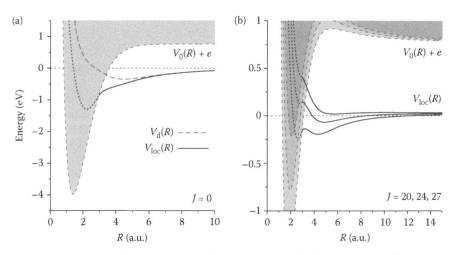

FIGURE 5.13 (**See color insert.**) Potential–energy curves (including the centrifugal term $J(J+1)/2\mu R^2$) of the $e^- + H_2$ system for angular momentum $J = 0$ (a) and $J = 20, 24, 27$ (b). The shaded area above $V_0(R)$ represents the electron continuum. $V_{loc}(R)$ corresponds to the energy of the (electronically) bound state (solid line) or to the pole of the K-matrix (dotted line).

TABLE 5.1

Theoretical and Experimental Results for Resonances with Lifetimes in the Microsecond to Millisecond Range

	J	v	E	τ (Theory)	τ (Experiment)
H_2^-	26	0	5 (−1)	0.25 (0.38)	8.2 ± 1.5
	27	0	28 (22)	0.002 (6.3)	
HD^-	31	0	22 (15)	23 (49)	50.7 ± 1
	30	0	1.7 (−5)	2 (4)	
	30	1	20 (17)	0.6 (0.8)	
D_2^-	38	0	19 (12)	2108 (3900)	1890 ± 80
	37	0	2 (−5)	61 (140)	84 ± 3
	37	1	18 (15)	16 (17)	23 ± 3
T_2^-	47	0	23	96 ms	
	46	0	9	21 ms	

Note: The rotational, J, and vibrational, v, quantum numbers are given for each resonance. The lifetimes τ are given in μs (unless stated otherwise). The energy E of each resonance relative to the dissociation threshold is given in meV. The values within parentheses illustrate the sensitivity of the energies and lifetimes with respect to small (10 meV) changes in the potential.

irregular patterns in the oscillatory structure due to the presence of the outer-well resonances was given by Houfek et al. (2008).

If the inner potential barrier between the autodetachment region and the outer well is high, the resonance states can live for a very long time as compared to the rotational or vibrational period of the molecule. Such states of H_2^- and D_2^- were actually observed experimentally and explained within the nonlocal theory by Golser et al. (2005). Using the projection-operator technique, the lifetimes of the H_2^-, HD^-, D_2^-, and T_2^- were determined from the nonlocal resonance model of Čížek et al. (1998). The longest lifetimes found for these anions by Čížek et al. (2007) are summarized in Table 5.1, see also Figure 5.14. The numbers collected in the table are in agreement with the measurement of the lifetimes in the ion beam trap (Heber et al. 2006), but it is necessary to take into account the sensitivity of the results to the small changes in the anion potential. The numbers in parentheses thus give the values of energies and lifetimes of the states when the potential is perturbed. All measured lifetimes are consistent with the data, provided that such a small perturbation is allowed. This is only indirect evidence that the observed negative ions are explained as the outer-well resonances. The direct evidence of high angular momentum of the observed anion states was provided by photofragment imaging of D_2^- ions (Lammich et al. 2009).

5.4 TEMPERATURE DEPENDENCE OF THE CROSS SECTIONS AND RATE CONSTANTS

Rate constants which can be determined from the temperature-dependent cross sections are of importance for the modeling of plasma processes, environmental

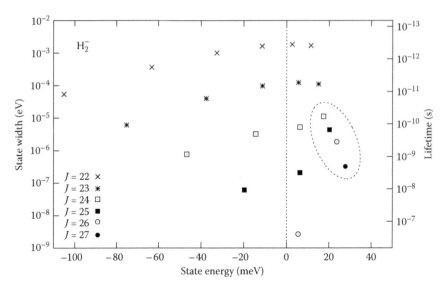

FIGURE 5.14 Summary of the energies and decay widths/lifetimes of outer-well resonance states in H_2^-. Energies are given relative to the $H + H^-$ threshold. The resonances with high dissociation rates are encircled.

chemistry, and so on. In this section, we summarize basic formulas defining rate constants for inelastic processes studied within the nonlocal theory of electron–molecule collisions. Because the rate constants at low temperatures are very sensitive to small changes in the threshold values of the cross sections which are on the other hand sensitive to the parameters of the nonlocal resonance model, the comparison of the calculated rate constants with the measured ones represents a severe test of the theory.

Generally, the rate constant for a reaction in the thermalized gas is defined as

$$k(T) = \int_0^\infty vf(v,T)\sigma(v,T)dv, \qquad (5.1)$$

where $f(v,T)$ is a normalized distribution of the relative velocities v at a given temperature T and $\sigma(v,T)$ is the temperature-averaged cross section of the reaction. For thermalized gas, the Maxwellian distribution is assumed which is given by

$$f(v,T) = 4\pi v^2 \left(\frac{\mu}{2\pi k_B T} \right)^{3/2} e^{-\frac{\mu v^2}{2k_B T}}, \qquad (5.2)$$

where μ is the reduced mass of the colliding particles and k_B is the Boltzmann constant.

Further we focus our attention to two processes, namely dissociative attachment (rate constants denoted by $\beta(T)$) and its reverse process associative detachment (rate constants denoted by $k(T)$). The principle of detailed balance can be used to obtain, for example, the AD cross section from the DA cross sections but in practice these cross sections are usually calculated separately because of contributions of many partial waves to the AD cross section. We also present examples of results for hydrogen halides as obtained using the nonlocal resonance models of Čížek et al. (1999) for HCl and DCl, of Čížek et al. (2001) for HBr and DBr, and of Horáček et al. (1997) for HI and DI.

The rate constants for hydrogen molecule are of great importance for applications in plasma physics and astrophysics. Therefore, the DA rate constants for many H_2 initial states were calculated using the NRM of Čížek et al. (1998) and published by Horáček et al. (2003). All data can be found in the database of National Institute for Fusion Science from Japan (http://dpc.nifs.ac.jp/DB/DA/). Here we show as an example the temperature dependence of the DA rate constant for H_2 calculated by Equation 5.2 in Figure 5.15. At room temperatures, the values are very small and unmeasurable (e.g., the value $\beta = 1.3 \times 10^{-12}$ cm^3 s^{-1} is reached for $T = 3000$ K).

5.4.1   DISSOCIATIVE ATTACHMENT

Since the mass of a molecule is much larger than the electron mass, that is, $m_e \ll m_{AB}$, we can write $\mu \simeq m_e$ and the velocity distribution of electrons in the DA process is practically equal to the distribution of the relative velocities given by Equation 5.2. The two-temperature rate constant for the process of dissociative attachment (as motivated by swarm experiments with different temperatures of electrons and of molecules) can be then defined as

$$\beta(T_e, T_m) = \int_0^\infty v f(v, T_e) \sigma_{DA}(v, T_m) \, dv, \tag{5.3}$$

where the temperature-averaged DA cross section $\sigma_{DA}(v, T_m)$ is defined as a weighted sum of cross sections over all vibrational and rotational states of the target molecules

$$\sigma_{DA}(v, T_m) = \sum_{v,J} c_v^J(T_m) \sigma_{v,DA}^J(v), \tag{5.4}$$

where σ_v^J denotes the DA cross section for the initial vibrational v and rotational J state of the molecule. The coefficients c_v^J are normalized by the condition $\sum_{v,J} c_v^J = 1$ and are proportional to the Maxwell–Boltzmann factor

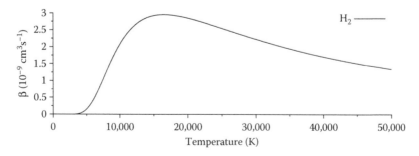

FIGURE 5.15   Temperature dependence of the dissociative attachment rate constant for H_2.

$$c_v^J(T_m) \propto (2J+1)e^{-(E_v^J - E_0^0)/kT_m}.$$ (5.5)

If thermalized mixture of electrons and molecules is considered, that is, $T_e = T_m$, Equation 5.3 leads to usual definition of the rate constant (now for convenience expressed in terms of energy $E = (1/2)mv^2$)

$$\beta(T) = \frac{8\pi}{m_e^2}\left(\frac{m_e}{2\pi kT}\right)^{3/2} \int_0^\infty E_e \sigma_{DA}(E_e, T)e^{E_e/kT}\, dE_e.$$ (5.6)

The temperature-averaged ($T = 1000$ K) DA cross sections for hydrogen halides (except for HF) and their deuterated analogs are shown in Figure 5.16a. The absolute values of the cross sections, and therefore also the rate constants shown in Figure 5.16b for thermalized mixture of electrons and molecules, increase significantly from HCl to HI. The two-temperature rate constant given by Equation 5.3 and a detailed discussion of the rate constants shown in Figures 5.16 and 5.17 can be found in Houfek et al. (2002).

5.4.2 Associative Detachment

For this process, it is assumed that the mixture of ions and neutral atoms is thermalized. Therefore, the rate constant $k(T)$ is given by (it follows directly from Equations 5.1 and 5.2, again expressed in terms of energy)

$$k(T) = \frac{8\pi}{\mu^2}\left(\frac{\mu}{2\pi kT}\right)^{3/2} \int_0^\infty E\sigma_{AD}(E)e^{E/kT}\, dE,$$ (5.7)

where σ_{AD} denotes the total cross section of associative detachment to all possible final states of the molecule

$$\sigma_{AD}(E) = \sum_{v,J}(2J+1)\sigma_{v,AD}^J(E).$$ (5.8)

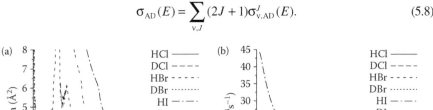
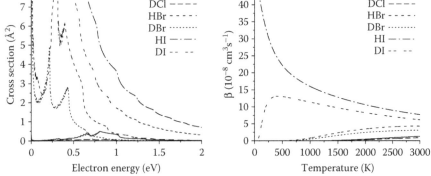

FIGURE 5.16 The temperature-averaged DA cross sections at $T = 1000$ K (a) and DA rate constants (b) for hydrogen halides and their deuterated analogs.

The calculated AD cross sections for HBr, HCl, and HI and their deuterated analogs are shown in Figure 5.17a. The AD cross sections for all molecules diverge as $E \to 0$ with an exception of the HI molecule. The AD process for HI is endothermic with the threshold at $E \sim 5$ meV. The AD rate constants are plotted in Figure 5.17a. The isotopic effect is very small for hydrogen halides, only for HI it is more pronounced at low temperatures due to different behavior of the cross section at small energies.

5.4.3 PRINCIPLE OF DETAILED BALANCE

The process of associative detachment is the reverse process to dissociative attachment and the DA and AD cross sections and rate constants are closely related by the principle of detailed balance. According to this principle, cross sections of dissociative attachment are related to the cross sections of associative detachment by

$$m_e E_e \sigma_{v,DA}^J(E_e) = \mu E \sigma_{v,AD}^J(E). \tag{5.9}$$

The energy E_e of an incident electron in the DA channel is connected with the collision energy E in the AD channel through the equation

$$E_e + E_v^J + E_a = E, \tag{5.10}$$

where E_a denotes the electron affinity and E_v^J is the rovibrational energy of the neutral molecule.

From Equation 5.9 and definitions (5.6) and (5.7), the relation between the AD and DA rate constants follows immediately:

$$k(T) = \left(\frac{m_e}{\mu}\right)^{3/2} Z(T)\beta(T)e^{-(E_a+E_0^0)/kT} \tag{5.11}$$

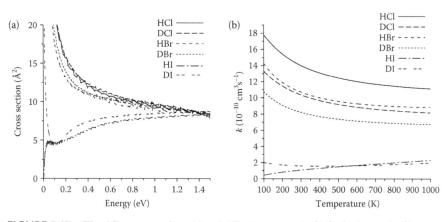

FIGURE 5.17   The AD cross sections (a) and AD rate constants (b) for hydrogen halides and their deuterated analogs.

TABLE 5.2

Comparison of the Calculated DA and AD Rate Constants for HCl, HBr, and HI with the Experimental Values of Smith and Adams (1987) and of Fehsenfeld (1975a)[a]

	T/K	HCl		HBr		HI	
		Theory	Expt.	Theory	Expt.	Theory	Expt.
DA	300	1.2^{-19}	–	8.0^{-13}	$<3^{-12}$	2.8^{-7}	3^{-7}
	515	3.1^{-14}	–	2.4^{-10}	3^{-10}	2.2^{-7}	–
AD	300	1.4^{-9}	9^{-10}	1.1^{-9}	7^{-10}	1.1^{-10}	3^{-10}
	300		9.6^{-10a}				$<6^{-11a}$
	515	1.2^{-9}	6^{-10}	9.5^{-10}	7^{-10}	1.5^{-10}	6^{-10}

Note: In $cm^3 s^{-1}$, the value 1.2^{-19}, means 1.2×10^{-19}. Electrons are thermalized, that is, $T_e = T_m$.

where

$$Z(T) = \sum_{v,J} (2J+1) e^{-(E_v^J - E_0^0)/kT} \tag{5.12}$$

is a partition sum of the Maxwell distribution of the rovibrational states of the molecules. Equation 5.11 can be used for testing of two independent codes for DA and AD calculations.

In Table 5.2, the calculated DA and AD rate constants for HCl, HBr, and HI molecules are compared with the experimental data of Smith and Adams (1987) and Fehsenfeld (1975). The calculated data are in very good agreement with the measured data both for DA and AD processes for HCl and HBr molecules. For hydrogen iodide, however, there is disagreement between the measured and the calculated data for the AD process. In spite of good agreement for the DA process, the calculated AD rate constant is smaller than the experimental value by a factor of 3 at $T = 300$ K and 4 at $T = 515$ K. This discrepancy is difficult to understand because both quantities are related to each other by the principle of detailed balance. A new theoretical and experimental work is obviously desirable for HI.

5.5 APPENDIX: DETAILS OF NONLOCAL RESONANCE MODELS FOR DIATOMIC MOLECULES

As we described earlier, all information about electronic degrees of freedom needed to calculate the resonant contribution to the VE, DA, and AD cross sections is contained in three basic quantities

- The ground-state potential $V_0 (R)$ of the neutral molecule,
- The discrete-state potential $V_d (R)$ of the temporary molecular anion, and

- The discrete-state-continuum coupling $V_{dE}(R)$ (or alternatively the resonance width $\Gamma(E,R) = 2\pi \mid V_{dE}(R) \mid^2$ and the shift $\Delta(E, R)$ can be given directly).

In this Appendix, we collect the nonlocal resonance models for most diatomic molecules for which the cross sections of low-energy resonant electron–molecule collisions were calculated during the last few decades. We should mention that the models for molecules with different isotopes (e.g., for H_2 and D_2, or HCl and DCl) are the same and differ only in the value of the molecular reduced mass.

We focus here especially on the hydrogen molecule and hydrogen halides. We summarize model functions and numerical values of all parameters which are needed for the calculations of the cross sections that were published in various papers. All values are given in atomic units unless stated otherwise.

In all the models (except for HF), it is possible to calculate the resonance shift

$$\Delta(\varepsilon) = \wp\!\int d\varepsilon' \mid V_{d\varepsilon'} \mid^2 /(\varepsilon - \varepsilon') \tag{5.13}$$

analytically using the formula (Domcke and Mündel 1985)

$$\int \frac{x^\alpha e^{-\beta x}}{\varepsilon - x}\,dx = \begin{cases} -\Gamma(1+\alpha)(-\varepsilon)^\alpha \Gamma(-\alpha, -\beta\varepsilon)e^{-\beta\varepsilon} & \varepsilon < 0, \\ \varepsilon^\alpha e^{-\beta x}[\pi\cot\alpha\pi - \Gamma(\alpha)(\beta\varepsilon)^{-\alpha}\,_1F_1(-\alpha, 1-\alpha; \beta\varepsilon)] & \varepsilon > 0, \end{cases} \tag{5.14}$$

where $\Gamma(\alpha, x)$ is the incomplete gamma function and $_1F_1(a,b,x)$ is the confluent hypergeometric function. These functions are conveniently evaluated via a continued fraction (Abramowitz and Stegun 1965, p. 263)

$$\Gamma(\alpha, x) = e^{-x}x^\alpha \left(\frac{1}{x+}\;\frac{1-\alpha}{1+}\;\frac{1}{x+}\;\frac{2-\alpha}{1+}\;\frac{2}{x+} \cdots \right) \tag{5.15}$$

and a series expansion (Abramowitz and Stegun 1965, p. 504)

$$_1F_1(a,b,x) = 1 + \frac{ax}{b} + \frac{a(a+1)}{b(b+1)}\cdot\frac{x^2}{2!} + \frac{a(a+1)(a+2)}{b(b+1)(b+2)}\cdot\frac{x^3}{3!} + \cdots \tag{5.16}$$

Input to all calculations includes the reduced mass of the molecule AB and the electron affinity of one of the atoms if its anion exists. Inaccuracy of these constants can make a significant influence on the results, particularly on the position of thresholds and sharp structures in the cross sections. Therefore, it is sometimes difficult to reproduce older calculations if these quantities were not specified in the papers. In Table 5.3, we summarize values of the reduced masses and electron affinities that were used in the most recent NRM calculations.

5.5.1 $H_2^-\ ^2\Sigma_u^+$ RESONANCE

The basic nonlocal resonance model for $H_2^-\ ^2\Sigma_u^+$ resonance (Mündel et al. 1985), referred to as the MBD model, was constructed from the fixed-nuclei *ab initio*

TABLE 5.3

Reduced Masses (in Atomic Units) and Electron Affinities Used in the Recent NRM Calculations for the Hydrogen Molecule and Hydrogen Halides and Their Deuterated Counterparts

X	μ (HX)	μ (DX)	E_a (eV)
H	918.576	1224.45	0.7542
D	—	1835.74	0.7546
F	1744.60	3319.56	3.4012
Cl	1785.69	3471.53	3.6127
Br	1813.99	3580.11	3.3636
I	1822.68	3614.12	3.0590

electron–molecule scattering calculations performed within the framework of the projection-operator formalism (Berman et al. 1985).

The function $V_0(R)$ in the MBD model is constructed as the spline interpolation of the data of Kołos and Wolniewicz (1965). The other model functions are given by

$$V_d(R) = D_1 \left\{ e^{-2\alpha_1(R-R_1)} - 2te^{-\alpha_1(R-R_1)} \right\} + E_a, \tag{5.17}$$

$$V_{de}(R) = \frac{1}{\sqrt{2\pi}} \sum_{i=1}^{3} f_i(\varepsilon) g_i(R), \tag{5.18}$$

where

$$f_i(E) = A_i E^{\frac{\alpha}{2}} e^{B_i E} \qquad i = 1, 2, 3$$
$$g_i(R) = \exp(-C_i^2 (R - R_0)^2) \quad i = 1, 2 \tag{5.19}$$
$$g_3(R) = \exp(-C_3 (R - R_0)).$$

The threshold exponent $\alpha = 3/2$ in Equation 5.19 is determined by the Wigner threshold law (4.114) for p-wave electron–molecule scattering. The values of the other constants are given as

$$A_1 = 1.6618 \quad A_2 = 1.3603 \quad A_3 = 1.0467$$
$$B_1 = 18.863 \quad B_2 = 4.6559 \quad B_3 = 1.4504$$
$$C_1 = 0.0 \quad\quad C_2 = 0.3302 \quad C_3 = 0.489$$
$$D_1 = 0.06247 \quad \alpha_1 = 1.19 \quad\quad t = 0.0$$

The equilibrium distance of the nuclei in the H_2 molecule is $R_0 = 1.4014$ a.u. (Kołos and Wolniewicz 1965).

The MBD model was later improved (Čížek et al. 1998) by changing the $V_d(R)$ when a new *ab initio* calculation of the long-range behavior of the H_2^--bound state (Senekowitsch et al. 1984) was taken into account and also by setting $C_1 = 0.2$ instead of an unphysical choice $C_1 = 0$ for which $V_{de}(R)$ does not vanish as $R \to \infty$. The new value of C_2 was obtained in the same way as the other parameters—from the best fit of the fixed-nuclei electron–molecule scattering data for internuclear distances $R = 1.4 - 2.75$ but this time the potential $V_d(R)$ was changed to be consistent with the data for H_2^--bound state. The resulting discrete-state potential curve is given by

$$V_d(R) = \begin{cases} 1.74e^{-2.37R} - \dfrac{94.4e^{-22.5/R}}{\left((R-2.54)^2 + 3.11\right)^2} & \text{for } R \leq 10.6 \\ -0.00845Re^{-0.35R} - \dfrac{2.25}{R^4} - \dfrac{97}{R^6} & \text{for } R > 10.6 \end{cases} \tag{5.20}$$

5.5.2 HYDROGEN HALIDES

5.5.2.1 HF

Description of the model construction from *ab initio* data is given in Čížek et al. (2003). The HF potential–energy curve and the discrete-state potential are given by

$$V_0(R) = D_0 e^{\alpha(R_0-R)}(e^{\alpha(R_0-R)} - 2) + E_a, \tag{5.21}$$

$$V_d(R) = -2.25\frac{R - v_1}{R^5 + v_2R^4 + v_3R^2} \quad \text{for } R > R_0, \tag{5.22}$$

$$= A_d e^{-2\alpha_d(R-R_0)} + p + qR \quad \text{for } R \leq R_0. \tag{5.23}$$

The parameters are listed in Table 5.4 except for p, q which are chosen to get a smooth function at $R_0 = 3.4$. The energy-dependent resonance width and shift were assumed in the form

$$\Gamma(E,R) = g(R)^2 \gamma(E), \tag{5.24}$$

$$\Delta(E,R) = g(R)^2 \delta(E), \tag{5.25}$$

TABLE 5.4
Model Parameters for HF

	$\gamma(\varepsilon > 0)$	$\delta(\varepsilon > 0)$	$\delta(\varepsilon < 0)$
a	0.227	0.734	−0.0677
b	0.00235	−0.177	0.00123
c	0.0590	1.528	−0.0104
d	−0.728	−19.4	−3.93
$g(R)$	$A_1 = 1.31$	$\alpha_1 = 0.1233$	$R_1 = 0.25$
$V_d(R)$	$v_1 = 2.99973$	$v_2 = -4.45722$	$v_3 = 32.6758$
	$A_d = 0.03307$	$\alpha_d = 1.8638$	$R_0 = 1.732517$
$V_0(r)$	$Q_0 = 0.125$	$D_0 = 0.2252$	$\alpha = 1.1735$

where $\gamma(E)$ and $\delta(E)$ are represented by the Padé approximant

$$\frac{a\varepsilon + b}{\varepsilon + c + d\varepsilon^2} \tag{5.26}$$

and

$$g(R) = A_1 e^{-\alpha_1 (R-R_1)^2}. \tag{5.27}$$

The parameters are also listed in Table 5.4.

5.5.2.2 HCl—DM Model

The first nonlocal resonance model for the molecule HCl whose parameters were determined by a least-squares fit of the fixed-nuclei eigenphase sum to *ab initio* electron–HCl scattering data of Padial and Norcross (1984) was constructed by Domcke and Mündel (1985). The potential for the neutral HCl molecule was taken in the Morse form

$$V_0(R) = D_0(e^{-2\alpha_0(R-R_0)} - 2e^{-\alpha_0(R-R_0)}) + E_a. \tag{5.28}$$

The free parameters were derived from spectroscopic data as $D_0 = 0.169414$, $\alpha_0 = 1.002$ (see references in Domcke and Mündel (1985)). The equilibrium distance in HCl is $R_0 = 2.4094$ and the electron affinity of Cl was taken as

$$E_a = 3.605\,\mathrm{eV} \doteq 0.132481\,\mathrm{a.u.}$$

The discrete state potential $V_d(R)$ and the coupling $V_{dE}(R)$ are given by the formulae

$$V_d(R) = 0.086361 e^{-1.74628(R-R_0)}, \tag{5.29}$$

$$V_{dE}(R) = a\left(\frac{E}{b}\right)^{\alpha/2} e^{-E/2b} g(R), \tag{5.30}$$

with $a = 0.428785$, $b = 0.199732$ and

$$g(R) = \begin{cases} e^{-0.031(R-R_0)} & R < 3 = R_1, \\ e^{-0.031(R-R_0)} e^{-0.78(R-R_1)} & R > R_1. \end{cases} \tag{5.31}$$

5.5.2.3 HCl—DMα(R) and DMHC Models

The DM model for HCl was later modified in two ways. First, because the threshold exponent α in the coupling (5.30) is determined by the electric dipole of the molecule which depends on the internuclear distance the threshold exponent was modified to be a function of R in a model of Horáček et al. (1998), the DMα (R) model. The R-dependence of $\alpha(R)$ is given by Equations 5.49 and 5.50.

Further improvement was achieved by taking into account a new *ab initio* calculation of the potential–energy curve of HCl⁻ for large internuclear distances. The resulting DMHC model was constructed by Čížek et al. (1999) using accurate data of

Åstrand and Karlström (1990). The model functions of the DMHC model are given by the formulas

$$V_{\mathrm{d}}(R) = 0.086361\,\mathrm{e}^{-1.74628(R-R_0)} - \frac{2.25}{((R-3.5)^2 + 51.37)^2 - 2500}, \tag{5.32}$$

$$V_{\mathrm{d}E}(R) = a\left(\frac{E}{b}\right)^{\alpha(R)/2} \mathrm{e}^{-E/2b}g(R), \tag{5.33}$$

where $a = 0.428785$, $b = 0.199732$ (the same values as in the DM model) and

$$g(R) = 1.07758(1 - 0.03101R)(1 + \mathrm{e}^{4.2(R-3.28)})^{-1}. \tag{5.34}$$

5.5.2.4 HCl—Fedor et al. Model

The most recent nonlocal resonance model for HCl constructed by Fedor et al. (2010) uses also the Morse function (5.28) as the HCl potential–energy curve and model functions

$$V_{\mathrm{d}}(R) = 0.0546277\,\mathrm{e}^{-2.45832(R-R_0)} - \frac{2.25}{((R-3.67061)^2 + 52.3152)^2 - 2500}, \tag{5.35}$$

$$V_{\mathrm{d}E}(R) = [\beta(R)E]^{\alpha(R)/2}\,\mathrm{e}^{-\beta(R)E/2}g(R), \tag{5.36}$$

where

$$\beta(R) = 0.43150R + 5.80996, \tag{5.37}$$

$$g(R) = 0.749083(1 - 0.166537R)(1 + \mathrm{e}^{4.2(R-3.18)})^{-1}. \tag{5.38}$$

5.5.2.5 HBr

The construction of the nonlocal resonance model of Čížek et al. (2001) is rather complicated and based only on *ab initio* calculations of the electron fixed-nuclei scattering and of the potential curve of the resonant complex HBr⁻. The potential $V_0(R)$ of the ground electronic state of HBr is described by the Morse potential

$$V_0(R) = 0.14406[\mathrm{e}^{-1.92(R-R_0)} - 2\mathrm{e}^{-0.96(R-R_0)}] + E_{\mathrm{a}}. \tag{5.39}$$

with the equilibrium distance $R_0 = 2.76$. The discrete state potential is given by

$$V_{\mathrm{d}}(R) = 9.934\mathrm{e}^{-1.535R} - \frac{2.25}{\left[(R-1.437)^2 + 2.884\right]^2}. \tag{5.40}$$

The coupling between the discrete state and the continuum $V_{\mathrm{d}E}(R)$ is given by the formulas

$$V_{\mathrm{d}E}(R) = \frac{1}{\sqrt{2\pi}}(\beta(R)E)^{\alpha(R)/2}\,\mathrm{e}^{-\beta(R)E/2}g(R), \tag{5.41}$$

where

$$g(R) = \begin{cases} 0.8688 - 0.1835R & R < 4.7345 \\ 0 & R > 4.7345 \end{cases}, \tag{5.42}$$

$$\beta(R) = 4.865\,R - 4.788 \tag{5.43}$$

and the threshold exponent $\alpha(R)$ is again given by Equations 5.49 and 5.50.

5.5.2.6 HI
In the case of HI, experimental data had to be used when the model of Horáček et al. (1997) was built because *ab initio* scattering calculations were not available for this molecule. The potential curves and the coupling are of the following forms

$$V_0(R) = 0.1175\left[e^{-1.86(R-R_0)} - 2e^{-0.93(R-R_0)}\right] + E_a, \tag{5.44}$$

$$V_d(R) = 0.0862e^{-2.219(R-R_0)} - 0.0124e^{-0.300(R-R_0)}, \tag{5.45}$$

$$V_{dE}(E,R) = f(E,R)g(R), \tag{5.46}$$

where the equilibrium distance $R_0 = 3.05$ and where

$$f(E,R) = 0.148\left(\frac{E}{0.158}\right)^{\alpha(R)/2} e^{E/0.316}, \tag{5.47}$$

$$g(R) = e^{-3.022(R-3.382)^2}. \tag{5.48}$$

Calculation of the threshold exponent $\alpha(R)$ is described next.

5.5.2.7 Threshold Exponent
The threshold exponent $\alpha(R)$ determines the behavior of the coupling $V_{dE}(R)$ at low energies. Its value for given R depends on the electric dipole of the neutral molecule and it is given by the following interpolation formula (Ogilvie et al. 1980)

$$\alpha(R) = \frac{1}{2} + a_1 M^2 + a_2 M^4 + a_3 M^6 + a_4 M^8, \tag{5.49}$$

where

$$a_1 = -0.101157, \quad a_2 = -0.014833, \quad a_3 = -0.007486, \quad a_4 = -0.003735$$

and

$$M(x) = M_0(1+x)^3(1+\Sigma e_i x^i)^{-1}, \quad x = \frac{R-R_0}{R_0}. \tag{5.50}$$

The parameters e_i and M_0 for molecules HCl, HBr, and HI are collated in Table 5.5.

TABLE 5.5

Parameters of Equation 5.50 for HCl, HBr, and HI

	HCl	HBr	HI
M_0	1.09333	0.81788	0.44722
e_1	1.897	2.199	3.166
e_2	0.871	0.808	2.393
e_3	1.465	1.483	2.243
e_4	1.829	3.868	11.794
e_5	−4.137	−2.612	10.380
e_6	13.886	13.209	7.423
e_7	0.416	0.255	−0.653

ACKNOWLEDGMENT

Support from the Czech Science Foundation (GAČR) by Grant No. 208/10/1281 and by Záměr MSM0021620860 of the Ministry of Education, Youth and Sports of the Czech Republic is gratefully acknowledged.

REFERENCES

Abouaf, R. and D. Teillet-Billy. 1977. Fine structure in dissociative-attachment cross sections for HCl and DCl. *J. Phys. B* *10*(11), 2261–2268.

Abouaf, R. and D. Teillet-Billy. 1980. Fine structure in the dissociative attachment cross sections for HBr and HF. *Chem. Phys. Lett.* *73*(1), 106–109.

Abramowitz, M. and I. A. Stegun (Eds.) 1965. *Handbook of Mathematical Functions with Formulas, Graphs, and Mathematical Tables.* New York: Dover Publications.

Allan, M. 1985. Experimental observation of structures in the energy dependence of vibrational excitation in H_2 by electron impact in the $^2\Sigma_u^+$ resonance region. *J. Phys. B* *18*(13), L451–L455.

Allan, M., M. Čížek, J. Horáček, and W. Domcke. 2000. Electron scattering in cooled HCl: Boomerang structures and outer-well resonances in elastic and vibrational excitation cross sections. *J. Phys. B* *33*, L209–L213.

Åstrand, P. and G. Karlström. 1990. A bound HCl⁻ species: An *ab initio* quantum-chemical study. *Chem. Phys. Lett.* *175*(6), 624–628.

Azria, R., R. Abouaf, and D. Teillet-Billy. 1982. Symmetry of Cl_2^- resonant states formed in dissociative electron attachment reaction on Cl_2. *J. Phys. B* *15*(16), L569–L574.

Bardsley, J. N. and J. M. Wadehra. 1979. Dissociative attachment and vibrational excitation in low-energy collisions of electrons with H_2 and D_2. *Phys. Rev. A* *20*(4), 1398–1404.

Barsotti, S., M.-W. Ruf, and H. Hotop. 2002. Clear experimental evidence for *p*-wave attachment-threshold behavior in electron attachment to chlorine molecules. *Phys. Rev. Lett.* *89*(8), 083201.

Berman, M. and W. Domcke. 1984. Projection-operator calculations for shape resonances: A new method based on the many-body optical-potential approach. *Phys. Rev. A* *29*(5), 2485–2496.

Berman, M., H. Estrada, L. S. Cederbaum, and W. Domcke. 1983. Nuclear dynamics in resonant electron-molecule scattering beyond the local approximation: The 2.3-eV shape resonance in N_2. *Phys. Rev. A* *28*(3), 1363–1381.

Berman, M., C. Mündel, and W. Domcke. 1985. Projection-operator calculations for molecular shape resonances: The $^2\Sigma_u^+$ resonance in electron-hydrogen scattering. *Phys. Rev. A 31*(2), 641–651.

Birtwistle, D. T. and A. Herzenberg. 1971. Vibrational excitation of N_2 by resonance scattering of electrons. *J. Phys. B 4*(1), 53–70.

Braun, M., M.-W. Ruf, I. I. Fabrikant, and H. Hotop. 2007. Observation of *p*-wave threshold behavior in electron attachment to F_2 molecules. *Phys. Rev. Lett. 99*(25), 253202.

Brems, V., T. Beyer, B. M. Nestmann, H.-D. Meyer, and L. S. Cederbaum. 2002. *Ab initio* study of the resonant electron attachment to the F_2 molecule. *J. Chem. Phys. 117*(23), 10635–10647.

Chandra, N. and A. Temkin. 1976. Hybrid theory and calculation of e-N_2 scattering. *Phys. Rev. A 13*(1), 188–203.

Chase, D. M. 1956. Adiabatic approximation for scattering processes. *Phys. Rev. 104*(3), 838–842.

Chutjian, A. and S. H. Alajajian. 1987. Electron attachment in F_2: Conclusive demonstration of nonresonant, s-wave coupling in the limit of zero electron energy. *Phys. Rev. A 35*(11), 4512–4514.

Čížek, M., J. Horáček, M. Allan, and W. Domcke. 2002. Resonances and threshold phenomena in low-energy electron collisions with hydrogen halides: New experimental and theoretical results. *Czech. J. Phys. 52*(9), 1057–1070.

Čížek, M., J. Horáček, M. Allan, I. I. Fabrikant, and W. Domcke. 2003. Vibrational excitation of hydrogen fluoride by low-energy electrons: Theory and experiment. *J. Phys. B 36*(13), 2837–2849.

Čížek, M., J. Horáček, and W. Domcke. 1998. Nuclear dynamics of the H_2^- collision complex beyond the local approximation: Associative detachment and dissociative attachment to rotationally and vibrationally excited molecules. *J. Phys. B 31*(11), 2571–2583.

Čížek, M., J. Horáček, and W. Domcke. 1999. Associative detachment, dissociative attachment and vibrational excitation of HCl by low-energy electrons. *Phys. Rev. A 60*(4), 2873–2881.

Čížek, M., J. Horáček, and W. Domcke. 2007. Long-lived anionic states of H_2, HD, D_2, and T_2. *Phys. Rev. A 75*(1), 012507.

Čížek, M., J. Horáček, A.-C. Sergenton, D. B. Popovič, M. Allan, W. Domcke, T. Leininger, and F. X. Gadea. 2001. Inelastic low-energy electron collisions with the HBr and DBr: Experiment and theory. *Phys. Rev. A 63*(6), 062710.

Čížek, M., J. Horáček, F. A. U. Thiel, and H. Hotop. 2001. Associative detachment in low-energy collisions between hydrogen atoms and atomic halogen anions. *J. Phys. B 34*(6), 983–1004.

Domcke, W. 1991. Theory of resonance and threshold effects in electron-molecule collisions: The projection-operator approach. *Phys. Rep. 208*(2), 97–188.

Domcke, W. and L. S. Cederbaum. 1980. On the interpretation of low-energy electron-HCl scattering phenomena. *J. Phys. B 14*(1), 149–173.

Domcke, W., L. S. Cederbaum, and F. Kaspar. 1979. Threshold phenomena in electron-molecule scattering: A non-adiabatic theory. *J. Phys. B 12*(12), L359–L364.

Domcke, W. and C. Mündel. 1985. Calculation of cross sections for vibrational excitation and dissociative attachment in HCl and DCl beyond the local-complex-potential approximation. *J. Phys. B 18*(22), 4491–4509.

Dubé, L. and A. Herzenberg. 1977. Vibrational excitation of polar molecules by slow electrons: HCl. *Phys. Rev. Lett. 38*(15), 820–823.

Dubé, L. and A. Herzenberg. 1979. Absolute cross sections from the boomerang model for resonant electron-molecule scattering. *Phys. Rev. A 20*(1), 194–213.

Fabrikant, I. I. 1977. Threshold behavior of the cross sections for scattering of electrons by polar molecules. *Sov. Phys. JETP 46*(4), 693–697.

Fabrikant, I. I., T. Leininger, and F. X. Gadéa. 2000. Low-energy dissociative electron attachment to Cl_2 molecules. *J. Phys. B 33*(21), 4575–4580.

Fabrikant, I. I., J. M. Wadehra, and Y. Xu. 2002. Resonance processes in e—H_2 collisions: Dissociative attachment and dissociation from vibrationally and rotationally excited states. *Phys. Scr. T96*, 45–51.

Fandreyer, R., P. G. Burke, L. A. Morgan, and C. J. Gillan. 1993. Low-energy electron scattering by HBr. *J. Phys. B 26*(20), 3625–3637.

Fedor, J., M. Cingel, J. D. Skalný, P. Scheier, T. D. Märk, M. Čížek, P. Kolorenč, and J. Horáček. 2007. Dissociative electron attachment to HBr: A temperature effect. *Phys. Rev. A 75*(2), 022703.

Fedor, J., O. May, and M. Allan. 2008. Absolute cross sections for dissociative electron attachment to HCl, HBr, and their deuterated analogs. *Phys. Rev. A 78*(3), 032701.

Fedor, J., C. Winstead, V. McKoy, M. Čížek, K. Houfek, P. Kolorenč, and J. Horáček. 2010. Electron scattering in HCl: An improved nonlocal resonance model. *Phys. Rev. A 81*(4), 042702.

Fehsenfeld, F. C. 1975. Associative detachment. In P. Ausloos (Ed.), *Interaction between Ions and Molecules*, pp. 387–412. New York: Plenum Press.

Gallup, G. A., Y. Xu, and I. I. Fabrikant. 1998. Nonlocal theory of dissociative electron attachment to H_2 and HF molecules. *Phys. Rev. A 57*(4), 2596–2607.

Gauyacq, J. P. 1985. Dissociative attachment in e^-–H_2 collisions. *J. Phys. B 18*(9), 1859–1872.

Gauyacq, J. P. and A. Herzenberg. 1982. Nuclear-excited Feshbach resonances in e+HCl scattering. *Phys. Rev. A 25*(6), 2959–2967.

Gertitschke, P. L. and W. Domcke. 1993. Time-dependent wave-packet description of dissociative electron attachment. *Phys. Rev. A 47*(2), 1031–1044.

Golser, R., H. Gnaser, W. Kutschera, A. Priller, P. Steier, A. Wallner, M. Čížek, J. Horáček, and W. Domcke. 2005. Experimental and theoretical evidence for long-lived molecular hydrogen anions H_2^- and D_2^-. *Phys. Rev. Lett. 94*(22), 223003.

Hazi, A. U., T. N. Rescigno, and M. Kurilla. 1981. Cross sections for resonant vibrational excitation of N_2 by electron impact. *Phys. Rev. A 23*(3), 1089–1099.

Heber, O., R. Golser, H. Gnaser, D. Berkovits, Y. Toker, M. Eritt, M. L. Rappaport, and D. Zajfman. 2006. Lifetimes of the negative molecular hydrogen ions: H_2^-, D_2^-, and HD^-. *Phys. Rev. A 73*(6), 060501.

Herzenberg, A. 1968. Oscillatory energy dependence of resonant electron-molecule scattering. *J. Phys. B 1*(4), 548–558.

Herzenberg, A. and F. Mandl. 1962. Vibrational excitation of molecules by resonance scattering of electrons. *Proc. R. Soc. London, Ser. A 270*(1340), 48–71.

Hickman, A. P. 1991. Dissociative attachment of electrons to vibrationally excited H_2. *Phys. Rev. A 43*(7), 3495–3502.

Horáček, J., M. Čížek, and W. Domcke. 1998. Generalization of the nonlocal resonance model for low-energy electron collisions with hydrogen halides: The variable threshold exponent. *Theor. Chem. Acc. 100*(1–4), 31–35.

Horáček, J., M. Čížek, K. Houfek, P. Kolorenč, and W. Domcke. 2004. Dissociative electron attachment and vibrational excitation of H_2 by low-energy electrons: Calculations based on an improved nonlocal resonance model. *Phys. Rev. A 70*(5), 052712.

Horáček, J., M. Čížek, K. Houfek, P. Kolorenč, and W. Domcke. 2006. Dissociative electron attachment and vibrational excitation of H_2 by low-energy electrons: Calculations based on an improved nonlocal resonance model. II. Vibrational excitation. *Phys. Rev. A 73*(2), 022701.

Horáček, J., M. Čížek, P. Kolorenč, and W. Domcke. 2005. Isotope effects in vibrational excitation and dissociative electron attachment of DCl and DBr. *Eur. Phys. J. D 35*(2), 225–230.

Horáček, J. and W. Domcke. 1996. Calculation of cross sections for vibrational excitation and dissociative attachment in electron collisions with HBr and DBr. *Phys. Rev. A 53*(4), 2262–2271.

Horáček, J., W. Domcke, and H. Nakamura. 1997. Electron attachment and vibrational excitation in hydrogen iodide: Calculations based on the nonlocal resonance model. *Z. Phys. D 42*(3), 181–185.

Horáček, J., K. Houfek, and M. Čížek. 2007. Giant structures in low-energy electron–deuterium-iodide elastic scattering cross section. *Phys. Rev. A 75*(2), 022719.

Horáček, J., K. Houfek, M. Čížek, I. Murakami, and T. Kato. 2003. Rate coefficients for low-energy electron dissociative attachment to molecular hydrogen. *NIFS-DATA 73*, 1–63.

Houfek, K., M. Čížek, and J. Horáček. 2002. Calculation of rate constants for dissociative attachment of low-energy electrons to hydrogen halides HCl, HBr and HI and their deuterated analogues. *Phys. Rev. A 66*(6), 062702.

Houfek, K., M. Čížek, and J. Horáček. 2008. On irregular oscillatory structures in resonant vibrational excitation cross-sections in diatomic molecules. *Chem. Phys 347*(1–3), 250–256.

Kazanskii, A. K. 1982. Antibonding intermediate state in the theory of vibrational excitation of diatomic molecules by slow electrons. *Sov. Phys. JETP 55*(5), 824–827.

Klar, D., B. Mirbach, H. J. Korsch, M.-W. Ruf, and H. Hotop. 1994. Comparison of rate coefficients for Rydberg electron and free electron attachment. *Z. Phys. D 31*(4), 235–244.

Kolorenč, P., V. Brems, and J. Horáček. 2005. Computing resonance positions, widths, and cross sections via the Feshbach-Fano R-matrix method: Application to potential scattering. *Phys. Rev. A 72*(1), 012708.

Kolorenč, P., M. Čížek, J. Horáček, G. Mil'nikov, and H. Nakamura. 2002. Study of dissociative electron attachment to HI molecule by using R-matrix representation for Green's function. *Phys. Scr. 65*(4), 328–335.

Kolorenč, P. and J. Horáček. 2006. Dissociative electron attachment and vibrational excitation of the chlorine molecule. *Phys. Rev. A 74*(6), 062703.

Kołos, W. and L. Wolniewicz. 1965. Potential–energy curves for the $X^1\Sigma_g^+$, $b^3\Sigma_u^+$ and $C^1\Pi_u$ states of the hydrogen molecule. *J. Chem. Phys. 43*(7), 2429–2441.

Kreckel, H., H. Bruhns, M. Čížek, S. C. O. Glover, K. A. Miller, X. Urbain, and D. W. Savin. 2010. Experimental results for H_2 formation from H^- and H and implications for first star formation. *Science 329*(5987), 69–71.

Lammich, L., L. H. Andersen, G. Aravind, and H. B. Pedersen. 2009. Experimental characterization of the metastable D_2^- ion by photofragment imaging. *Phys. Rev. A 80*(2), 023413.

Launay, J. M., M. Le Dourneuf, and C. J. Zeippen. 1991. The $H + H^- \leftrightarrow H_2(v,j) + e^-$ reaction: A consistent description of the associative detachment and dissociative attachment processes using the resonant scattering theory. *Astron. Astrophys. 252*(2), 842–852.

Mazevet, S., M. A. Morrison, O. Boydstun, and R. K. Nesbet. 1999. Adiabatic treatments of vibrational dynamics in low-energy electron-molecule scattering. *J. Phys. B 32*(5), 1269–1294.

Mündel, C., M. Berman, and W. Domcke. 1985. Nuclear dynamics in resonant electron-molecule scattering beyond the local approximation: Vibrational excitation and dissociative attachment in H_2 and D_2. *Phys. Rev. A 32*(1), 181–193.

Narevicius, E. and N. Moiseyev. 1998. Fingerprints of broad overlapping resonances in the e+H_2 cross section. *Phys. Rev. Lett. 81*(11), 2221–2224.

Narevicius, E. and N. Moiseyev. 2000. Trapping of an electron due to molecular vibrations. *Phys. Rev. Lett. 84*(8), 1681–1684.

Nestmann, B. M. 1998. Characterization of metastable anionic states within the R-matrix approach. *J. Phys. B 31*(17), 3929–3948.

Ogilvie, J. F., W. R. Rodwell, and R. H. Tipping. 1980. Dipole moment functions of the hydrogen halides. *J. Chem. Phys. 73*(10), 5221–5229.

Padial, N. T. and D. W. Norcross. 1984. Ro-vibrational exccitation of HCl by electron impact. *Phys. Rev. A 29*(3), 1590–1593.

Rapp, D., T. E. Sharp, and D. D. Briglia. 1965. Large isotope effect in the formation of H⁻ or D⁻ by electron impact on H_2, HD, and D_2. *Phys. Rev. Lett. 14*(14), 533–535.

Rohr, K. 1978. Interaction mechanisms and cross sections for the scattering of low-energy electrons from HBr. *J. Phys. B 11*(10), 1849–1860.

Rohr, K. and F. Linder. 1976. Vibrational excitation of polar molecules by electron impact I. Threshold resonances in HF and HCl. *J. Phys. B 9*(14), 2521–2537.

Ruf, M.-W., S. Barsotti, M. Braun, H. Hotop, and I. I. Fabrikant. 2004. Dissociative attachment and vibrational excitation in low-energy electron collisions with chlorine molecules. *J. Phys. B 37*(1), 41–62.

Schafer, O. and M. Allan. 1991. Measurement of near-threshold vibrational excitation of HCl by electron impact. *J. Phys. B 24*(13), 3069–3076.

Schneider, B. I., M. Le Dourneuf, and Vo Ky Lan. 1979. Resonant vibrational excitation of N_2 by low-energy electrons: An *ab initio* R-matrix calculation. *Phys. Rev. Lett. 43*(26), 1926–1929.

Schulz, G. J. 1959. Formation of H⁻ ions by electron impact on H_2. *Phys. Rev. 113*(3), 816–819.

Schulz, G. J. 1962. Vibrational excitation of nitrogen by electron impact. *Phys. Rev. 125*(1), 229–232.

Schulz, G. J. 1964. Vibrational excitation of N_2, CO, and H_2 by electron impact. *Phys. Rev. 135*(4A), A988–A994.

Schulz, G. J. and R. K. Asundi. 1965. Formation of H⁻ by electron impact on H_2 at low energy. *Phys. Rev. Lett. 15*(25), 946–949.

Schulz, G. J. and J. T. Dowell. 1962. Excitation of vibrational and electronic levels in O_2 by electron impact. *Phys. Rev. 128*(1), 174–177.

Senekowitsch, J., P. Rosmus, W. Domcke, and H. Werner. 1984. An accurate potential energy function of the H_3^- ion at large internuclear distances. *Chem. Phys. Lett. 111*(3), 211–214.

Seng, G. and F. Linder. 1976. Vibrational excitation of polar molecules by electron impact. II. Direct and resonant excitation in H_2O. *J. Phys. B 9*(14), 2539–2551.

Smith, D. and N. G. Adams. 1987. Studies of the reactions HBr(HI) + e ↔ Br⁻(I⁻) + H using the FALP and SIFT techniques. *J. Phys. B 20*(18), 4903–4913.

Tam, W.-C. and S. F. Wong. 1978. Dissociative attachment of halogen molecules by 0–8 eV electrons. *J. Chem. Phys. 68*(12), 5626–5630.

Teillet-Billy, D. and J. P. Gauyacq. 1984. Dissociative attachment in e⁻-HCl, DCl collisions. *J. Phys. B 17*(19), 4041–4058.

Wadehra, J. M. and J. N. Bardsley. 1978. Vibrational- and rotational-state dependence of dissociative attachment in e-H_2 collisions. *Phys. Rev. Lett. 41*(26), 1795–1798.

Wigner, E. P. 1948. On the behavior of cross sections near thresholds. *Phys. Rev. 73*(9), 1002–1009.

Živanov, S., M. Allan, M. Čížek, J. Horáček, F. A. U. Thiel, and H. Hotop. 2002. Effects of interchannel coupling in associative detachment: Electron spectra for H+Cl⁻ and H+Br⁻ collisions. *Phys. Rev. Lett. 89*(7), 073201.

6 Theoretical Studies of Electron Interactions with DNA and Its Subunits

From Tetrahydrofuran to Plasmid DNA

Laurent G. Caron and Léon Sanche

CONTENTS

6.1 INTRODUCTION

Biomolecules are quite sensitive to damage caused by high-energy radiation of natural or man-made origin. There is a sequence of three major groups of initial events that can lead to damage: primary, secondary, and reactive (Sanche 2009). The first two are considered in Figure 6.1 for the case of a two-component biomolecules AB. Multiply charged products are omitted from this figure. The direct, primary processes are ionization and excitation which can lead to dissociation. In cells, single and multiple ionization leads to the production of copious amounts of low-energy secondary electrons of the order 3×10^4/MeV of deposited energy. These electrons thus carry a large fraction of the energy of the impinging radiation. Their most probable energy lies between 9 and 10 eV (Pimblott and LaVerne 2007). These low-energy electrons (LEEs) can interact resonantly or directly with the irradiated biomolecules; the latter interaction is depicted by the elastic channel and left vertical arrow in Figure 6.1, which represents inelastic channels. These latter channels, along with electron recombination with ions, can produce excited neutral species. The resonance processes involve the formation of a transient negative ion (TNI) having a lifetime lying between 10^{-15} and 10^{-3} s. The extra electron in the TNI can occupy a usually empty orbital of the ground state molecule (i.e., a shape resonance) or of an electronically singly excited state of the molecule (i.e., a core-excited resonance). If the energy of the core-excited resonance is less than that of the parent electronically excited state of the neutral molecule, the incoming electron is captured essentially by the electron affinity of that state; in this case, it falls in the category of Feshbach resonances which usually have a fairly long

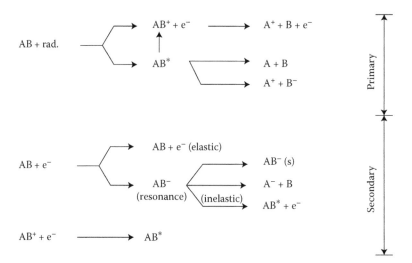

FIGURE 6.1 Reactions induced by primary ionizing radiation and secondary electrons on a molecule AB. (Adapted from Sanche, L. 2009. *Low-Energy Electron Interaction with DNA: Bond Dissociation and Formation of Transient Anions*, Radicals, and Radical Anions, pp. 239–293. John Wiley & Sons, Inc.; Sanche, L./Greenberg, M. (Ed.), *Radical and Radical Ion Reactivity in Nucleic Acid Chemistry*, pp. 239–293, 2010. Copyright Wiley-VCH Verlag GmbH & Co. KGaA. Reprinted with permission.)

lifetime, of the order of 10^{-12} to 10^{-14} s. When the TNI state lies above the parent electronically excited state, it is called a core-excited shape resonance. The TNI can decay via resonance stabilization, dissociation, or inelastic scattering leaving behind an excited molecule. The excited molecules that are produced by secondary processes can also ionize or dissociate as shown in the upper part of Figure 6.1. Finally, the radicals and ions produced by primary and secondary processes proceed through reactive scattering and chemical reactions with nearby biomolecules.

It is relevant to discuss at this point the different environments that are encountered in various experiments and the consequences for theory. In electron scattering experiments with gas-phase molecules under vacuum conditions, the molecule densities are small and the incident electrons undergo spatially isolated single-collision processes with these molecules. Thus, electrons scatter away from the incident beam. This electron absorption can be used to obtain total scattering cross sections. Whereas in vacuum there are no available states below 0 eV for electrons, in the solid and liquid phases the situation is usually different. In molecular films, for instance, the molecules are tightly packed and multiple collisions can occur. In thick films, the electrons see a polarizable medium having some band structure. As a consequence, the wavenumber k of the incident electrons changes after penetration such that, in a first approximation, $\hbar^2 k^2 / (2m^*) + E_c = E_i$ where E_i is the incident energy in vacuum, m^* is the electron's effective mass in the conduction band of the film, and E_c is the energy at the bottom of this band. Here, E_c can be positive or negative depending on bulk medium composition (Bader et al. 1984). In the latter situation, there will be electron states below vacuum. The mechanisms of Figure 6.1 still prevail as in the gas phase although most often they occur at incident electron energy slightly different than that in the gas phase, owing to the band structure or metallic substrate image potential in thin films deposited on metallic substrates. Films can also trap electrons and care must be taken by the experimentalists to monitor charging. Dissociation data obtained from measurements of desorbed products in vacuum will mostly be sensitive to the molecules on the film's surface as fragments generated in the bulk have a large probability to be trapped.

The motion of the electrons in the bulk of the film is more complicated to describe due to strong elastic and inelastic events. Correspondingly, cross sections are more difficult to measure. For example, the quantum states of an electron in water are multifaceted. There are extended states, which form a conduction band, whose bottom is likely (Bernas et al. 1997) to lie around 0.75 eV below vacuum and localized states forming an Urbach tail extending to some 0.5 eV below the conduction band. The solvation or hydration energy is estimated (Jortner and Noyes 1966, Hickel and Sehested 1985, Schwarz 1991) to be in the range −1.4 to −1.6 eV. The physical time scale of events (Kumar and Sevilla 2010) and energy loss processes are also quite different, particularly in biological cells. LEE interaction with the complex water environment in cells results in lossy propagation in the conduction band, which terminates with the formation of presolvated electron states, followed by rapid solvation at a timescale less than a picosecond. Attachment of a thermalized electron on a molecule is characterized by its affinity which, when positive, signals the stability of the electronic ground-state anion as compared with separated electron and neutral molecule in the surrounding medium. The reference energies for the free electron are thus the vacuum energy *in vacuo* or the solvated electron energy in water. The calculation of affinities is quite delicate as discussed in Kumar and Sevilla (2008a). It is sensitive to the choice of basis set and to environment such as in base pairs or in solution. The affinity has likewise been shown to be responsive to π-stacking (Kobylecka et al. 2008a, 2009) sequence in double-stranded DNA. In a homogeneous film, the notion of affinity transforms into the energy of the bottom of the conduction band relative to vacuum. If impurities (i.e., different molecules) are added to the film, the gas-phase notion may still be valid, when a localized trapped electron state is created below the conduction band. Thus, the reference energy for an electron on a substituted molecule, within an otherwise homogeneous film, is the conduction band minimum of this homogeneous medium.

In this chapter, we focus on the interaction of LEEs with DNA and RNA, their constituent subunits, and related molecules. Figure 6.2 shows an exploded view of a DNA sequence with the bases adenine, cytosine, guanine, and thymine, with the

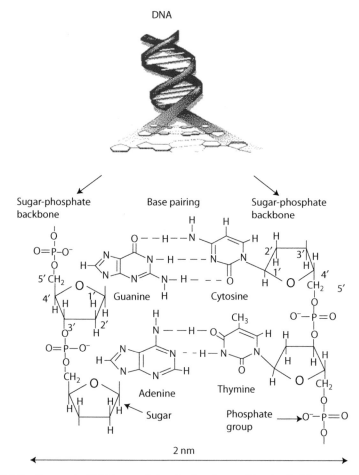

FIGURE 6.2 Segment of DNA containing the four bases. (Sanche, L./Greenberg, M. (Ed.), *Radical and Radical Ion Reactivity in Nucleic Acid Chemistry*, pp. 239–293, 2010. Copyright Wiley-VCH Verlag GmbH & Co. KGaA. Reprinted with permission.)

backbone sugar unit, deoxyribose, and the phosphate group. In RNA, thymine is replaced by uracil and the sugar is ribose.

If one were to single out the experiment that initially drew attention to the interaction of LEEs with DNA, one would point to the work of Boudaïffa et al. (2000) in which strand breaks were measured in supercoiled plasmid DNA deposited on a tantalum substrate in the incident electron energy range 3–20 eV. The measurements were performed in a clean ultrahigh vacuum environment with just the presence of structural water molecules within DNA. The results are shown in Figure 6.3. They provide clear evidence that resonance mechanisms are operative in the strand-breaking process. It took a few years for Martin et al. (2004) to develop a more efficient purification technique for DNA and reinvestigate LEE damage in the energy range below 4 eV. They found two peaks at 0.8 and 2.2 eV, which they unequivocally associated with the known shape resonances of the DNA bases (Aflatooni et al. 1998). Their result is shown in Figure 6.4. In the meantime, Pan et al. (2003) investigated the desorption of H⁻, O⁻, and OH⁻ anions from films of DNA and subunits. The yield functions exhibit resonances in the same energy range as those of Figure 6.3. The reported results indicate that H⁻ produced by dissociative electron attachment (DEA) to DNA arises principally

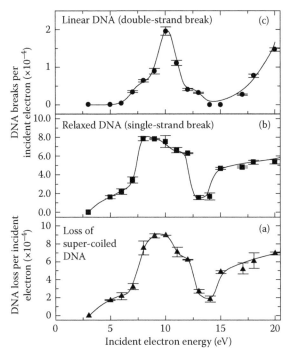

FIGURE 6.3   Measured yields, per incident electron of 520 eV, for the loss of the supercoiled DNA form (a), induction of SSBs (b), and DSBs (c) in dry DNA films. The error bars correspond to one standard deviation from six measurements. (From Boudaïffa, B. et al. 2000. *Science* (Washington, DC, U.S.) *287*(5458), 1658–1660, Copyright 2000. Reprinted with permission of American Association for the Advancement of Science.)

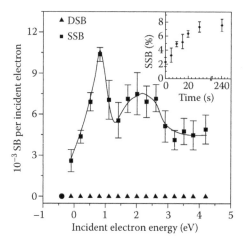

FIGURE 6.4 Quantum yield of DNA single-strand breaks (SSBs) and double-strand breaks (DSBs) vs incident electron energy from Martin et al. (2004). The inset shows the dependence of the percentage of circular DNA (i.e., SSBs) on irradiation time for 0.6 eV electrons. (Reprinted with permission from Martin, F. et al., *Phys. Rev. Lett. 93*(6), 068101, 2004. Copyright 2004 by the American Physical Society.)

from the bases and, to a lesser extent, from the sugar ring of the backbone, whereas O^- is produced from fragmentation of the phosphate group in the backbone. DEA is thus involved in LEE-induced DNA damage. The subsequent work of Zheng et al. (2005) on solid films of oligonucleotide tetramers (CGTA and GCAT) shed light on the pathway to damage by LEEs. The observed distribution of nonmodified products suggests a mechanism of damage involving initial electron attachment to nucleobase moieties, followed by electron transfer to the sugar–phosphate backbone, and subsequent dissociation of the phosphodiester bond which primarily takes place via C–O bond rather than the P–O bond within the entire 4- to 15-eV range. Yield functions for the nucleobase and oligonucleotide fragments (Zheng et al. 2006a) in self-assembled monolayer films of the tetramers showed broad peaks at 6 eV and 10–12 eV, corresponding to those in the yield functions for damage to plasmid DNA, which confirmed the mechanism leading to phosphodiester bond cleavage and further showed the possibility of a direct rupture of the *N*-glycosidic bond from the sugar. More details on these key experiments and many others can be found in Sanche (2008, 2009).

The offshoot is that damage by LEEs proceeds preferentially through DEA. The important quantities influencing DEA can be deduced from the analysis of O'Malley and Taylor (1968). They write the nuclear coordinate ($\mathbf{R}$) dependent *electronic* matrix element coupling an incident electron plane wave of wave vector $\mathbf{k}$ and a resonance wave function as

$$V_{\mathbf{k}}(\mathbf{R}) = \sqrt{4\pi} \sum_{L} V_L(R) Y_L^*(\mathbf{k}) \qquad (6.1)$$

in which Y_L is a spherical harmonic. In their one-center approximation, only the dominant contribution $V(L_0)$ associated with the resonance is retained reducing

Equation 6.1 to $V_k(\mathbf{R}) \approx \sqrt{4\pi} V_{L_0}(R) Y_{L_0}^*(\mathbf{k})$. The corresponding T-matrix for DEA is thus proportional to $\sqrt{4\pi} V_{L_0}(R) Y_{L_0}^*(\mathbf{k})$.

The DEA cross section therefore depends on the nuclear degrees of freedom, the nature of the resonant state, and the incident wave function. In DNA, the total incident wave function on a subunit is not only the incident electron plane wave, but also includes the scattered waves from all other subunits. We shall be examining, herein, these direct and indirect contributing factors to DEA in this chapter. Elastic and inelastic scattering information is germane to the total wave function on a subunit and carries the imprint of resonances. *Ab initio* calculations, density functional theory (DFT), and time-dependent DFT (TDDFT) are relevant to attachment, excited states, and dissociation of subunits. Multiple-scattering theory is at the core of theoretical treatments of the total incident wave function on subunits of long oligomers. The abbreviations used in this chapter are collected at the end of the text.

6.2   MOLECULAR UNITS

6.2.1   Scattering from Subunits

In this section, we examine several theoretical *ab initio*-based methods that were applied to calculations of electron scattering from DNA and RNA subunits (i.e., the bases, sugar, and phosphate units) and some closely related molecules. The *R*-matrix and variational methods that are referred to in this section and discussed in Chapter 2 are capable of doing elastic as well as inelastic calculations. We focus on the usual LEE energy range 0–20 eV. A fair number of the experimental results and some theoretical ones cited in this section pertain to DEA and will be revisited in the next section.

Let us first briefly mention two methods that are quite efficient at higher energies. First, there is the independent atom model (IAM) as used by Możejko and Sanche (2005) and references therein to earlier works. It was applied to a large number of biomolecules and is reputed excellent for energies above 50 eV. Then there is the work of Blanco and García (2007) and later Milosavljević et al. (2008a), and references therein to earlier works, which combine a screening-corrected additivity-rule (SCAR) with *ab initio* quasifree inelastic atomic potential. It improves over the independent atom model (IAM) by taking overlapping of atom pairs into account. The results are satisfactory above 10 eV but show little structure in the integral cross sections (ICS) at smaller energies. They can, however, provide information on inelastic processes.

6.2.1.1   Sugar and Analogs

The deoxyribose analog tetrahydrofuran (THF) was very early thoroughly examined using various state of the art theoretical methods. It is chemically and electronically much simpler than deoxyribose and all other DNA subunits. Prior to 2006, several LEE collision experiments had been performed on the THF molecule. These are documented in Table 6.1. Note that the results of Lepage et al. (1998) clearly show that resonances in THF films are some 1 eV lower than in the gas phase. THF is thus rich in resonances, although their identification is often uncertain, and the molecule offers a good testing ground for theory.

TABLE 6.1

Major Experimental Resonances Observed in THF Prior to 2006

Source	Method	E_r [eV]	Nature
	Gas phase		
Bremmer et al. (1991)	EEL	6.4	$^1n_0$3s core-excited
Lepage et al. (1998)	EEL excitation functions	8.4	$^1n_0-1 \rightarrow$ 3s or $^1n_0$4p core-excited
Zecca et al. (2005)	Transmission	7.5	Shape
	Films		
Lepage et al. (1998)	EEL excitation functions	4	σ^* shape
Lepage et al. (1998)	EEL excitation functions	7.5	$^1n_0-1 \rightarrow$ 3s or $^1n_0$4p core-excited
Lepage et al. (1998)	EEL excitation functions	10	
Antic et al. (1999)	ESD of H$^-$	10	Hole below HOMO
Breton et al. (2004)	DEA yield of aldehydes	3.5	Shape
Breton et al. (2004)	DEA yield of aldehydes	12.5	Hole in σ_{CO} or $\sigma_{CO\perp}$, 2e$^-$ in σ^*_{CO}
Breton et al. (2004)	DEA yield of aldehydes	15	Hole in π co or π co$\perp$ 2e$^-$ in σ^*_{CO}
Breton et al. (2004)	DEA yield of aldehydes	18	Hole in σ_{CH}, 2e$^-$ in σ^*_{CO}

Note: ESD refers to electron stimulated desorption and EEL to electron energy loss. HOMO refers to the highest occupied molecular orbital.

Bouchiha et al. (2006) were the first to tackle the calculation of elastic and inelastic scattering from THF. They used the UK molecular R-matrix codes (Burke and Berrington 1993, Gillan et al. 1995), which are amenable to calculation of the T-matrix for elastic and inelastic scattering and are well suited to core-excited states. They focused on a simple planar form having C_{2v} symmetry. The Rydberg character of some of the excited states of THF made it necessary to use diffuse functions and a bigger R-matrix sphere of 15a_0. The calculation included the double zeta plus polarization (DZP) basis set with diffuse functions on the heavy atoms (see Davidson and Feller (1986) for an introduction to basis sets). To improve the representation of the excited states and thus of the inelastic cross section, they built state-averaged pseudonatural orbitals (NOs are discussed in Springborg (2000).) from the molecular orbitals (MOs) obtained from a Hartree–Fock self-consistent field (HF-SCF) calculation. Two models were employed for these NOs using differing weights for selected target states. The target ground state and excited states were then obtained using a complete active space configuration interaction model (CASCI) and 26 orbitals. For the scattering calculations, two different computations were performed using an active space including the ground state plus the 7 or 14 lowest singly excited states. The results for the inelastic cross section are shown in Figure 6.5. Note the elastic cross section (ECS) is over 100 times larger. Its revised form will be shown a

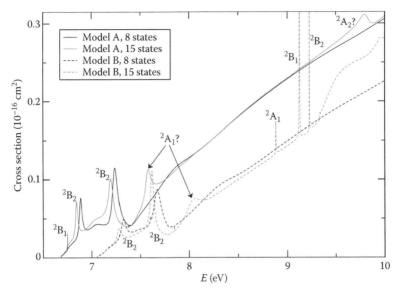

FIGURE 6.5 (**See color insert**.) Inelastic cross sections for different models used in Bouchiha et al. (2006). The symmetry of the various resonances is also indicated; those with a question mark are features that cannot be clearly identified as resonances. (Reprinted with permission from Bouchiha, D. et al. 2006. *J. Phys. B: At. Mol. Opt. Phys. 39*(4), 975–986. Copyright 2006 IOP Publishing Ltd.)

bit farther down! Feshbach resonances were observed below the first excitation threshold in the interval 6.2–6.9 eV, correlating well with the 6.4 eV resonance of Bremmer et al. (1991), and core-excited ones around 7.5 eV, in the region of the broad maximum of Zecca et al. (2005) and of Lepage et al. (1998), the values being sensitive to the exactitude of the model. This is the only calculation explicitly exhibiting such resonances. THF confines the *R*-matrix codes to considerable stress, owing to the large *R*-matrix sphere and the large number of NOs needed in the configuration interaction calculations. As a consequence, the dipole moment was too large and no shape resonance was found. This latter irritating result led to a flurry of calculations using other techniques.

Trevisan et al. (2006) performed an *ab initio* calculation of THF. With their complex Kohn variational method, they found a broad-shaped resonance behavior around 9 eV in the momentum-transfer cross section (MTCS). The Schwinger multichannel method in the static exchange (SE) approximation was chosen by Winstead and McKoy (2006c) in their calculations on various molecular symmetries. They found significant difference in the ICS between the planar and nonplanar forms of THF; particularly shifts in peak energies by 1–2 eV around 10 and 15 eV. When static exchange plus polarization was used (SEP), they obtained a broad peak in the MTCS around 8 eV, close to the position found by Trevisan et al. (2006), and a hump around 13 eV. In the same time period, Tonzani and Greene (2006b) applied their own original 3D finite element approach of the *R*-matrix (Tonzani and Greene 2005). It is not strictly speaking an *ab initio* calculation since they use a one-electron potential including exchange and polarization for the interaction with the molecule. The electrostatic part of the electronic potential is, however, derived from the *ab initio* treatment of the molecule. By its nature, the method is reliable for treating shape resonances but cannot handle core-excited anion transient states. These authors find two shape resonances at 8.6 and 14.1 eV for a nonplanar molecule, in reasonable agreement with the results of Winstead and McKoy (2006c).

So the question is what went wrong with the calculations of Bouchiha et al.? In their effort to get a good representation of the excited states of THF, they used NOs instead of the standard molecular orbitals. In her thesis work (Bouchiha 2007), Bouchiha reexamined THF with emphasis on a better description of the electronic ground state incorporating the MOs in her calculations. She found two shape resonances at 8 and 10 eV. As it so turns out, these resonances were actually faintly present above 10 eV in the original work but were not previously seen as they lay outside the energy range in the calculations of Bouchiha et al. (2006) (see Figure 6.6 for energies extended to 14 eV). Table 6.2 and Figure 6.6 present a comparison of Bouchiha's new calculations with other calculated shape resonances and measured values. The new cross section of Bouchiha is now seen to be quite consistent with measured and calculated values. All predicted resonance positions lie at higher energy than most experimental values probably owing to the limited description of the effects of polarization. It is puzzling to realize that the peaks found below 4 eV cannot be duplicated by theory.

Theory can also, when the computational expense allows it, give information on the resonant wave function. For instance, this can be achieved from the *R*-matrix

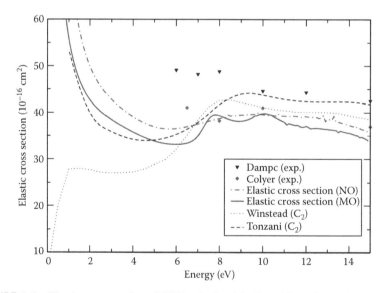

FIGURE 6.6 Elastic cross section of THF calculated by Bouchiha using MOs (Bouchiha 2007) and NOs (Bouchiha et al. 2006) compared to the experimental results of Dampc et al. (2007) and of (Colyer et al. 2007) as well as the theoretical computations of Winstead and McKoy (2006c) and Tonzani and Greene (2006b).

TABLE 6.2

Energy (eV) of the Shape Resonances of THF from Some Elastic Scattering Calculations or Post-2006 Measurements

Source	Method	Res. 1	Res. 2	Res. 3
Tonzani and Greene (2006b)	R-matrix, one-electron potential		8.6	14.1
Trevisan et al. (2006)	Kohn variational		8.6	9
Winstead and McKoy (2006c) ($C_{S/2}$)	Schwinger multichannel		8.3	13.5
Winstead and McKoy (2006c) (C_{2V})	Schwinger multichannel		9	16
Allan (2007)	Gas phase scattering	2.6	6.2	10.8
Dampc et al. (2007)	Gas phase scattering		6	9
Colyer et al. (2007)	Gas phase scattering		6.5	
Możejko et al. (2006)	Transmission in gas phase	3.5	6.5	8.5
Sulzer et al. (2006)	DEA in gas phase		7.5	
Aflatooni et al. (2006)	DEA in gas phase		6.2	8
Milosavljević et al. (2008a)	EEL in gas phase		7.5	9.5
Bouchiha (2007) (C_{2V})	R-matrix		8	10

approach of Tonzani and Greene (2006b) through the time-delay eigenfunction which corresponds to the maximum eigenvalue of the Hermitian matrix

$$Q = iS \frac{dS^\dagger}{dE}, \tag{6.2}$$

where S is the scattering matrix and E is the energy (Stibbe and Tennyson 1996). As illustrated in Figure 6.7 for the 8.6 eV resonance, through this wave function one can see a possible antibonding character of C_1–H and C_4–H bonds.

Winstead and McKoy (2006c) also calculated the ICS of deoxyribose. It is some 30% larger than that of THF mostly due to a larger background. They observe a broader maximum than in THF and argue this likely arises from the contribution of additional resonances including C–O and O–H σ^* resonances arising from the hydroxyl groups. Using the 2 eV downward energy shift due to polarization deduced from their THF calculation, they predict the maxima to be at 8 and 13 eV, in quite good agreement with the experiments of Ptasińska et al. (2004). In the same publication, the authors also looked at scattering on deoxyribose monophosphate (dMP). Scattering off of the larger moiety increases globally the ICS while the broad peak around 10 eV (8 eV with downward shift) is noticeably more pronounced. The small hump for deoxyribose at 13 eV has however disappeared, probably widened and overlapping with the 8 eV one. The phosphate group thus participates significantly in the deoxyribose resonances at those energies. Baccarelli et al. (2008a) carried out a single-center expansion study of ribose and deoxyribose in both the furanose and pyranose forms. This approach is well suited for the study of shape resonances. They identified a series of overlapping resonances in the range 6–13 and 8–15 eV for the furanose and pyranose forms, respectively, consistent with the findings of Winstead and McKoy (2006c) and

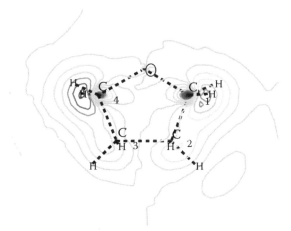

FIGURE 6.7 (**See color insert**.) Slice of the three-dimensional time-delay eigenfunction for THF on the plane that contains OC_1C_4, while C_3 is above the plane and C_2 below, at the 8.6 eV resonance. (Reprinted with permission from Tonzani, S. and C. H. Greene 2006. *J. Chem. Phys. 125*, 094504-1-7. Copyright 2006, American Institute of Physics.)

Ptasińska et al. (2004). The calculated ICS of the more relevant deoxyribofuranose form shows peaks at 3, 5, 8, 10, and 16 eV. It is worth mentioning that the 5 eV structure in the ICS has not been seen by McKoy (2006c). The peaks in the ICS of ribofuranose are at about 5, 7, 9.5, 13.5, and 18.5 eV. The pyranose forms show fewer peaks: 4 and 6 eV for deoxyribopyranose and 4, 6.5, and 15.5 eV for ribopyranose. Note, however, Bald et al. (2006) caution that, in gas-phase experiments, it was demonstrated that ribose and deoxyribose exist in the six-membered ring form (pyranose form), which behaves differently from the furanose form (Baccarelli et al. 2008a).

Another analog of deoxyribose, hydroxytetrahydrofuran (HTHF), was also theoretically investigated. Milosavljević et al. (2008b) observed EEL peaks at 7.5 and 10 eV, at similar positions to those of THF. Vizcaino et al. (2008) measured the differential cross section (DCS) at 6.5, 8, 10, 15, and 20 eV which they complemented with a Schwinger multichannel calculation. The theoretical and experimental DCS compared favorably whereas the ICS and MTCS differed substantially in magnitude and trend. The HTHF ICS and MTCS were very similar to those of THF. A peak was seen in the DCS at 120° and at an energy of 6.5 eV, similar to THF. They associated this feature to shape resonances of large angular momentum, much like those found by Tonzani and Greene (2006b) in THF. But then, the DEA measurements of Aflatooni et al. (2006) and those of Bremner et al. (1991) revealed a core-excited resonance of HTHF at 6.12 and 6.4 eV, respectively, which may well correspond to the same resonance. The DEA cross section measured by Aflatooni et al. (2006) in HTHF is some 30 times larger than that for THF. These authors claim it is the presence of the OH group, which is responsible for the huge increase. Consequently, they argue that a molecule such as deoxyribose, which contains three OH groups, will have an even larger DEA cross section. According to this interpretation, the molecule is not a suitable model for the dissociative response of the sugar ring in DNA, which does not possess these groups.

We can summarize by saying that the theoretical calculations are, not surprisingly, sensitive to the choices made in the ingredients of the model which, in turn, determine the energy and shape of the features in the calculated cross sections. These choices include basis set, NOs versus MOs, level of quantum treatment of the wave functions, one-electron potential components, treatment of polarization, and so on. There is nevertheless good overall correspondence between the different theoretical models and, generally, with experimental data for the five-membered furanose ring. Shape and core-excited resonances were found in the energy range of interest to DNA strand breaks. As discussed in Section 6.1, the direct role of deoxyribose in strand breaks is probably minimal. But it seems to be an actor in electron transfer from the bases to the phosphate group and its resonances may play a role in this channeling function.

6.2.1.2 Phosphate

Little theoretical work has been done on LEE elastic scattering off from the phosphate group. The *R*-matrix calculations of Tonzani and Greene (2006b) for phosphoric acid show two shape resonances at 7.7 and 12.5 eV of width 2 and 1.5 eV, respectively. The experimental evidence from the cousin molecule trimethylphosphate (Aflatooni et al. 2006) reveals a peak in the DEA cross section at 7.4 eV which they, however, assign to a core-excited resonance. Burrow et al. (2008) later came to the conclusion that trimethylphosphate does not exhibit properties that suggest a local π^* orbital associated with the P=O double bond. They further argue that there is little reason therefore to expect or invoke a π^* resonance resident on the phosphate group, as traditionally understood. This surely is relevant to the dissociation of the phosphodiester bond discussed in Section 6.1 or to direct attachment to the phosphate group. Measurements on dihydrogenphosphate (Pan and Sanche 2006) reveal broad peaks in the interval 7–9 eV while those on dibutyl phosphate (König et al. 2006) exhibit a low-energy resonant feature below 1 eV and a much weaker one at 8 eV. There is concurrence for the ~8 eV resonance, which could be a shape resonance in view of the above discussion. It should be realized, however, that above the first electronic excitation threshold, DEA occurs principally via core-excited resonances, which, despite the increase in decay channels, often have a sufficiently long lifetime to retain the electron in the molecule up to the time of its stabilization on a fragment.

6.2.1.3 Bases

We begin by looking at the RNA base uracil. Three π^* resonances have been found experimentally by Aflatooni et al. (1998) at 0.22, 1.58, and 3.83 eV. du Penhoat et al. (2001) observed an H$^-$ peak at 8.7 eV in the electron-stimulated anion desorption from uracil films. Hanel et al. (2003), Denifl et al. (2004a), and Abouaf and Dunet (2005) find the dominant DEA peaks in the interval 0.8–2.2 eV. The former two references also see less intense DEA activity around 5, 6, 7, 9, and 10 eV, depending on the anion fragment.

Very early in the history of LEE scattering calculations on DNA and RNA subunits, Gianturco and Lucchese (2004) calculated TNIs of A′ (planar symmetric) at 0.012 and 10.4 eV and A″ symmetry at 2.27, 3.51, and 6.50 eV. They associated the two lower-energy A″ peaks to the π_2^* and π_3^* resonances of Aflatooni et al. (1998) and the lower A to a σ* state. This was challenged by Burrow (2005) who contends that their wave function maps clearly identify the three A″ peaks as the π_1^*, π_2^*, and π_3^* resonances of Aflatooni et al. (1998). Consequently, all computed energies of Gianturco and Lucchese are approximately 2 eV too large (a number, i.e., in common with R-matrix and Schwinger calculations without polarization). Burrow also places the σ* resonance above 2 eV. Figure 6.8 shows a wave function map for the 10.4 eV resonance of Gianturco and Lucchese. The authors pointed out the antibonding nature at the C_5–H bond and proposed that this may be a precursor state for the TNI structure responsible for the excision of H⁻ observed around 9 eV (du Penhoat et al. 2001). This conclusion was found premature by Burrow (2005) who pointed out that in the absence of bond strengths, electron affinities of the fragments, and the potential surface gradients, one cannot make a strong case for the end products. Grandi et al. (2004) added a correlation and polarization correction to their one-electron potential and computed the partial cross sections of uracil and found a shape resonance of b1g symmetry at 9.07 eV. More recently, Gianturco et al. (2008) reexamined uracil also adding the correlation and polarization correction. Their new resonance energies, shown in Figure 6.9, are slightly different. Even though they now claim four π* resonances, they stick with their assignment of their lowest three being those observed experimentally. This interpretation was disputed by Winstead and McKoy (2008a), who restate Burrow's assignment and argue that the lowest-energy peak is likely a computational artifact of the strong dipole potential of uracil. Burrow's assignment has been corroborated by the authors (Gianturco et al. 2009) who realized that (Gianturco et al. 2008) had cutoff the dipole field of uracil at too short a distance in their original publication. The low-energy peak was indeed an artifact that disappeared when increasing the cutoff distance. Gianturco et al. (2008) also find a σ* resonance near 8.5 eV for which they have, most interestingly, studied the energy dependence

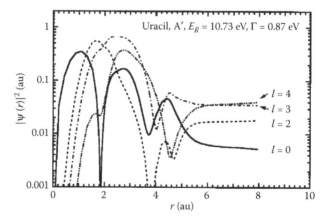

FIGURE 6.8 Computed map of the real part of the A′ symmetry resonant wave function at energy 10.73 eV calculated by Gianturco and Lucchese (2004). Changes of sign are given by solid and dotted lines. The molecular frame is also shown. (Reprinted with permission from Gianturco, F. and R. Lucchese. 2004. *J. Chem. Phys.* 120(16), 7446–7455. Copyright 2004, American Institute of Physics.)

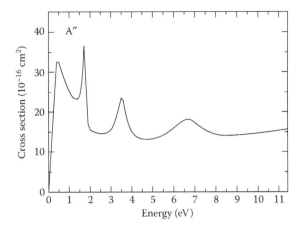

FIGURE 6.9 Single-center expansion A″ component of the ECS for gaseous uracil at its equilibrium geometry. (Reprinted with permission from Gianturco, F. A. et al. 2008. *J. Chem. Phys. 128*(17), 174302-1-8. Copyright 2008, American Institute of Physics.)

of the resonances with respect to C_3–N_5 and C_4–N_2 bond stretching. We shall come back to this bond stretching study in the next section.

The *R*-matrix calculations of Tonzani and Greene (2006a) identified three shape resonances at 2.16 eV, 5.16 eV, and a very broad one at 7.8 eV in uracil. They plotted the total time delay, given by the sum of the eigenvalues of Equation 6.2, which is quite effective in locating resonances. It is shown in the lower panel of Figure 6.12. These presumably correspond to the three higher π^* resonances of Gianturco and Lucchese (2004) but oddly quite differently spaced. Tonzani and Greene (2006a) also fail to see a fourth A′ resonance around 10 eV as do the other calculations. This proves somewhat puzzling. When down shifting by some 2 eV, these three resonances tentatively correlate with the three experimentally reported ones in Aflatooni et al. (1998) with a larger spread. Winstead and McKoy (2006b) found, with their Schwinger multichannel calculations at the SEP level, three π^* resonances of A″ symmetry at 2.08, 4.2, and 8.2 eV (upper panel of Figure 6.13) and a very broad A′ symmetry peak at 10.5 eV plus a presumably reliable one at 1.75 eV. Here also, a downward energy shift of order 2 eV occurs when polarization is taken into account at the SE plus polarization level. The two A′ peaks shift to 1.45 and 8.5 eV while the A″ contributions show a very large dipolar background between 0.1 and 0.4 eV and peaks at 0.32, 1.91, and 5.08 eV. Interestingly, Winstead and McKoy also looked at the ICS for electron-impact excitation of the HOMO → LUMO transition in uracil. This is their only inelastic scattering calculation on the bases. Their threshold energies are 4.15 eV for ^{3}A′ and 7.73 eV for ^{1}A′ transitions.

Quite recently, Dora et al. (2009) performed an *R*-matrix calculation on uracil. This is quite a feat considering uracil has 58 electrons, much more than THF. These authors used a state-averaged complete-active space self-consistent field method (CASSCF) with several basis sets and different active spaces for the target. Vertical excitation energies for 14 active electrons in 10 orbitals using 6–31G basis set are shown in Table 6.3. The electron impact excitation cross sections are shown in Figure 6.10, where they are compared to those of Winstead and McKoy (2006b). In their scattering calculations, Dora et al. experimented with various approximations: SE, SEP, and close-coupling (CC) by varying the number of frozen electrons and the distribution of electrons in valence and virtual states. Their A″ symmetry contribution to the ECS is shown in Figure 6.11. It shows the three π* shape resonances at 0.134, 1.94, and 4.95 eV. Note that the low-energy part in the insert differs markedly from the results of Winstead and McKoy. They have also found three A′ resonances at 6.17, 7.62, and 8.12 eV.

A novel and relatively simple approach was also proposed by Yalunin and Leble (2007) for uracil. They use a muffin-tin multiple scattering method in which phase

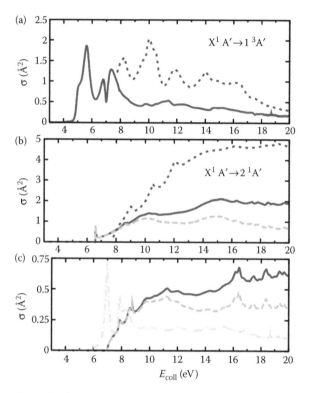

FIGURE 6.10 (**See color insert.**) Low-energy electron impact electronic excitation cross sections of Dora et al. (2009) for uracil in comparison with the calculations of Winstead and McKoy (2006b): Only excitations to the four lowest A′ symmetry states are shown since the cross sections for excitation to the corresponding A″ states are very small. (a) Solid curve: Dora et al. calculation, short-dashed curve: Winstead and McKoy. (b) Solid curve: Dora et al. calculation with Born correction, long-dashed curve: Dora et al. calculation without Born correction, short-dashed curve: Winstead and McKoy. (c) Solid curve: Dora et al. calculation for $X^1A′ \rightarrow 3^1A′$ excitation with Born correction, long-dashed curve: Dora et al. calculation for $X^1A′ \rightarrow 3^1A′$ excitation without Born correction, chain curve: Dora et al. calculation for $X^1A′ \rightarrow 2^3A′$ excitation. (Reprinted with permission from Dora, A. et al. 2009. *J. Chem. Phys.* 130, 164307-1-8. Copyright 2009, American Institute of Physics.)

TABLE 6.3
Summary of Measured and Calculated Resonances or Threshold Excitations for Uracil

Source	Method	Energies in eV								
		π_1^*	π_2^*	π_3^*						
Aflatooni et al. (1998)	ETS gas phase	0.22	1.58	3.83						
Abdoul-Carime et al. (2001)	DEA H⁻ on films									
Isaacson (1972)	EEL for high energy electrons in films				4.7	5.93	6.93	8.45		10.6
Abouaf and Dunet (2005)	DEA in gas and ECS (a)	0.27 (a), 0.64, 0.95	1.22, 1.41, ≈1.6, ≈1.9							
du Penhoat et al. (2001)	H⁻ desorption from film								8.7	
Hanel et al. (2003)	DEA in gas	0.7	1, 1.4, 1.7–3		≈5		≈7		≈9	≈10
Denifl et al. (2004a)	DEA in gas	0.69	1.0, 1.5, 1.7, 1.9		≈5	≈6	≈7		≈9	≈10
Shukla and Leszczynski (2004)	TDDFT (basis set C, oscillator strength ≥0.01) singlet excitations				$\pi\pi^*$ 5.17	$\pi\pi^*$ 5.89, 6.42		$\pi\pi^*$ 7.38		
Gianturco and Lucchese (2004)	Single-center expansion	2.27	3.51	6.5						10.4
Gianturco et al. (2008)	Single-center expansion	1.7	3.5	6.5					σ^* 8.5	
Grandi et al. (2004)	Single-center expansion							9.07		
Tonzani and Greene (2006a)	R-matrix, one-electron potential	2.16	5.16	7.8						
Winstead and McKoy (2006b)	Threshold excitation				3A′ 4.15			1A′ 7.73		
Dora et al. (2009)	Schwinger variational, SEP level	0.32	1.91	5.08					8.5	
	Threshold excitation				13A′ 3.8	23A′ 5.4	21A′ 6.6	31A′ 7.0		
	R-matrix resonances	0.134	1.94	4.95		6.17	7.62	8.12		

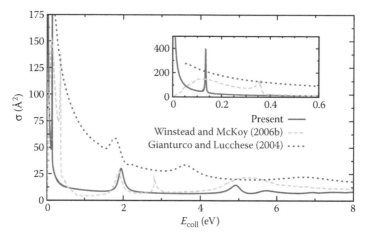

FIGURE 6.11 **(See color insert.)** A″ symmetry contribution to the elastic cross section for low-energy electron collisions with uracil by Dora et al. (2009). The inset gives the low-energy behavior showing the very sharp resonance feature. Comparison is made with the calculations of Winstead and McKoy (2006b) and Gianturco and Lucchese (2004). (Reprinted with permission from Dora et al. 2009. *J. Chem. Phys. 130*(16), 164307-1-8. Copyright 2009, American Institute of Physics.)

shifts are calculated for each "atom" using a muffin potential generated from *ab initio* calculations. They find A′ symmetry peaks at 1.2, 2.2, and 6.7 eV and A″ ones at 1.9, 5.6, 6.1 eV. Quite obviously, the A′ symmetry has been contaminated by A″ symmetry. The A″ energies are quite competitive.

We have summarized in Table 6.3 a large number of experimental and theoretical results for the uracil resonances. Most were discussed above, and some, like the TDDFT calculations that we shall meet again in the next section, are there to provide complementary information on the electronic single excitations which may serve as reference parent states to potential core-excited resonances. We present TDDFT results for uracil and the DNA bases from Shukla and Leszczynski (2004), Tsolakidis and Kaxiras (2005), and Sobolewski and Domcke (2002). There are others as well and many *ab initio* calculations of single excitations at the CASSCF or Complete Active Space with Second-order Perturbation Theory (CASPT2) level that are cited in these three last references. See also Nguyen et al. (2004).

The four DNA bases have been looked at by Tonzani and Greene (2006a), see Figure 6.12, and by Winstead and McKoy (2006a), Winstead et al. (2007), see Figure 6.13. One observes that the purines adenine and guanine show a triplet of resonances in the range 3–5 eV as compared to two for the pyrimidines cytosine, thymine, and uracil. Note that the two visible resonances for guanine in the results of Winstead and McKoy (2006a) without polarization are split into three when polarization is included. The polarization corrections are somewhat sensitive to the basis set used, especially at low energy. Tables 6.4 through Table 6.7 summarize a number of experimental and theoretical results for the DNA bases.

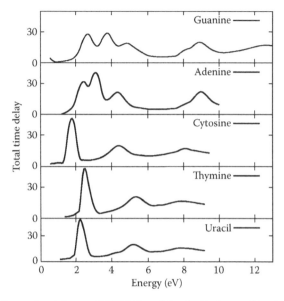

FIGURE 6.12   Tonzani and Greene (2006a) time delay for the bases. (Reprinted with permission from Tonzani, S. and C. H.Greene. 2006. *J. Chem. Phys. 124*, 054312-1-11. Copyright 2006, American Institute of Physics.)

6.2.1.4   Larger Moieties

The Schwinger multichannel approach is thus far the only method reporting scattering calculations on such large units as the nucleosides and nucleotides (Winstead and McKoy 2006a, Winstead et al. 2007, Winstead and McKoy 2008b). This sets it apart from the other approaches. The general finding is that the cross section of the nucleosides is larger, as to be expected by a quantity of order the cross section of deoxyribose, with the peak modulations of the π^* resonances having similar amplitudes but with a slight upward shift in energy. When adding a phosphate group to deoxyadenosine at the $5'$ position to get dAMP, the ICS globally increases by a quantity of order the ICS of the phosphate group. The π^* resonances are unaltered, indicating that the fairly distant phosphate group has negligible influence on the base's resonances, but a small hump appears around 8 eV (SE level), which is perhaps due to the phosphate group not unlike in dMP.

6.2.1.5   Concluding Remarks on Scattering from Subunits

What can be learned from the many tables summarizing resonance information for the bases? On the experimental side, there are many scattering or dissociative peaks distributed fairly evenly over the energy range 0–10 eV that denote the presence of numerous TNIs. The three lower-energy π^* shape resonances are calibration checkpoints for the theoretical methods. The higher energies ones are mostly in the domain of core-excited resonances. There is a fair concordance between the TDDFT excitation energies and most of the experimental resonances. There is thus no great difficulty in finding core-excited resonances of proper energy that might contribute to strand breaks in DNA, as proposed in Section 6.1. The problem with the theoretical scattering calculations is

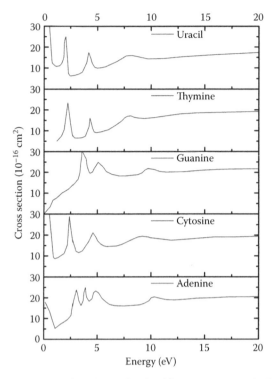

FIGURE 6.13 Cross section of the bases for the A″ symmetry as calculated by Winstead and McKoy (2006b) without polarization.

getting good resonance energies and properly handling these core-excited resonances. The calculated energies for the π* shape resonances are usually too large. The one-electron potential methods have larger errors by a factor of order 2 eV; surely, because of the nature of the local one-electron exchange or polarization potentials used which seem unable to provide an efficient description of the interaction with the neutral molecule. The calculations of Tonzani and Greene (2006a) for uracil, cytosine, and thymine furthermore fail to show the higher energy A′ resonances near 10 eV. The Schwinger variational approach fares better at the SEP level. The R-matrix calculations, at least judging from its application to THF (using MOs) and uracil, seem to be more precise and able to reproduce most of the experimentally observed resonances. The other theoretical treatments ignore the intermediate energy range 5–8 eV range, in which a number of experimental resonances have been found experimentally. Threshold excitation calculations were quite successful for uracil. It is unfortunate that such calculations were not repeated for the other bases. At this point in time, the Schwinger variational method can be considered as the most useful for large systems.

6.2.2 DISSOCIATIVE ELECTRON ATTACHMENT

The previous section presented various theoretical methods with which to calculate cross sections and detect resonances. But it proves somewhat difficult, using these

TABLE 6.4
Summary of Measured and Calculated Resonances for Adenine

Resonance Energies in eV (columns π₁* through σσ*)

Source	Method	π_1^*	π_2^*	π_3^*			n→3s		$\pi\pi^*$	$\pi\pi^*$	$\sigma\sigma^*$
Dillon et al. (1989)	EEL at small-scattering angle in gas*				$\pi\sigma^*$ 4.96		6.15	$\pi\sigma^*$ 6.5	7.28	8.56	11.9
Panajotović et al. (2007)	ICS for excitation of vibrational states in ≈ monolayer film				4.7	5	5.5	6.1	6.6		
Isaacson (1972)	EEL for high-energy electrons in films				4.53			5.84	6.5	7.71	13
Aflatooni et al. (1998)	ETS gas phase	0.54	1.36	2.17							
Aflatooni et al. (2006)	Total dissociation in gas phase		1.18	2.17				6.28			
Abdoul-Carime et al. (2001)	DEA H⁻ on films									9.5	
Shukla and Domcke (2002)	TDDFT (basis set C; oscillator strength ≥0.01) singlet excitations				$n\pi^*$ 4.86	$\pi\pi^*$ 4.97, 5.20; $\sigma\pi^*$ 5.23		$\pi\sigma^*$ 6.03–6.16			
Sobolewski and Domcke (2002)	TDDFT				$n\pi^*$ 4.75	$\pi\sigma^*$ 4.99; $\pi\pi^*$ 5.05					
Tsolakidis and Kaxiras (2005)	TDDFT				4.51	4.95, 5.58		5.79, 6.28, 6.63	6.92, 7.47	7.81	
Tonzani and Greene (2006a)	R-matrix, one-electron potential	2.4	3.2	4.4						9	
Winstead and McKoy (2006a)	Schwinger variational, SEP (SE) level	1.1	1.8	4.1						SE π^*10	SE σ^*11

TABLE 6.5

Summary of Measured and Calculated Resonances for Cytosine

Source	Method	Resonance Energies in eV						
		π_1^* 0.32	π_2^* 1.53	π_3^* 4.5				
Aflatooni et al. (1998)	ETS gas phase							
Aflatooni et al. (2006)	Total dissociation in gas phase		1.54		5.48			
Huels et al. (1998)	DEA H⁻ gas phase		1.4		5.6		9.5	
Abdoul-Carime et al. (2001)	DEA H⁻ on films						8.5	10.3
Denifl et al. (2004b)	DEA O⁻/NH2⁻ gas phase				5.74		9.64	
Huels et al. (1998)	DEA O⁻/NH2⁻ gas phase		2.3	3.8	5.2	7.4	9.2	
Denifl et al. (2003)	EA in gas phase	0.04, 1.1	1.54		5.2	6.7		
Crewe et al. (1971)	EEL in films			4.7		6.5	8.3	
Shukla and Leszczynski (2004)	TDDFT (basis set C; oscillator strength ≥0.01) singlet excitations			$\pi\pi^*$ 4.62	$\pi\pi^*$ 5.44 $\pi\sigma^*$ 5.66	$\pi\pi^*$ 6.17, 6.36, 6.54		
Tsolakidis and Kaxiras (2005)	TDDFT			4.1	4.9	5.92, 6.39, 6.48, 6.88, 7.16		
Tonzani and Greene (2006a)	R-matrix one-electron potential	1.7	4.3	8.1				
Winstead et al. (2007)	Schwinger variational SEP (SE) level	0.5	2.4	6.3			SE π^*9.5	SE σ^*11

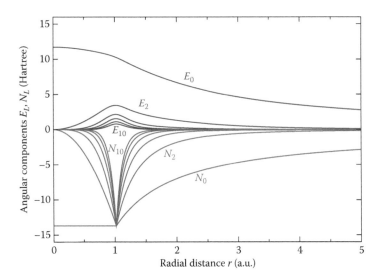

FIGURE 2.1 The angular expansion (2.9) of the nuclear attractive potential (red lines) and the electronic repulsive part (blue lines). Their sum forms the electrostatic interaction between scattered electron and N_2 molecule. The partial contributions are shown for even L-values because the odd L-values do not contribute in case of the homonuclear diatomic molecule. The contributions up to $L = 10$ are displayed. The target electron density was obtained by the Hartree–Fock method.

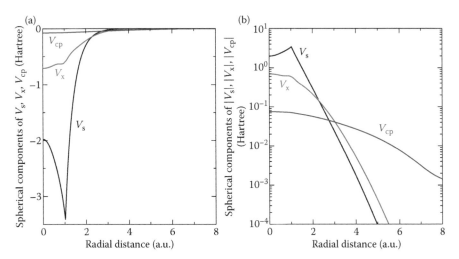

FIGURE 2.2 (a) Spherical components of static V_s, exchange V_x and correlation–polarization V_{cp} (LSD model) potentials for nitrogen molecule. (b) Absolute values of the potentials on a logarithmic scale.

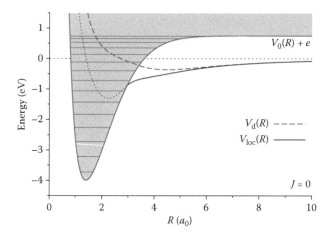

FIGURE 4.1 Typical potential–energy curves for the electron collisions with diatomic molecules.

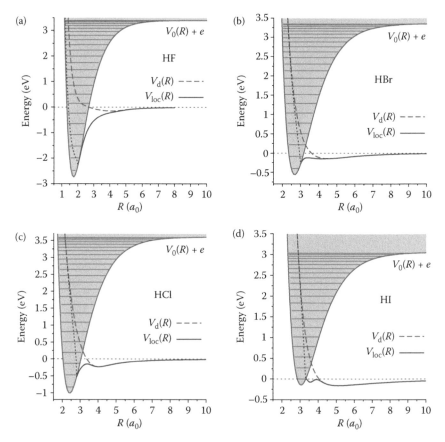

FIGURE 5.5 Potential–energy curves of the nonlocal resonance models for the following hydrogen halides: (a) HF molecule, (b) HBr molecule, (c) HCl molecule, and (d) HI molecule.

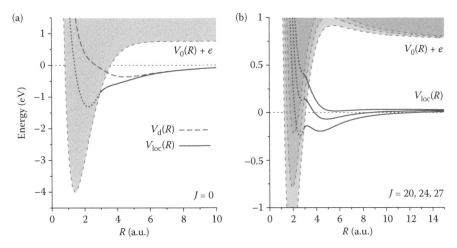

FIGURE 5.13 Potential–energy curves (including the centrifugal term $J(J + 1)/2\mu R^2$) of the $e^- + H_2$ system for angular momentum $J = 0$ (a) and $J = 20, 24, 27$ (b). The shaded area above $V_0(R)$ represents the electron continuum. $V_{loc}(R)$ corresponds to the energy of the (electronically) bound state (solid line) or to the pole of the K-matrix (dotted line).

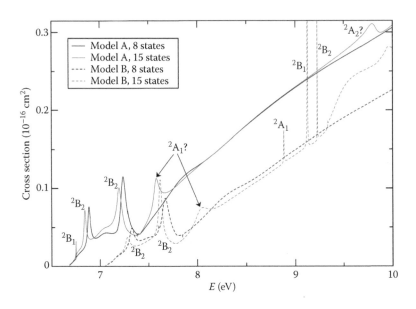

FIGURE 6.5 Inelastic cross sections for different models used in Bouchiha et al. (2006). The symmetry of the various resonances is also indicated; those with a question mark are features that cannot be clearly identified as resonances. (Reprinted with permission from Bouchiha, D. et al. 2006. *J. Phys. B: At. Mol. Opt. Phys.* *39*(4), 975–986. Copyright 2006 IOP Publishing Ltd.)

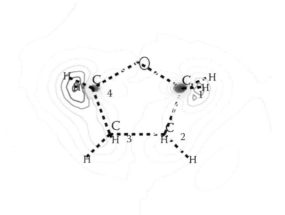

FIGURE 6.7 Slice of the three-dimensional time-delay eigenfunction for THF on the plane that contains OC_1C_4, while C_3 is above the plane and C_2 below, at the 8.6 eV resonance. (Reprinted with permission from Tonzani, S. and C. H. Greene 2006. *J. Chem. Phys. 125*, 094504-1-7. Copyright 2006, American Institute of Physics.)

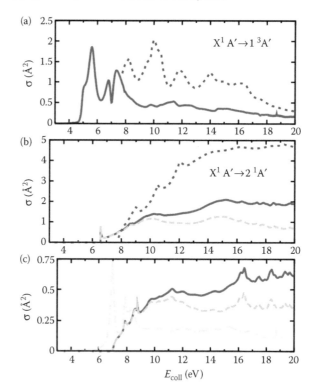

FIGURE 6.10 Low-energy electron impact electronic excitation cross sections of Dora et al. (2009) for uracil in comparison with the calculations of Winstead and McKoy (2006b): Only excitations to the four lowest A′ symmetry states are shown since the cross sections for excitation to the corresponding A″ states are very small. (a) Solid curve: Dora et al. calculation, short-dashed curve: Winstead and McKoy. (b) Solid curve: Dora et al. calculation with Born correction, long-dashed curve: Dora et al. calculation without Born correction, short-dashed curve: Winstead and McKoy. (c) Solid curve: Dora et al. calculation for $X^1A' \rightarrow 3^1A'$ excitation with Born correction, long-dashed curve: Dora et al. calculation for $X^1A' \rightarrow 3^1A'$ excitation without Born correction, chain curve: Dora et al. calculation for $X^1A' \rightarrow 2^3A'$ excitation. (Reprinted with permission from Dora, A. et al. 2009. *J. Chem. Phys. 130*, 164307-1-8. Copyright 2009 American Institute of Physics.)

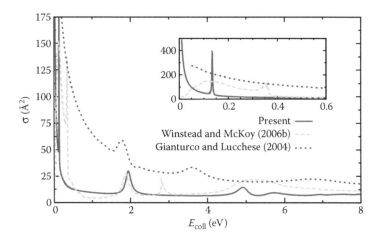

FIGURE 6.11 A″ symmetry contribution to the elastic cross section for low-energy electron collisions with uracil by Dora et al. (2009). The inset gives the low-energy behavior showing the very sharp resonance feature. Comparison is made with the calculations of Winstead and McKoy (2006b) and Gianturco and Lucchese (2004). (Reprinted with permission from Dora, A. et al. 2009. *J. Chem. Phys. 130*(16), 164307-1-8. Copyright 2009, American Institute of Physics.)

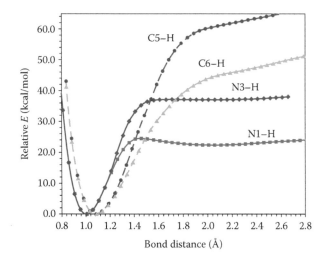

FIGURE 6.15 Adiabatic potential energy surfaces of anionic uracil along each N–H or C–H coordinate, calculated by Li et al. (2004) at the B3LYP/6-31+G(d) level. Energy relative to that of the optimized anion in the equilibrium state. Zero-point energy is not included. (Reprinted with permission from Li, X., L. Sanche, and M. D. Sevilla. 2004. *J. Phys. Chem. B 108*(17), 5472–5476. Copyright 2004, American Institute of Physics.)

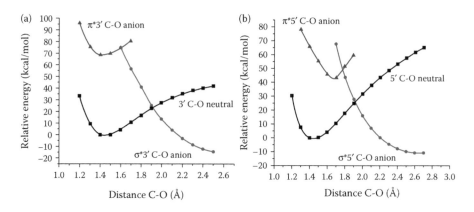

FIGURE 6.20 Energies calculated by Berdys et al. (2004a) of neutral dPd, π^* anion, and σ^* anion for $C_3'-O$ (a) and $C_5'-O$ (b) bond rupture versus C–O bond length.

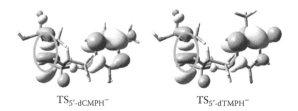

FIGURE 6.23 The SOMOs of the transition states of the 5′-dCMPH and 5′-dTMPH radical anions calculated by ref108. The typical characteristics of the σ antibonding orbital along the C5′–O5′ bond are shown in the orange circle. (Reprinted with permission from Bao, X. et al. 2006. *Proc. Natl. Acad. Sci. U. S. A. 103*(15), 5658–5663. Copyright 2006 National Academy of Sciences, USA.)

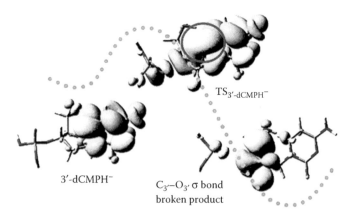

FIGURE 6.24 The SOMO of the electron attached state, the transition state, and the C3′–O3′ σ-bond broken product of 3′-dCMPH radical anion as calculated by Gu et al. (2006). The migration of the excessive negative charge through the atomic orbital overlap between the C6 of pyrimidine and the C3′ of deoxyribose is shown in the orange circle. (Reprinted with permission from Gu, J., J. Wang, and J. Leszczynski. 2006. *J. Am. Chem. Soc. 128*(29), 9322–9323. PMID: 16848454. Copyright 2006 American Chemical Society.)

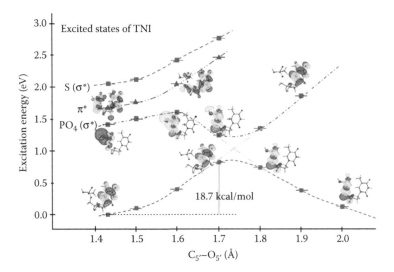

FIGURE 6.25 Lower curve: PES of the 5′-dTMPH transient negative ion (TNI); calculated in the neutral optimized geometry of 5′-dTMPH with C5′–O5′ bond elongation. SOMO is shown at selected points. Upper curves: Calculated vertical excitation energies of the radical anion at each point along the PES, MOs involved in excitations are also shown. Energies and distances are given in eV and Å, respectively. The lowest $\pi\pi^*$ state (triangles) and lowest $\pi\sigma^*$ states (square) are shown. (Reprinted with permission from Kumar, A. and M. D. Sevilla. 2008b. *J. Am. Chem. Soc. 130*(7), 2130–2131. Copyright 2008 American Chemical Society.)

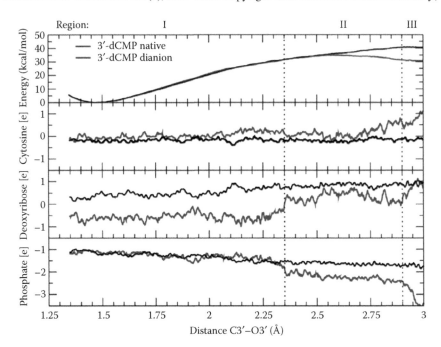

FIGURE 6.26 The uppermost graph shows the energy when breaking the C3′–O3′ bond in 3′-dCMP. The three lower graphs show the charges for each moiety of the 3′-dCMP nucleotide while rupturing the bond. (Reprinted from *Chem. Phys. Lett. 462*(4–6), Schyman, P., A. Laaksonen, and H. W. Hugosson, Phosphodiester bond rupture in 5′ and 3′ cytosine monophosphate in aqueous environment and the effect of low-energy electron attachment: A Car-Parrinello QM/MM molecular dynamics study. 289–294. Copyright 2008, with permission from Elsevier.)

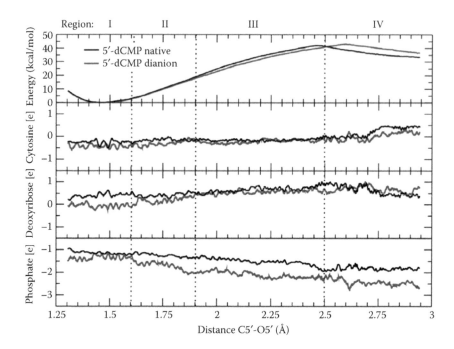

FIGURE 6.27 The uppermost graph shows the energy when breaking the C5′–O5′ bond in 5′-dCMP. The three lower graphs show the charges for each moiety of the 5′-dCMP nucleotide while rupturing the bond. (Reprinted from *Chem. Phys. Lett.* 462(4–6), Schyman, P., A. Laaksonen, and H. W. Hugosson, Phosphodiester bond rupture in 5′ and 3′ cytosine monophosphate in aqueous environment and the effect of low-energy electron attachment: A Car-Parrinello QM/MM molecular dynamics study. 289–294. Copyright 2008, with permission from Elsevier.)

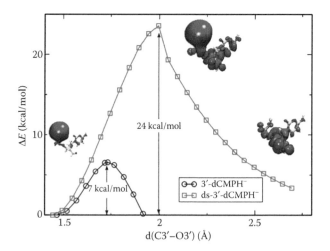

FIGURE 6.28 PES of Loos et al. (2009) for the C3′–O3′ bond in ss-3′-dCMPH⁻ and ds-3′-dCMPH⁻. The SOMOs corresponding to the minimum and maximum along the bond cleavage coordinates as well as the dissociated product are depicted for ds-3′-dCMPH⁻. (Reprinted from *Chem. Phys. Lett.*, 475(1–3), Loos, P.-F., E. Dumont, A. D. Laurent, and X. Assfeld. Important effects of neighbouring nucleotides on electron induced DNA single-strand breaks. 120–123. Copyright 2009, with permission from Elsevier.)

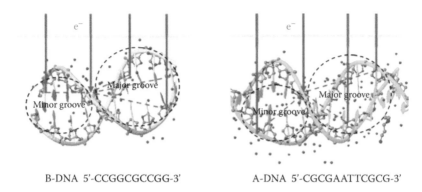

B-DNA 5'-CCGGCGCCGG-3'   A-DNA 5'-CGCGAATTCGCG-3'

FIGURE 6.38 The two DNA targets used in the calculations of Orlando et al. (2008) are illustrated in Figure 1 of their paper. The detailed structural information was obtained from the protein data bank (PDB). The dots in and around target represent the structural water positions extracted from the database. For B-DNA 5'-CCGGCGCCGG-3' and A-DNA 5'-CGCGAATTCGCG-3', the minor and major grooves are indicated by the dashed-line circles. DNA strands are aligned with the symmetry axis parallel to the surface. Bold vertical lines describe the incident electrons whereas dashed lines are the first-order scattered components. (Reprinted with permission from Orlando, T. M. et al. 2008. *J. Chem. Phys. 128*(19), 195102-1-7. Copyright 2008 American Institute of Physics.)

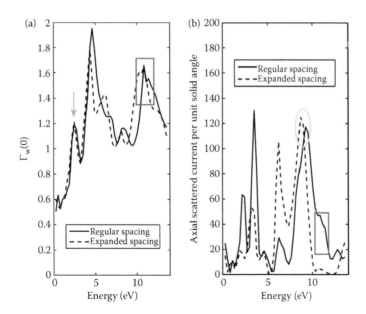

FIGURE 6.41 (a) Interpolated value of the weighted partial capture factors for the $L = 0$ component, and (b) Scattered current in a.u. in the $+ \hat{z}$ (axial) direction of the poly(A) •poly(T) decamer as a function of incident electron energy: For normal spacing and for an expanded spacing 1.05 times the regular one, from Caron et al. (2008). The encircled and boxed regions are discussed in the text. (Reprinted with permission from Caron, L. et al. *Phys. Rev. A 78*(4), 042710-1-13, 2008. Copyright 2008 by the American Physical Society.)

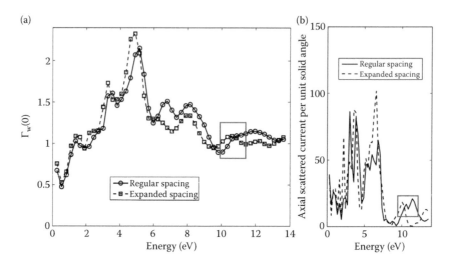

FIGURE 6.42 (a) Interpolated value of the weighted partial capture factors for the $L = 0$ component, and (b) Scattered current in a.u. in the $+\ \hat{z}$ (axial) direction of the B-form GCGAATTGGC decamer as a function of incident electron energy: for normal spacing and for an expanded spacing 1.05 times the regular one, from Caron et al. (2008). The boxed regions are discussed in the text. (Reprinted with permission from Caron, L., L. Sanche, S. Tonzani, and C. H. Greene (2008). Diffraction in low-energy electron scattering from DNA: Bridging gas-phase and solid-state theory. *Phys. Rev. A 78*(4), 042710-1-13. Copyright 2008 American Physical Society.)

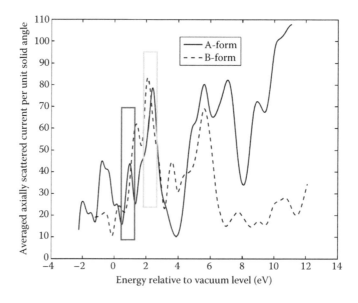

FIGURE 6.43 Average of the square of the electron wave function on the bases over two incident and mutually orthogonal incident directions on the GCGAATTGGC A-form and B-form decamers as a function of incident electron energy corrected for the estimated polarization energy of each form from Caron et al. (2008, 2009). The two rectangular boxes are discussed in the text. (Reprinted with permission from Caron, L. et al. *Phys. Rev. A 78*(4), 042710-1-13, 2008. Copyright 2008 by the American Physical Society.)

FIGURE 6.44 Side view of the decamer studied by Caron et al. (2009). The oxygen atom of the retained water molecules appear as isolated spheres. (Reprinted with permission from Caron, L. et al. *Phys. Rev. A 80*(1), 012705-1-6, 2009. Copyright 2009 by the American Physical Society.)

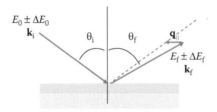

FIGURE 7.1 Schematic plot of the electron-scattering geometry in the plane of incidence. k_i and k_f are the initial and final electron wave vectors, respectively. The angles θ are defined with respect to the normal to the surface. $q_\parallel$ is the surface parallel excitation wave vector.

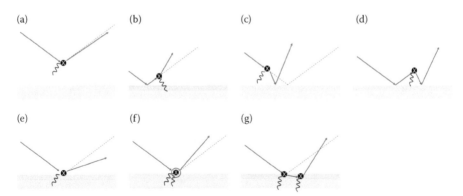

FIGURE 7.2 Schematic representation of seven possible scattering scenarios for an incoming electron at a substrate surface. Crosses denote the points where inelastic-scattering events take place. Schemes (a)–(d) all correspond to dipolar scattering and result in a single energy loss event taking place at large distance. Schemes (b)–(d) picture the involvement of elastic scattering away from the substrate. (Adapted from Thiry, P. A. et al. 1987. *Phys. Scripta 35*(3), 368–379.) Scheme (e) pictures one possible scenario of impact scattering. Further scenarios involving elastic scattering can be derived from it. Schemes (f) and (g) picture multiple energy loss events. In scheme (f), the multiple energy loss proceeds through a single-scattering event, involving the temporary trapping of the electron (resonant scattering), and in scheme (g) it proceeds through successive inelastic impact-scattering events.

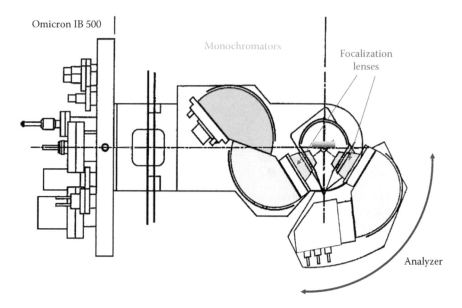

FIGURE 7.3 Schematic representation of the Omicron IB500 HREEL spectrometer. (Adapted from Omicron. 2000. *IB 500 User's Guide, version 1.0.*)

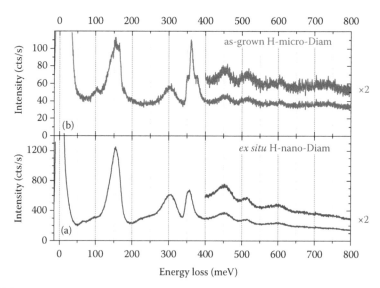

FIGURE 7.4 Energy loss spectra measured at $E_0 = 5$ eV for (a) an *ex situ* MW-hydrogenated nanocrystalline (5 nm) dc GD CVD diamond film with a typical resolution of $\Delta E_{FWHM} \sim 12$ meV; and (b) an as-grown-hydrogenated microcrystalline (300 nm) HF CVD diamond film with a typical resolution of $\Delta E_{FWHM} \sim 5$ meV.

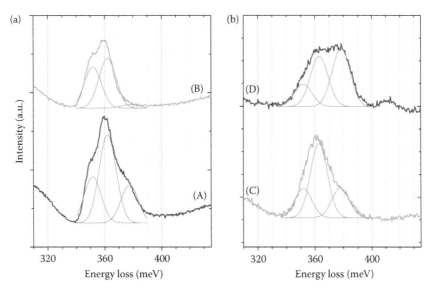

FIGURE 7.5 (a): Hydrogen termination-stretching band v(sp^m-CH$_x$) of as-grown hydrogenated HF CVD microcrystalline diamond films after annealing to 350°C (A), and after subsequent *ex situ* MW-hydrogenation and annealing to 350°C (B) (Adapted from Lafosse, A. et al. 2008. *Diam. Relat. Mater. 17*(6), 949–953.) (b) Hydrogen termination-stretching band v(sp^m-CH$_x$) of *in situ*-hydrogenated microcrystalline diamond films annealed to 600°C (C) and after further Ar$^+$ ion irradiation ($E = 1$ keV, ~10^{15} cm^{-2}) (D) (Adapted from Bertin, M. et al. 2007. *Phys. Lett. 90*(6), 061918.) The represented Gaussians are only introduced to guide the eye, and are indicative of the different main sp^m-CH$_x$ contributions. All spectra were recorded at $E_0 = 4$ eV and with a resolution ΔE_{FWHM} ~ 9–14 meV.

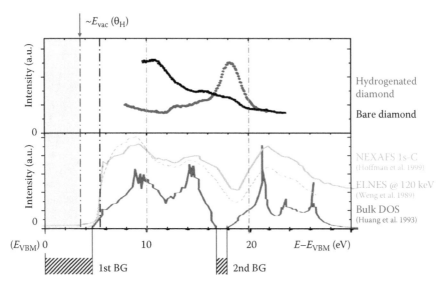

FIGURE 7.6 Upper panel: Elastic reflectivity curves measured for the *in situ* hydrogenated microcrystalline diamond surface (in blue) and bare microcrystalline diamond surface (in black). Lower panel: Comparison of the bulk diamond calculated DOS (Huang and Ching 1993) with the diamond C(1s) ELNES (Weng et al. 1989) (dashed line) and NEXAFS (Hoffman et al. 1999) (full line) spectra. The five spectra are drawn using a common energy scale, whose origin was chosen as the VBM of the substrate ($E_{VBM} = 0$ eV). The relative intensities of the spectra have been arbitrarily fixed. (Adapted from Lafosse, A. et al. 2005. *Surf. Sci. 587*(1–2), 134–141.)

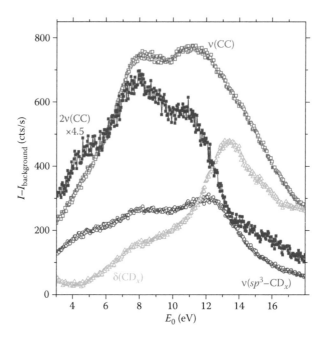

FIGURE 7.7 Selected vibrational excitation functions for deuterated nanocrystalline dc GD CVD diamond, after subtraction of the background: ν(CC) at 152 meV (red), δ(CD$_x$) at 107 meV (green), ν(sp^3-CD$_x$) at 272 meV (violet), and 2 ν(CC) at 300 meV (blue, intensity multiplied by a factor 4.5). (Adapted from Amiaud, L. et al. 2011. *Phys. Chem. Chem. Phys. 13(24)*, 11495–11502.)

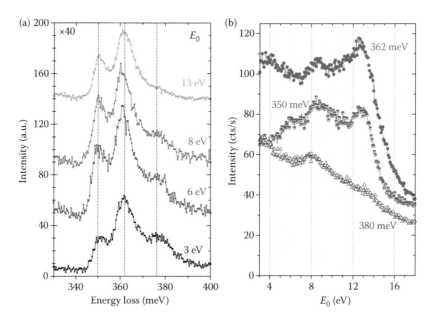

FIGURE 7.8 (a) Hydrogen termination-stretching band ν(sp^m-CH$_x$) of as-grown-hydrogenated HF CVD microcrystalline diamond films recorded at increasing incident electron energy. (b) Specular vibrational excitation functions recorded for the energy losses 362, 350, and 380 meV, respectively, attributed to sp^3-CH$_x$, sp^3-CH$_x$, and sp^2-CH$_x$ groups. (Adapted from Lafosse, A. et al. 2006. *Phys. Rev. B 73(19)*, 195308.)

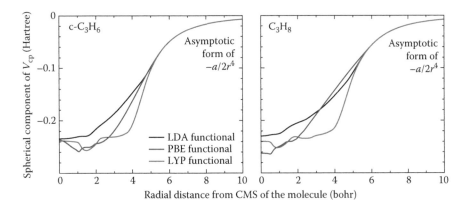

FIGURE 8.2 Visual comparison of different correlation functionals used in the present study. For simplicity, only the dominant spherical components $u_0^0(r)$ of Equation 8.47 are displayed. Left panel shows the data for the cyclopropane molecule, right panel for the propane molecule. (Reprinted with permission from Čurík, R. and M. Šulc 2010. *J. Phys. B: At. Mol. Opt. Phys. 43*, 175205. Copyright 2010 IOP Publishing Ltd.)

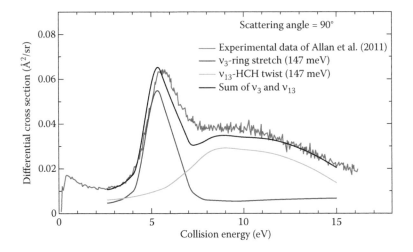

FIGURE 8.6 Rotationally summed $0 \rightarrow 1$ differential cross section of cyclopropane as a function of the collision energy. Scattering angle is fixed at $\vartheta = 90°$. Blue line shows present calculations of the resonant vibrational excitation of v_3 mode while the green line displays the excitation of HCH twisting mode v_{13}. The sum of these two cross sections (black line) is compared to experimental data. (Adapted from Čurík, R., P. Čársky, and M. Allan. 2011. In preparation.)

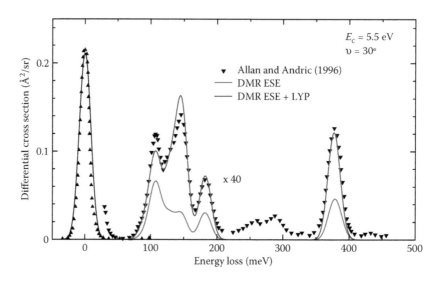

FIGURE 8.7 Computed and measured electron energy-loss spectra of cyclopropane for the scattering angle $\vartheta = 30°$ at the resonant energy of 5.5 eV. The red line displays exact static-exchange approximation while the blue lines include also polarization effects modeled by LYP functional. Bands observed at 250–300 meV are not due to fundamental vibrational transitions. Triangles denote experimental spectra. (Adapted from Allan, M. and L. Andrič. 1996. *J. Chem. Phys. 105*, 3559–3568.)

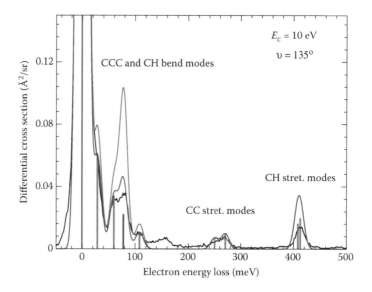

FIGURE 8.8 Computed and measured electron energy-loss spectra of diacetylene for the scattering angle $\vartheta = 135°$ and collision energy of 10 eV. Vertical bars represent the actual calculated value of the differential cross section for the particular mode. Assuming the Gaussian band shape, the calculated SEP results (blue line) are compared with the experimental spectra (black curve). For completeness, we also present static-exchange calculations denoted by the red line. (Adapted from Allan, M. et al. 2011. *Phys. Rev. A, 83*, 052701.)

TABLE 6.6
Summary of Measured and Calculated Resonances for Guanine

Source	Method	Resonance Energies in eV						
Aflatooni et al. (1998)	ETS gas phase (enol tautomer?)	π_1^* 0.462	π_2^* 1.237	π_3^* 2.36				10.6
Abdoul-Carime et al. (2001)	DEA H- on films							
Crewe et al. (1971)	EEL in films				4.80,	5.3	7	8.8
Shukla and Leszczynski (2004)	TDDFT (basis set C; oscillator strength ≥0.01) singlet excitations				$\pi\pi^*$ 4.64, 4.86	$\pi\sigma^*$ 5.04; $\pi\pi^*$ 5.17		
Tsolakidis and Kaxiras (2005)	TDDFT				4.46, 4.71,	5.04, 5.64	6.23–7.26	
Tonzani and Greene (2006a)	R-matrix one-electron potential	2.4	3.8	4.8			8.9	12
Winstead and McKoy (2006a)	Schwinger variational SEP or (SE) level	1.55	2.4	3.75				SE π^*, σ^* 10

TABLE 6.7
Summary of Measured and Calculated Resonances for Thymine

Source	Method	π_1^*	π_2^*	π_3^*			
				Resonance Energies in eV			
Aflatooni et al. (1998)	ETS gas phase	0.29	1.71	4.05			
Aflatooni et al. (2006)	Total dissociation in gas phase	1.01	1.7			6.63	
Huels et al. (1998)	DEA H⁻ gas phase			4	5.2	6.6	8
Abdoul-Carime et al. (2001)	DEA H⁻ on films						9.3
Huels et al. (1998)	DEA O⁻ gas phase		1.8	3.2, 4.5		6.6	8.7
Denifl et al. (2004b)	DEA O⁻ gas phase						9.8
Denifl et al. (2003)	EA in gas phase	0.04–1.48	1.75				
Isaacson (1972)	EEL for high energy electrons in films				5.5	6.7	7.8
Abouaf and Dunet (2005)	DEA in gas	0.67–1.45		4.66	5.94	7.8	8.82
Levesque et al. (2005)	Cross section in films			3.7, 4.0, 4.9	6.0, 6.3	7.3	9
Ptasińska et al. (2005)	DEA in gas	0.01–1.64			6.03, 6.23	7.1, 7.4	8.4
du Penhoat et al. (2001)	H⁻ desorption from film			4.75	5.9	7.45	8.7
Shukla and Leszczynski (2004)	TDDFT(basis set C, oscillator strength ≥0.01) singlet excitations			$\pi\pi^*$ 4.96	$\pi\pi^*$ 5.95, 6.19	$\pi\pi^*$ 6.50, 7.36	
Tsolakidis and Kaxiras (2005)	TDDFT			4.45	5.24, 5.68	6.86, 7.07, 7.52	
Tonzani and Greene (2006a)	R-matrix, one-electron potential	2.4	5.5	7.9			
Winstead et al. (2007)	Schwinger variational, SEP (SE) level	0.3	1.9	5.7			SE σ* 10

numerical techniques, to make a strong case for probable dissociation products fathered by these resonances. Commendable analyses were made, however, by Baccarelli et al. (2007, 2008b) on ribose and deoxyribose. These authors calculated the shape resonance energies and their lifetimes as well as the nuclear vibrational modes. By analyzing a TNI's electron density and slices of the wave function, these authors were able to correlate the mode displacements with those bonds showing strong antibonding character and hence be in a position to propose likely dissociative paths and fragment products. According to the comment of Burrow (2005), in the context of uracil, more quantitative information is needed in order to make such predictions. But Baccarelli et al. discuss conditions that are perhaps necessary if not sufficient for dissociation. The most interesting study in this context is to be found in the recent work of Gianturco et al. (2008) who improved on their earlier study of (Gianturco and Lucchese 2004) and studied the bond-stretching response in uracil. Figure 6.14 illustrates their results for the energy dependence of various resonances on two-bond distances. Their choice of bonds to stretch was inspired by the strong antibonding character of the two σ^* wave functions. Stretching lowers the energy of the σ^* TNI having the antibonding character at the stretched distance. They argue that a sizeable energy would be released to the nuclei with a large stretch resulting in ring breaking and fragmentation. They predict likely fragment compositions based on the wave function.

DFT quickly imposed itself as an alternative approach, which could elegantly address the dissociation conundrum. (An introduction to DFT can be found in Springborg (2000), a more detailed description in Koch and Holthausen (2001), while the popular hybrid functionals and their use are discussed in Rienstra-Kiracofe et al. (2002).) In DFT, one searches for the ground state of the anion. One can then follow the potential energy curves for any bond stretching to assess the possibility of fragmentation.

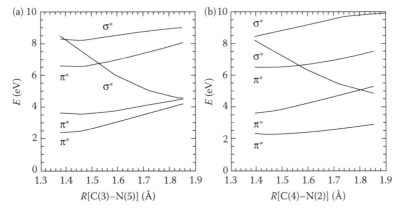

FIGURE 6.14 Computed real parts of the A′ and A″ resonance energies as a function of ring-breaking deformations in the uracil ring from Gianturco et al. (2008). Two specific bond-stretching effects are reported. Figure (a) deals with the C_3–N_5 stretch while (b) reports the C_4–N_2 stretch. (Reproduced with permission from Gianturco, F. A. et al. 2008. *J. Chem. Phys.* *128*(17), 174302-1-8. Copyright 2008, American Institute of Physics.)

Barrier heights are deduced and bond breaking is confirmed for positive energy release; the nature and charge of the fragments are determined. These calculations address the situation of thermalized electrons since they produce a minimal energy state of the anion. The method is thus informative only in cases where the TNI of a resonance has similarities with the wave function of the extra electron on the anion. This can occur for very low-energy resonances. Let us illustrate this in the case of uracil. Li et al. (2004a) studied the valence-bound anion states of this RNA base and followed the adiabatic anionic potential energy surfaces (PESs) for C–H and N–H bond dissociation. Their calculated PESs are shown in Figure 6.15. All the N–H and C–H bond dissociations are endothermic. The singly occupied molecular orbital (SOMO) of the anion is dominantly a π^* state.

The extracted hydrogen is neutral while the remaining U–yl$^-$ part retains the extra electron. The adiabatic electron affinities of U–yl radicals are quite large, of the order of 2–4 eV depending on extraction site, thus helping in lowering the energy barrier. The adiabatic PESs suggest that the energy threshold for the formation of hydrogen from N–H and C–H bonds are in the order: 0.78 (N_1–H)<1.3 (N_3–H)<2.2 (C_6–H)<2.7 (C_5–H) eV. Interestingly, these energies are within the range of the three π^* shape resonances observed by Aflatooni et al. (1998). The SOMO in Li et al.'s analysis is also π^* and most likely has a strong component of these three π^* shape resonances. An electron in these shape resonances has enough energy to break C–H and N–H bonds while in a TNI, is closely related to the anion's SOMO and thus following energetics similar to those described by Li et al. In contrast with this, a later study of uracil by Takayanagi et al. (2009) focused on the dipole-bound state. By comparing

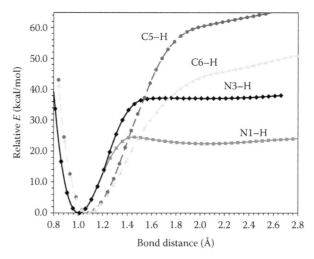

FIGURE 6.15 (**See color insert.**) Adiabatic potential energy surfaces of anionic uracil along each N–H or C–H coordinate, calculated by Li et al. (2004a) at the B3LYP/6-31+G(d) level. Energy relative to that of the optimized anion in the equilibrium state. Zero-point energy is not included. (Reprinted with permission from Li, X., L. Sanche, and M. D. Sevilla. 2004. *J. Phys. Chem. B 108*(17), 5472–5476. Copyright 2004, American Institute of Physics.)

calculations using various hybrid functionals with the 6–311++G(2d,p) basis set, they were able to continuously go from the dipole bound to the valence anion state. The motivation for this particular course of action came from Scheer et al. (2004, 2005) who found vibrational Feshbach resonances below 1 eV, that is, excited vibrational levels of the dipole-bound anion states (DBS) of uracil. Figure 6.16 shows their proposed avoided crossing between the σ^* and the DBS anion states. Takayanagi et al. (2009) did a very systematic survey of one- and two-dimensional PES having for variables the N_1–H bond distance and out of plane angle, and the C_6–H out of plane angle. The authors determined energy minima and saddle point energies and plotting wave functions for which they could monitor DBS and π^* character. They predicted a barrier height for N_1–H bond breaking of order 1 eV, in line with the experimental results of Scheer et al. (2004, 2005). This work stands as an example of clear success for DFT.

5-Halouracils, whose radiosensitizing properties have been well recognized and useful in radiation therapy, offer another nice example of DEA calculation. Li et al. (2002) examined the PESs for dehalogenation. The procedure is similar to the one on dehydrogenation of uracil described earlier. The PESs for stretching of the C_5–X bond reveal very low barriers of 1.88 and 3.99 kcal/mol for X=Br, Cl, respectively,

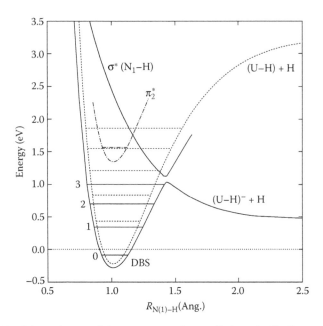

FIGURE 6.16 Schematic potential energy curves for uracil along the N_1–H stretching coordinate from Scheer et al. (2004). The neutral molecule and its vibrational levels are shown as dashed lines, the dipole bound (DBS) and σ^*(N1–H) anion states as solid lines, and the π_2^* anion state as a dot-dashed line. At N_1–H distances greater than $\approx 1/4$ Å, the fate of the upper resonance curve is uncertain. It may connect with the unobserved (U–H) + H$^-$ asymptote. (Reprinted with permission from Scheer, A. M. et al. *Phys. Rev. Lett.* 92(6), 068102-1-4, 2004. Copyright 2004 by the American Physical Society.)

and a much larger one of 20.80 kcal/mol for X=F. Dehalogenation of BrU and ClU anion radicals most likely leads to formation of the U–yl radical and halogen anion. For the FU anion radical, however, the fluorine leaves with the C_6 proton as a HF molecule, which explains its larger barrier height. The same authors later examined the changes brought about by pairing 5-halouracils with adenine (Li et al. 2003b). They found that the barriers for dehalogenation increased to values of 3.29, 5.62, 25.6 kcal/mol for X=Br, Cl, F, respectively. The authors concluded that the competitiveness of halo-uracils in base pairs for excess electrons is significantly reduced relative to nonbase-paired halouracils and thus the radiosensitivity of halouracils is likely to be reduced in double-stranded DNA in which the halouracils are paired with adenine. This pre-diction was later verified experimentally with double-stranded DNA and its impor-tance for radiotherapy discussed (Cecchini et al. 2005).

The DFT calculations are quite valuable at very low-impinging energies. At higher-impact energies, c.a., larger than 3 eV, one should consider if the resonance's TNI state has any chance of looking anything like the SOMO of the extra electron on the anion. This is unlikely and especially so for core-excited resonances for which TDDFT can be more expedient. In the rest of this section, we shall see how DFT and TDDFT were applied to study DEA, on small subunits to very large units. We shall also present a few alternate approaches which stand out by their ingenuity.

6.2.2.1 DNA Bases

Li et al. (2004b) also examined cytosine and thymine. They used the same approach as for uracil. In both bases, the N_1–H bond is the weakest. The PES of thymine is nearly identical to the one of uracil, while that of cytosine has a barrier height larger by some 5 kcal/mol, which might explain why it is less prone to dehydrogenation. The next weak bond is the amino N–H bond in cytosine or the N_3–H bond in thymine, the latter having a barrier height of 37 kcal/mol. The strongest bonds are C_4–NH_2 in cytosine and C_7–H in the methyl group of thymine. Breaking these bonds leaves anion radical fragments with affinity of the order of 1 eV, much smaller than that for those anion radical fragments resulting from other broken bonds. The authors interest-ingly conclude that since N_1 is the site of the glycosidic bond between the deoxyribose and the base in DNA, the vulnerable nature of this site toward bond rupture suggests that LEEs are likely to induce considerable pyrimidine base release in DNA, as verified experimentally (Zheng et al. 2005).

Théodore et al. (2006) proposed an alternate *ab initio* approach to thymine. This paper is the last in a significant series by Simons and collaborators dealing with cyto-sine and thymine nucleotides which we shall report on when covering larger subunits. What is remarkable is that this group of people focused on excited states of their molecular units instead of the ground anion state as in DFT. By doing so, they can go beyond DFT in their ability to treat LEE resonance states that are metastable and thus of direct experimental relevance. In this paper, they utilized a SCF calculation with a modest 6–31+G* basis set which they could fine tune to produce an anion with an electron attached to a thymine π^* state at a targeted energy. This is achieved by scaling the exponents of the most diffuse p-type basis functions on the atoms within the base ring. They address the metastability of the ensuing anion state by using a stabilizing method in which an external potential λV, especially crafted for the unit studied and the targeted anion states, is applied and can be made sufficiently strong to confine the extra electron of the anion in its orbital. This makes the calculations amenable to conventional quantum chemistry treatment. By extrapolating the anion and neutral state energies to $\lambda = 0$, they obtain the stabilization estimate for the anion energy. Théodore et al. focused on the N_3–H bond on the premise that the release of this hydrogen on bond rupture could cause damage to DNA. The authors tuned the lowest π^* state of the neutral molecule to an energy of 1 eV. The stabilization of the meta-stable anion states was then achieved by a V that increased the nuclear charge on N_3 and its partner H. These authors were able to follow the energy of the anion, having its extra electron on the thymine π^* state or the N_3–H σ^* anion state, as the N_3–H bond was stretched. Their results are shown in Figure 6.17. They find a barrier height for N_3–H bond rupture of 27.7 kcal/mol for an electron in the π^* resonance which is appreciably smaller than the DFT value of 37 kcal/mol taken from the work of Li et al. (2004b) for a thermalized electron. An electron of energy 2.2 eV is likely to break the bond directly or tunnel through the barrier if it has slightly lower energy.

Xie and Cao (2007) calculated the barriers for dehydrogenation of the purine bases adenine and guanine. Their PESs are shown in Figure 6.18. In the purines also, the site of the glycosidic bond N9 is very vulnerable to bond rupture although there is strong competition, in guanine, from the N_1–H and amino N_{10}–H bonds. Guanine should thus be more prone to hydrogen release at very low energy.

In an attempt to go beyond the single bases, various studies on the affinity of base pairs and more complex arrangements were made. These offer only partial information relevant to DEA. They are amply described in Kumar and Sevilla (2008a). We shall

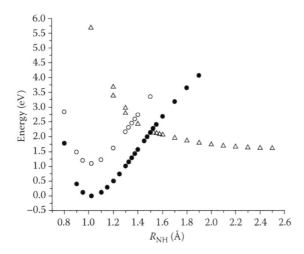

FIGURE 6.17   Théodore et al. (2006) calculated energies (eV relative to neutral thymine at its equilibrium geometry) of neutral thymine (filled circles), π^*-attached anion (open circles), and σ^*-attached anion (triangles) as functions of the N_3–H bond length (Å). (Reprinted from *Chem. Phys.* 329(1–3), Théodore, M., M. Sobczyk, and J. Simons, Cleavage of thymine N_3–H bonds by low-energy electrons attached to base π^* orbitals. 139–147. Copyright 2010, with permission from Elsevier.)

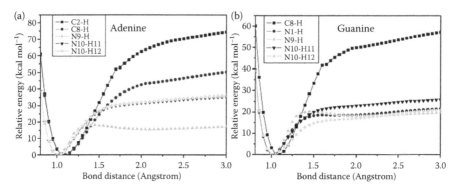

FIGURE 6.18 Potential energy surface profiles calculated by Xie and Cao (2007) along the N–H and C–H bond dissociations in the anionic adenine (a) and guanine (b) by B3LYP/6–31+G(d). (Xie, H. and Z. Cao. Electron attachment to the DNA bases adenine and guanine and dehydrogenation of their anionic derivatives: Density functional study. *Int. J. Quantum Chem.*, 2007, 107(5), 1261–1269. Copyright Wiley-VCH Verlag GmbH & Co. KGaA. Reprinted with permission.)

focus mostly on studies involving glycosidic and phosphodiester bond stretching in larger units.

6.2.2.2 Sugar–Phosphate–Sugar

Very early in the quest for answers to strand breaking in DNA, the choice of a sugar-phosphate-sugar model by Li et al. (2003a) proved simple but yet creative. They used a multilayered integrated molecular orbital plus molecular mechanics method, called ONIUM (Svensson et al. 1996), which simplifies the calculations. The phosphate group and one deoxyribose were targeted for DFT calculations at the B3LYP with 6–31+G basis set while the other deoxyribose was treated by less accurate semiempirical AM1 computations. Their model and PESs are shown in Figure 6.19. The energy barriers for breaking the C–O bonds are of order 10 kcal/mol and easily accessible to very LEEs landing on the group. This is precisely the interesting aspect of the simulation. No matter where the extra electron comes from, directly from vacuum (König et al. 2006) or by transfer from a core-excited resonance on a base. For instance, in the case where decay of the resonance leaves behind the core excitation, as proposed in Section 6.1. As long as it is nearly thermalized on the sugar-phosphate-sugar group, then the DFT SOMO should be a proper wave function and the results quite credible. The predicted phosphodiester bond breaking would lead to strand breaks. The authors' proposition that an electron of near-zero energy could attach to the phosphate group has, however, been challenged in the later work of Berdys et al. (2004a).

These authors focused on the phosphate group to see if an electron might not directly attach to the P=O bond, and used the stabilization procedure that has already been described for thymine (Théodore et al. 2006 in Section 6.2.2.1). The external potential λV employed for dPd increased the nuclear charge on the phosphorus atom to render the anionic states electronically bound. By extrapolating λ to zero value, they could then follow the metastable anion's states (electron attached to a π^* on the PO

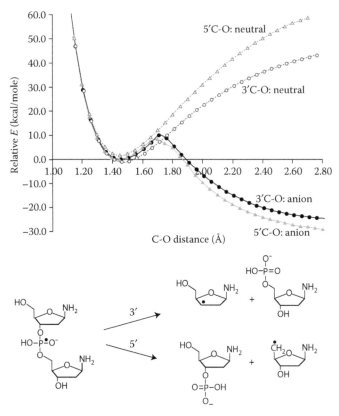

FIGURE 6.19 Sugar-phosphate-sugar model representing a section of DNA backbone. Electron-induced bond dissociations at the 3′ and 5′ ends of the model as investigated by Li et al. (2003a) along with the adiabatic potential energy surfaces for C_3'–O and C_5'–O bond rupture. All energies are relative to the energies of the anions at equilibrium. (Reprinted with permission from Li, X., M. D. Sevilla, and L. Sanche. 2003. *J. Am. Chem. Soc.* 125, 13668–13669. Copyright 2003 American Chemical Society.)

bond state or a σ* state on the CO bond) as they stretched the C_3'–O or C_5'–O bonds. It is quite obvious from Figure 6.20 that near zero energy electrons cannot directly attach to the phosphate group. It takes at least 2.5–3 eV electrons to be able to vertically attach to a π* orbital on the phosphate. Rupture would then be thermodynamically assured for the C_5'–O bond while the barrier for the C_3'–O bond is quite small, of order 5–6 kcal/mol, and might even be smaller due to avoided crossing. It should be noted, however, that Burrow et al. (2008) questions the existence of π* states on the phosphate group. Furthermore, in TDDFT calculations by Kumar and collaborators (Kumar and Sevilla 2008b), that will be presented further on, PO_4 σ* resonances were found in nucleotides at the smaller C–O bond lengths. Perhaps the stabilization procedure used by Berdys et al. generates incorrect orbitals on the phosphate group or, as Burrow suggests, the orbital might have been incorrectly identified.

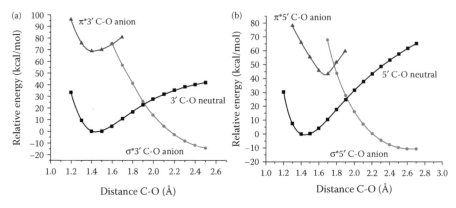

FIGURE 6.20 (**See color insert.**) Energies calculated by Berdys et al. (2004a) of neutral dPd, π^* anion, and σ^* anion for C'_3–O (a) and C'_5–O (b) bond rupture versus C–O bond length.

6.2.2.3 Nucleosides

The DFT calculations on nucleosides of Gu et al. (2005) at the B3LYP with DZP++ basis set were the first of two studies we shall report on. The authors wanted to examine the glycosidic bond-breaking process of pyrimidine nucleosides. The deoxyribose was terminated at the C'_5 position by a hydroxyl group. This group will be seen to play an important part in the predicted energetics of the N1–C1' glycosidic bond-stretching process. The anion's SOMO is located on the base and sugar with antibonding character at the glycosidic bond. The adiabatic energy barrier at the transition state (TS) is found to be of order 18.9 and 21.6 kcal/mol for the dT and dC anions, respectively, a relatively high value compared to the phosphodiester barrier found by Li et al. (2003a). The TS is characterized by a hydrogen bonding between O5' and N1. The tail end around C5' is quite flexible and can easily bend in a scorpion-like fashion such that O5' gets close to N1. At further stretching, the hydrogen bonding persists between the anion base and the radical sugar. This is an interesting study although one might think that it is perhaps a bit far from the situation encountered in a DNA strand in which the terminal hydroxyl group to the sugar is replaced by the phosphate group and which has a much more rigid backbone. We shall see, however, that these differences seem to have little impact on the energy barrier for rupture.

Li et al. (2006) later investigated thymidine (dT), deoxycytidine (dC), and deoxyadenosine (dA) using B3LYP functionals with the 6–31+G(d) basis set. In view of the results of the preceding work of Gu et al. (2005) regarding the hydrogen bridge between O5' and N1, the authors examined two geometries for dA, one with a forced bridge (dA1) and another without (dA2). The SOMO of the vertically attached extra electron has a heavy diffuse character which it loses to become a π^* valence state on the base after molecular relaxation. The PESs for N–C bond stretching are shown in Figure 6.21. The TS wave function is spread out throughout the nucleoside. Beyond the maximum, it transforms into a dissociative σ^* state with an anionic base and a free radical sugar. Note the discontinuity in the dA2 PES for an N9–C1' bond length around 2.4 Å. A clear anomaly can also be seen in the derivative of the PESs of all nucleosides. This

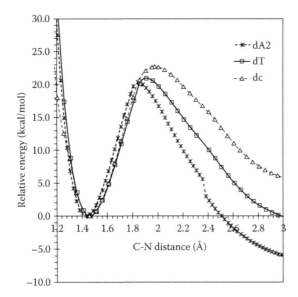

FIGURE 6.21   Anionic N–C potential energy surfaces of dA2, dT, and dC as calculated by Li et al. (2006). dA1 has a similar profile to dA2. (Reprinted with permission from Barrios, R., P. Skurski, and J. Simons 2002. *J. Phys. Chem. B 106*(33), 7991–7994. Copyright 2002 American Chemical Society.)

is where the O5′–H–N1 hydrogen bond kicks in. Li et al. (2006) thus find that the H-bonding does not significantly alter the low-energy profile for fragmentation in vacuum and raises the barrier by 5 kcal/mol. The authors find, however, a slight effect on the low-energy profile in aqueous solution, using a polarizable conductor model (CPCM, Tomasi et al. 2005). The energy release on bond rupture is also larger in water. The PESs barrier heights are compatible with those of Gu et al. (2005).

6.2.2.4   Nucleotides

The nucleotides have been extensively studied. The earliest studies came from Simons' group. First Barrios et al. (2002) who examined protonated 2′-deoxycytidine-3′-monophosphate (3′-dCMPH) using an *ab initio* approach to probe its sensitivity to C3′–O phosphodiester bond stretching. The calculations were performed using a simple 6–31+G* basis set. This basis set produced a LUMO on cytosine in vacuum having an energy around 1 eV, that is, near the experimental findings of a shape-resonance state of cytosine. The authors could thus simulate the behavior of a π* state formed by attaching a 1 eV electron on cytosine. Another attractive feature of this basis set is that the π* anion was electronically stable at the Koopmans' theorem level, even though it is energetically unstable with respect to the neutral unit. The energy of the neutral and anionic units was determined from the RHF and UHF methods, respectively. The calculations are vertical in the optimized neutral nucleotide geometry at all C3′–O bond lengths. The authors also considered the situation in aqueous solution using the polarizable continuum model (PCM, Tomasi et al. 2005). The energy curves are reproduced in Figure 6.22. An analysis of the SOMO of the anion reveals a transfer of

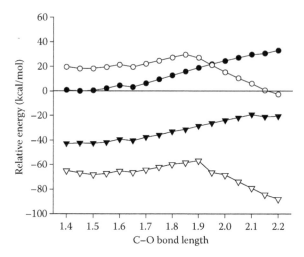

FIGURE 6.22 Barrios et al. (2002) plots SCF energies (kcal/mol) of unsolvated neutral dCMPH (filled circle) and anionic (empty circle) species as well as the fully solvated neutral (filled triangle) and anion (empty triangle) vs the C–O bond length (Å). (Reprinted with permission from Berdys, J. et al. 2004a. *J. Am. Chem. Soc. 126*, 6441–6447. Copyright 2004 American Chemical Society.)

the extra electron from the base to the sugar-phosphate region beyond the maxima in the energy curves. The barrier for bond rupture is estimated to be 13 kcal/mol in vacuum and drops slightly to 12 kcal/mol in solution. The rupture is exothermic and the π^* anion state is stable in water. The authors go on to estimate the rate of bond rupture through thermal processes to be some 10^4 s^{-1} in vacuum, which yields a negligible probability of bond rupture within the shape-resonance lifetime of 10^{-13} to 10^{-14} s. The situation is quite different in water since the π^* anion state is stable. The rupture rate of $6 \cdot 10^{-4}$ s^{-1} would then be fully operational.

Berdys et al. (2004b) extended the preceding study of 3′-dCMPH to other resonant energies in order to cover energies accessible through the Heizenberg width of a 0.5 eV π^* shape resonance. The authors also considered external media of varying dielectric constant using PCM. They used the same approach as used by Barrios et al. Their results are summarized in Table 6.8.

Solvents ($\varepsilon > 1$) are seen to first increase the barrier height as the dielectric constant ε increases, it then reaches a maximum, and decreases thereon. The reduction is especially significant at the higher energies. What the authors also found is that the π^* anion state is stabilized for all solvent strengths considered. Moreover, they estimated the through-bond transfer rate of the electron to be some 10^{12}–10^{14} s^{-1} using the calculated avoided crossing gap between π^* an σ^* states. These values are however similar to the estimated autodetachment rate of 10^{13} s^{-1} in vacuum ($\varepsilon = 1$). This through bond transfer is much faster than the bond rupture time through thermal processes, which varies from 10^{-9} to 10^5 s in their calculations covering the parameters in Table 6.8, and so is not the rate-limiting factor for bond breaking.

Afterwards, Berdys et al. (2004c) studied the nucleotide with cytosine replaced by thymine. The target π^* orbital for thymine was set at 0.3 eV. The incident electron

TABLE 6.8

Barriers kcal/mol along the C3′–O Bond Length for Various Electron Kinetic Energies (eV) and Various Solvent Dielectric Constants ε as Calculated by Berdys et al. (2004b)

Electron energy	0.2	0.3	0.8	1.0	1.3	1.5
Barrier ($\varepsilon = 1.0$)	15.6	15.1	12.1	11.25	9.0	8.38
Barrier ($\varepsilon = 4.9$)	18.3	18.5	13.1	10.47	10.2	7.95
Barrier ($\varepsilon = 10.4$)	19.0	19.8	13.7	10.51	10.5	8.38
Barrier ($\varepsilon = 78$)	28.1	21.8	11.3	9.5	5.3	5.1

Source: Reprinted with permission from Berdys, J. et al. 2004b. *J. Phys. Chem. A 108*, 2999–3005. Copyright 2004 American Chemical Society.

energy was chosen in the range 0.25–1.0 eV and same solvents were used. The results for dTMPH were quite similar to those for dCMPH which is not surprising between siblings in the pyrimidine family.

Bao et al. (2006) investigated 2′-deoxycytidine-5′-monophosphate (5′-dCMPH) and 2′-deoxythymidine-5′-monophosphate (5′-dTMPH) in their protonated form at the B3LYP DZP ++ level. The authors followed the wave function as the C5′–O5 phosphodiester bond is stretched. The C5′ tail is here also flexible and bends toward the base in the relaxed configurations. The energy barriers are 14.27 and 13.84 kcal/mol for 5′-dCMPH and 5′-dTMPH, respectively. The TSs in Figure 6.23 illustrate the electron transfer to the sugar and phosphate group and the antibonding character at the C5′–O5′ bond. These values increase to 17.97 and 17.86 kcal/mol, respectively, in aqueous solution as calculated using PCM. They are appreciably lower than corresponding values in the nucleoside calculations. The chemical nature of the tail end, OH versus protonated phosphate, thus has significant effect of the energetics of the C5′–O5′ bond rupture in these calculations.

Shortly thereafter, Gu et al. (2006) looked at anionic 3′-dCMPH and 3′-dTMPH using a B3LYP with DZP ++ basis set. The PES barrier heights during C3′–O3′ bond stretching are surprisingly quite small, 6.17 and 7.06 kcal/mol for 3′-dCMPH

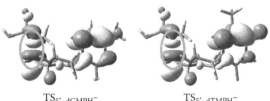

$\text{TS}_{5'\text{-dCMPH}^-}$ $\text{TS}_{5'\text{-dTMPH}^-}$

FIGURE 6.23 (**See color insert.**) The SOMOs of the transition states of the 5′-dCMPH and 5′-dTMPH radical anions calculated by ref108. The typical characteristics of the σ antibonding orbital along the C5′–O5′ bond are shown in the orange circle. (Reprinted with permission from Bao, X. et al. 2006. *Proc. Natl. Acad. Sci. U. S. A. 103*(15), 5658–5663. Copyright 2006 National Academy of Sciences, USA.)

and 3′-dTMPH, respectively. These increase to 12.82 and 13.73 kcal/mol in aqueous solution using the PCM. The reason for these smaller numbers is that the extra electron is going directly from the base to C3′ of the sugar as shown in Figure 6.24. This contact transfer is easier than going through the glycosidic bond. This C3′–O3′ bond might well be the weak link involved in strand breaks provided that this contact transfer in preserved in the DNA structure.

Kumar and Sevilla (2007) looked at 5′-dTMPH and compared the vertical and adiabatic PES when stretching the C5′–O5′ bond. They used the B3LYP functional and 6–31G* and 6–31 ++ G** basis sets. An examination of the SOMO using the 6–31 ++ G** basis set revealed an appreciable diffuse character and thus mixing with the dipolar state. The authors reasoned that the 6–31G* basis set would provide more reliable results. The calculated adiabatic path barrier height for this set was 14.8 kcal/mol. This compares favorably with the results of Bao et al. (2006). The vertical path goes from 4.8 kcal/mol to a TS potential energy of 25.5 kcal/mol for a net barrier height of 9 kcal/mol. The authors argued that transiently bound electrons to the virtual molecular orbitals of the neutral molecule, the base in particular, likely play a key role in the cleavage of the sugar-phosphate C5′–O5′ bond in DNA resulting in the direct formation of single-strand breaks without significant molecular relaxation. The same authors further considered the effects of an aqueous environment by examining 5′-dTMP surrounded by 5 or 11 water molecules with Na+ as a counterion. In the optimized geometries, the water molecules are concentrated around the phosphate group with less influence on thymine. This is a much more microscopic approach to salvation. The barrier heights were found to be 26.0 and 28.9 kcal/mol for 5 and 11 water molecules, respectively. These values are surprisingly close to the

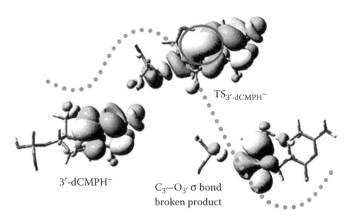

$TS_{3'-dCMPH^-}$

$3'-dCMPH^-$

$C_{3'}-O_{3'}$ σ bond
broken product

FIGURE 6.24 (See color insert.) The SOMO of the electron attached state, the transition state, and the C3′–O3′ σ-bond broken product of 3′-dCMPH radical anion as calculated by Gu et al. (2006). The migration of the excessive negative charge through the atomic orbital overlap between the C6 of pyrimidine and the C3′ of deoxyribose is shown in the orange circle. (Reprinted with permission from Gu, J., J. Wang, and J. Leszczynski. 2006. *J. Am. Chem. Soc. 128*(29), 9322–9323. PMID: 16848454. Copyright 2006 American Chemical Society.)

other and to the value of 30.0 kcal/mol found by the same authors using PCM, which turns out to be a fairly accurate method after all.

Kumar and Sevilla (2008b) also did impressive work on 5′-dTMPH using TDDFT (Marques and Gross 2003). They mapped the electronic excitations of anionic 5′-dTMPH as they calculated the PES of the anion under C5′–O5′ bond stretching. The results of their TD-BH&HLYP/6–31G* calculations are shown in Figure 6.25.

The avoided crossing between the σ* curve with its extra electron on the PO$_4$ with the ground π* state very nicely illustrates the charge transfer that occurs in the bond breaking. They propose that an LEE having some 2 eV energy could produce a π → σ* or π → π* transition. Since the π → σ* is difficult to observe, a π → π* transition could be followed by a π* → σ* transfer due to π* and σ* coupling. This would lead to rapid dissociation.

Xie et al. (2008) have explored the C5′–O5′ and C1′–N1 bond-stretching behavior of anionic 5′-dCMPH and 5′-dTMPH using B3LYP with a DZP ++ basis set. They have done so in the gas phase and in water solution using CPCM. Their results for the barrier heights are shown in Table 6.9. The barriers for the glycosidic bond in the gas phase are similar to those found in the nucleosides. They are quite reduced in aqueous solution in contrast to the general trend for the phosphodiester bond. But although the activation energy for the phosphodiester bond is similar to the values

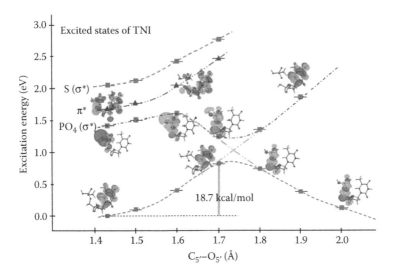

FIGURE 6.25 (**See color insert.**) Lower curve: PES of the 5′-dTMPH transient negative ion (TNI); calculated in the neutral optimized geometry of 5′-dTMPH with C5′–O5′ bond elongation. SOMO is shown at selected points. Upper curves: Calculated vertical excitation energies of the radical anion at each point along the PES, MOs involved in excitations are also shown. Energies and distances are given in eV and Å, respectively. The lowest ππ* state (triangles) and lowest πσ* states (square) are shown. (Reprinted with permission from Kumar, A. and M. D. Sevilla. 2008b. *J. Am. Chem. Soc. 130*(7), 2130–2131. Copyright 2008 American Chemical Society.)

TABLE 6.9

Energy Barriers in kcal/mol Calculated by Xie et al. (2008) Without Zero-Point Energy Correction

Medium	$C_5'-O_5'$ Bond		$C_1'-N_1$ Bond	
	dCMPH	dTMPH	dCMPH	dTMPH
Gas phase	8.7	9.8	25.3	22.8
Aqueous solution	6.6	14.1	12.1	17.5

Source: Xie, H., R. Wu, F. Xia, and Z. Cao. Effects of electron attachment on $C_5'-O_5'$ and $C_1'-N_1$ bond cleavages of pyrimidine nucleotides: A theoretical study. *J. Comput. Chem.*, 2008, 29(12), 2025–2032. Copyright Wiley-VCH Verlag GmbH & Co. KGaA. Reprinted with permission.

calculated by Li et al. (2003a), for the sugar-phosphate-sugar unit, they are substantially smaller than those values obtained by Bao et al. (2006) or by Kumar and Sevilla (2007). In solution, the barrier of dTMPH increases but, surprisingly, decreases for dCMPH. The authors found that the relatively low activation energies result from the intramolecular hydrogen bond between the phosphate and sugar subunits (O3'–H ⋯ O=P). One may wonder at this point if the diffuse component of the DZP ++ basis set used by Xie et al. might not the responsible for this, inasmuch as the 6–31 ++ G** basis set used by Kumar and Sevilla (2007) led to suspiciously small results for the vertical PES calculation

Schyman and Laaksonen (2008) studied anionic protonated 3'-guanosine monophosphate (3'-GMPH) using a B3LYP with DZP ++ basis set. They calculated the PES during C3'–O3' bond stretching. The gas-phase barrier height was found to be 10.28 kcal/mol whereas the value in aqueous solution was 5.25 kcal/mol using the integral equation formalism-polarized continuum model (IEF–PCM) (Tomasi et al. 2005) a variant of PCM, and 6.54 kcal/mol when doing a calculation with 21 explicit water molecules around the GMP unit. This contrasts with the computations of Gu et al. (2006) which showed an opposite trend from gas phase to solution for the pyrimidines. The very important diffuse component of the SOMO calculated by Schyman and Laaksonen for GMPH also contrasts with the quite regular SOMO of Gu et al. for dCMPH and DTMPH using the same basis set.

Schyman et al. (2008) also examined 5'-dCMP and 3'-dCMP (nonprotonated) in an aqueous environment using a mixed Car–Parrinello (Car and Parrinello 1985, Springborg 2000) QM/MM molecular dynamics approach. In the Car–Parrinello conceptualization, the electrons and the ions are treated in the Born–Oppenheimer approximation and obey time-dependent-coupled equations, quantum for the electrons, and classical for the ions. The quantum treatment is DFT-like using a plane wave basis set, a pseudopotential interaction with the ions and exchange, correlation, and polarization contributions. One interesting aspect is that the atomic kinetic energy can serve as a measure of temperature and annealing procedures can be implemented. The calculations implemented by Shyman et al. included some 1150 water molecules. The latter were handled by a classical force field at the molecular mechanics (MM) level, whereas the nucleotide was treated at the quantum mechanics (QM) level in a

$32 \times 34 \times 32$ Å³ box. The energy profile of the bond-breaking reaction was calculated by constrained dynamics, in which the phosphodiester bond C3′–O3′/C5′–O5′ was slowly extended. The results for native dCMP having an negatively charged phosphate group or with an extra electron are shown in Figures 6.26 and 6.27. The energy barrier for rupture is very high for both bond types, much more so than for anionic dCMPH.

We now present an innovative two-step process proposed by Dabkowska et al. (2005) for 3′dCMPH, involving either an H radical and an LEE or two LEEs interacting with the same nucleotide, which proceeds through bound anionic states, not through TNIs. In the TS, a proton transfers from the C2′ atom of the sugar to C6 of (Cy+H)⁻ with a barrier height of 3.6 or 5.0 kcal/mol depending on choice of basis set. This is followed by spontaneous phosphodiester bond rupture. The two-step process was invoked in later works by Zhang and Eriksson (2006), Zhang et al. (2007), Gu et al. (2007), and Kobylecka et al. (2008b).

The final nucleotide calculation we will present is on the anions of the RNA nucleotide 3′-Uridine Monophosphate (3′-UMP and protonated 3′-UMPH) made by Zhang et al. using B3LYP with DZP ++ basis set. Their results for barrier heights are summarized in Table 6.10. Their values for UMPH are consistent with those on the

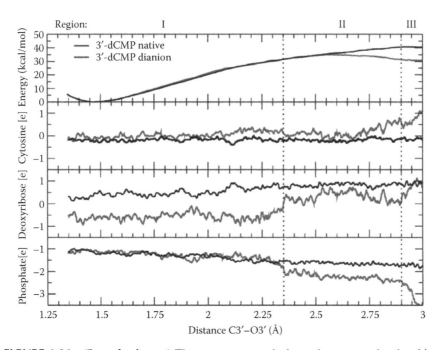

FIGURE 6.26 (**See color insert.**) The uppermost graph shows the energy when breaking the C3′–O3′ bond in 3′-dCMP. The three lower graphs show the charges for each moiety of the 3′-dCMP nucleotide while rupturing the bond. (Reprinted from *Chem. Phys. Lett.* 462(4–6), Schyman, P., A. Laaksonen, and H. W. Hugosson, Phosphodiester bond rupture in 5′ and 3′ cytosine monophosphate in aqueous environment and the effect of low-energy electron attachment: A Car-Parrinello QM/MM molecular dynamics study. 289–294. Copyright 2008, with permission from Elsevier.)

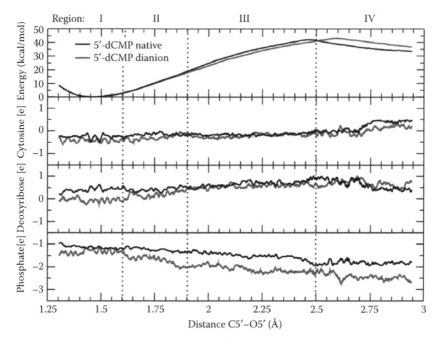

FIGURE 6.27 (**See color insert.**) The uppermost graph shows the energy when breaking the C5′–O5′ bond in 5′-dCMP. The three lower graphs show the charges for each moiety of the 5′-dCMP nucleotide while rupturing the bond. (Reprinted from *Chem. Phys. Lett.* 462(4–6), Schyman, P., A. Laaksonen, and H. W. Hugosson, Phosphodiester bond rupture in 5′ and 3′ cytosine monophosphate in aqueous environment and the effect of low-energy electron attachment: A Car-Parrinello QM/MM molecular dynamics study. 289–294. Copyright 2008, with permission from Elsevier.)

pyrimidine nucleotides. One should note the very important decrease of barrier height of the C3′–O3′ bond for UMP under solvation. This contrasts with the calculations of Schyman et al. (2008) for dCMP in aqueous solution which predict a much larger barrier.

6.2.2.5 Larger Moieties

Anusiewicz et al. (2004) finalized their series on dCMPH by considering a more complex single-strand trimer with the objective of examining the effect of stacking on the C3′–O3′ and C1′–N1 bonds of the central nucleotide. The authors applied a two-layered ONIOM method to carry out a series of *ab initio* electronic structure calculations on the neutral and anion trimer formed by adding an electron to the π* LUMO of the central cytosine unit. The central nucleotide was treated within the SCF method using a 6–311 ++ G* basis sets while a 4-31G basis set was used on the two terminal "low layer" nucleotides. Incident electron energies were considered in the range 0.3–2.0 eV. The barrier heights in vacuum for C3′–O3′ bond rupture are some 70% larger than those of the single nucleotide results of Berdys et al. (2004b) mainly due to π stacking. The fragment separation is exothermic. The barrier height for C1′–N1 bond breaking is very large at 43 kcal/mol at an incident electron energy

TABLE 6.10

Barrier Heights in kcal/mol Calculated by Zhang et al. (2008) with Zero-Point Energy Included

	$C_3'-O_3'$ Bond		$C_1'-N_1$ Bond	
Medium	UMP	UMPH	UMP	UMPH
Gas phase	29.6	8.2	21.0	23.6
Aqueous solution	14.7	10.3	19.1	20.0

Source: Zhang, R. B., K. Zhang, and L. A. Eriksson. Theoretical studies of damage to 3'-uridine monophosphate induced by electron attachment. *Chem. Eur. J.*, 2008, 14(9), 2850–2856. Copyright Wiley-VCH Verlag GmbH & Co. KGaA. Reprinted with permission.

of 0.8 eV. Furthermore, the rupture is endothermic. The authors conclude that this last route is not operative for single-strand breaks at such energies.

Loos et al. did a challenging QM/MM study of C3'–O3' bond cleavage in single-stranded (ss) and double-stranded (ds) trimers of 3'-dCMPH (although technically speaking the article's Figure 1 indicates a Na^+ counterion on the phosphate group and thus one could write 3'-dCMPNa). The QM part consists of the central 3'-dCMPH nucleotide. The B3LYP functional with 6–311 + G^* basis set was used. Interestingly, these authors find that the effect of the stacking leads to an anion's SOMO that relocates the excess electron from the π^* orbital on the base to a diffuse phosphate-centered orbital leading to the formation of a dipole-bound state. One might wonder, however, how the predicted diffuse orbital on the phosphate is affected by the presence of structural water molecules or by a more aggressive water environment. Their PES is shown in Figure 6.28. The excess electron is transferred in the process from a diffuse phosphate-centered orbital to the π^*orbital of the C3'–O3' bond. What is especially significant in these results is the dramatic increase in barrier height from 7 kcal/mol in ss-3'-dCMPH, a value consistent with previous mentioned results, to 24 kcal/mol in ds-3'-dCMPH. The extra rigidity of the double-stranded trimer is surely responsible for this. This supports the view we have taken throughout this section as well as the one expressed in the papers from Simons' group about the lack of rigidity in the smaller units.

We end this subsection with the superb TDDFT calculations at the BH and HLYP/6–31G* level of Kumar and Sevilla (2009) on protonated 2'-deoxy-X-3',5'-diphosphate(3',5'-dXDP), where X = G, A, T, and C, radical anions in their TNI and adiabatic states, in the gas phase and in solution. Although there is no bond stretching involved, the exploration of the electronic excitations is very informative on the sensitivity to molecular shape and environment as well as on the capabilities of TDDFT. We will only present the authors' results on adenine. Figure 6.29 shows both the shape and core excitations, their energies, and their wave functions. Figure 6.30 follows the $\pi \rightarrow \sigma(5'-PO_4)^*$ excitation energy as a function of the extent of solvation. There are similar results for the $\pi \rightarrow \sigma(3'-PO_4)^*$ transition. This shows that in the gas phase LEEs with energies <4 eV can access PO_4 (σ^*) states which lead to strand break formation as discussed by Kumar and Sevilla (2008b). Water increases the energy of these states which should render DNA in living systems less vulnerable to

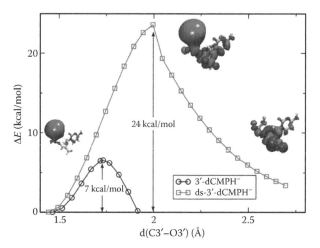

FIGURE 6.28 (**See color insert**.) PES of Loos et al. (2009) for the C3′–O3′ bond in ss-3′-dCMPH⁻ and ds-3′-dCMPH⁻. The SOMOs corresponding to the minimum and maximum along the bond cleavage coordinates as well as the dissociated product are depicted for ds-3′-dCMPH⁻. (Reprinted from *Chem. Phys. Lett.*, 475(1–3), Loos, P.-F., E. Dumont, A. D. Laurent, and X. Assfeld. Important effects of neighbouring nucleotides on electron induced DNA single-strand breaks. 120–123. Copyright 2009, with permission from Elsevier.)

very LEEs. Interestingly, this study shows the capability of TDDFT to provide considerable information on the excited states accessible to LEEs. Even the core-excited states mentioned in Section 6.1 can be addressed.

6.2.2.6 Concluding Remarks on Dissociative Electron Attachment

The DEA studies that were presented examined the strength of various bonds in diverse subunits. A summary is presented in Table 6.11. Of these, the phosphodiester and glycosidic bonds are surely the most relevant to the stability of DNA and their likelihood of being ruptured by LEEs has been intensely scrutinized. There is a global consensus that the phosphodiester bond is the weakest with barrier height estimates varying between 7 and 15 kcal/mol for the C3′–O3′ bond and 9–14 kcal/mol for the C5′–O5′ bond. The calculated glycosidic barrier heights lie in the energy range 19–30 kcal/mol for the pyrimidines and around 20 kcal/mol for the purines. The calculations are seen to be quite dependent on subunit size and computational level used (hybrid functional for DFT and basis set). The glycosidic bond is, however, clearly the strongest. This conclusion is corroborated by the experiment of Zheng et al. (2005), Zheng et al. (2006a), and Zheng et al. (2006b). The effect of a water solvent is extremely model dependent with predictions indicating heightening of the barrier while others pretending to the contrary. Calculations on ss or ds trimers are especially valuable even though some conclusions are conflicting, for the ss trimer barrier height for instance. Examination of larger units should be continued.

We have seen that although DFT has been the preferred approach to look at bond breaking, there are some original alternatives such as the Car–Parrinello treatment and those that target resonances instead of the ground state SOMO, which more often than not shows attachment to a base π* orbital. We believe that TDDFT provides

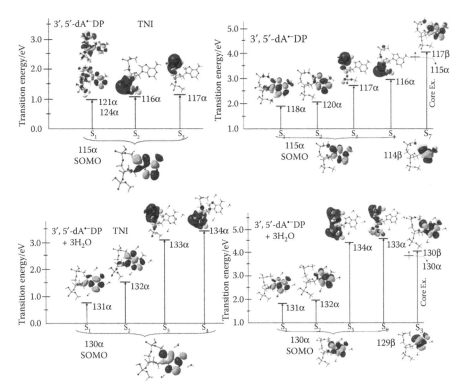

FIGURE 6.29 Transition energies computed by Kumar and Sevilla (2009) for the TNI state of 3′,5′-dA·⁻DP (left) and for the adiabatic state of 3′,5′-dA·⁻DP (right) in the gas phase (top) and aqueous solution (bottom). The effect of bulk water solvent was considered using the IEF–PCM model on the trihydrated 3′,5′-dA·⁻DP system. Transitions from the SOMO to different MOs (shape resonance) and from an inner-shell MO to higher MOs (core excitation) are shown. (Kumar, A. and M. D. Sevilla. Role of excited states in Low-Energy Electron (LEE) induced strand breaks in DNA model systems: Influence of aqueous environment. *Chem. Phys. Chem.*, 2009, 10(Special Issue 9–10), 1426–1430. Copyright Wiley-VCH Verlag GmbH & Co. KGaA. Reprinted with permission.)

a reliable theoretical environment with which to explore core-excited subunits of DNA. Combining this method with bond stretching is invaluable.

6.3   LARGE DNA MOLECULAR STRUCTURES

We have seen how fundamental units of DNA have thus far been characterized with regard to their LEE scattering properties or dissociative electron attachment behavior. What remains to be examined is how these different pieces of data can be packaged together to calculate the LEE response of larger molecular structures, oligomers for instance. This is a difficult problem which has barely started to be addressed. The first timid approach dealt with elastic scattering dates back to a period in time prior to the availability of reliable scattering computations on DNA subunits. An interesting observation can be made about the electron energy typically 5–15 eV leading to

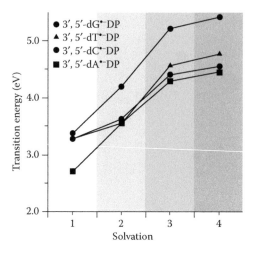

FIGURE 6.30 Variation of transition energies of the $\pi \rightarrow \sigma(5'-PO_4)^*$ excited state of nucleotides in adiabatic state with increasing solvation. Different solvation levels labeled on the X-axis are (1) gas phase, (2) three discrete water molecules, (3) three water molecules and a continuous dielectric ($\varepsilon = 7.0$), and (4) three water molecules and a continuous dielectric ($\varepsilon = 78.4$). (Kumar, A. and M. D. Sevilla. Role of excited states in Low-Energy Electron (LEE) induced strand breaks in DNA model systems: Influence of aqueous environment. *Chem. Phys. Chem.*, 2009, 10(Special Issue 9–10), 1426–1430. Copyright Wiley-VCH Verlag GmbH & Co. KGaA. Reprinted with permission.)

strand breaking (Boudaïffa et al. 2000) in Figure 6.3. At this energy, the electron has a wavelength of the order of sizes of molecular subunits and intermolecular distances in DNA. So, might not the quasi-periodic stacking of the base pairs within the DNA strand produce quantum interference of scattered electron waves, which could cause visible structure in Figure 6.3? This led Caron and Sanche (2003, 2004) to propose a toy model of scattering on DNA based on multiple scattering (MS) and having as first objective to elucidate any role the stacking periodicity of the bases might have on the interaction of LEEs with the DNA structure. Searching for physically instructive scattering information on the bases, these authors conjured up substitute pseudo-molecules (PMOLs) constructed with Argon atoms, whose phase shifts are known, mimicking the heavier elements (C, O, N) of the bases. Figure 6.31 shows the construction for the cytosine and guanine substitutes and their repeated distribution in a B-type DNA segment. Interestingly, the elastic cross section for the PMOLs is of the order of that of the six-membered ring of the benzene molecule. Arguably, this structure does reproduce the quasiperiodic stacking of base pairs and as such can shed light on interference effects within the segment. But it can in no way emulate molecular shape or core-excited resonances. MS is implemented through an expansion of the total wave function $\psi_{k,i}(r)$ incident on each of the N_s scatterers (a superposition of an external plane wave of wave vector $\mathbf{k}$ with the outgoing contribution of all other scatterers) in angular momentum components taking the form

$$\psi_{k,i}(\mathbf{r_n}) = 4\pi e^{i\mathbf{k}\cdot\mathbf{R_n}} \sum_L i^l B_{kL}^{(n)} Y_L(\Omega_{\mathbf{r_n}}) j_l(kr_n) \quad (6.3)$$

TABLE 6.11
Summary of Studies on DEA

Author	Molecule	Stretched Bond	Barrier kcal/mol	Barrier in H_2O kcal/mol	Notes
Li	U	N1–H	25		e⁻ in U(π*) → (U–H)⁻ + H
Takayanagi	U	N1–H	23		e⁻ in DBS → (U–H)⁻ + H
Li	Halouracil	C5–Br	1.88	3.29	e⁻ in U(π*)→(U–yl)• + Br⁻
		C5–Cl	3.99	5.62	e⁻ in U(π*)→(U–yl)• + Cl⁻
		C5–F	20.8	25.6	e⁻ in U(π*) → (U–yl)⁻ + HF
Li	C	N1–H	30		e⁻ in C(π*) → (C–H)⁻ + H
	T	N1–H	25		e⁻ in T(π*) → (T–H)⁻ + H
	T	N3–H	37		e⁻ in T(π*) → (T–H)⁻ + H
Théodore	T	N3–H	27.7		1 eV e⁻ in T(π*) → (T–H)⁻ + H
Xie	A	N9–H glyco	19		e⁻ in A(π*) → (A–H)⁻ + H
	G	N9–H glyco	21		e⁻ in DBS → (G–H)⁻ + H
Li	dPd	C3'–O	10		Very LEE on PO4
	dPd	C5'–O	9		Very LEE on PO4
Berdys	dPd	C3'–O	5–6		E≥2.5 3 e⁻ in P=O(π*) → C–O(σ*)
	dPd	C5'–O	Rupture		E≥2.5 3 e⁻ in P=O(π*) → C–O(σ*)
Gu	dC; dT	N1–Cl'	21.6; 18.9		SOMO on base and sugar with antibonding character on N1–Cl' bond → anion base + sugar radical
Li	dC	N1–Cl'	22.7		Vertical SOMO has strong diffuse character, relaxed π* SOMO on base → base anion + sugar radical
	dT	N1–Cl'	20.9		
	dA1	N9–Cl'	20.3		
	dA2	N9–Cl'	19.5		

continued

TABLE 6.11 (continued)
Summary of Studies on DEA

Author	Molecule	Stretched Bond	Barrier kcal/mol	Barrier in H_2O kcal/mol	Notes
Barrios	3'-dCMPH	C3'–O	13	12	1 eV e in $C(\pi^*) \to$ sugar phosphate
Berdys	3'-dCMPH	C3'–O	15–8	28–5	0.2 eV < E < 1.5 eV e⁻ in $C(\pi^*)$
Berdys	5'-dTMPH	C5'–O	≈like ref. 106		0.25 eV;E;1.0 eV e⁻ in $T(\pi^*)$
Bao	5'-dCMPH	C5'–O5	14.3	18	Very LEE in base(π^*) $\to$ sugar phosphate
	5'-dTMPH	C5'–O5	13.8	18	Very LEE in base(π^*) $\to$ sugar phosphate
Gu	3'-dCMPH	C3'–O3'	6.2	12.8	Very LEE in base(π^*) $\to$ sugar phosphate
	3'-dTMPH	C3'–O3'	7	13.7	Very LEE in base(π^*) $\to$ sugar phosphate
Kumar	5'-dTMPH	C5'–O5'	14.8 adiabatic		Very LEE in base(π^*) $\to$ sugar phosphate; 6–31G* basis
			9 vertical		0.6 eV e⁻ in $T(\pi^*) \to$ sugar phosphate; 6–31G* basis
Kumar	5'-dTMPNa	C5'–O5'	≈4	28.9	11 water molecules; 6–31G** basis
	5'-dTMPH	C5'–O5'	18.7		E=2 eV PO4(σ^*) vertical state using TDDFT
			23.3		SOMO of TNI in $T(\pi^*) \to$ sugar phosphate Adiabatic
Xie	5'-dCMPH	C5'–O5'	8.7	6.6	Very LEE in base(π^*) $\to$ sugar phosphate
		N1–C1'	25.3	12.1	Somo mostly on base $\to$ phosphate anion + base-sugar radical
	5'-dTMPH	C5'–O5'	9.8	14.1	Very LEE in base(π^*) $\to$ sugar phosphate
		N1–C1'	22.8	17.5	SOMO mostly on base $\to$ phosphate anion + base-sugar radical
Schyman	3'-GMPH	C3'–O3'	10.3	5.25–6.5	Very diffuse SOMO on base
Schyman	5'-dCMP	C5'–O5'		40	Car–Parrinello

Author	System	Bond	Value	Description	Ref
Dabkowska	3'-dCMP	C3'–O3'		Car–Parrinello	35
	3'-dCMPH	C3'–O3'	3.6–5.0 for proton transfer	Two-step process with proton transfer followed by spontaneous phosphodiester bond cleavage	
Anusiewicz	ss dCMPH trimer	C3'–O3'	25–11	0.3 eV;E¡2.0 eV e⁻ in C(π^*)→sugar phosphate	
Loos	ss 3'-dCMPH	N1–C1'	43	E=0.8 e⁻ in C(π^*)	
	ds 3'-dCMPH	C3'–O3'	7	SOMO in base(π^*) relocated to very diffuse phosphate state due to surrounding nucleotides	
Kumar	3',5'-dXDP		24	TDDFT shape and core-excited anion states	

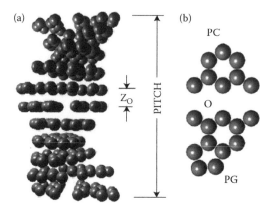

FIGURE 6.31 The basic pseudomolecular PC–PG unit, defining a base pair in Caron and Sanche (2003), is shown on (b). The distance between Ar scatterers (balls) is 0.14 nm. The open circle marks the position of the helix's rotation axis. The helical structure obtained by applying the screw operation of B-type DNA to the PC–PG unit appears on (a) (not to scale) showing parameters associated with a helical structure. (Reprinted with permission from Caron, L. G. and L. Sanche. *Phys. Rev. Lett.* *91*(11), 113201-1-4, 2003. Copyright 2003 by the American Physical Society.)

in which L is short for the angular momentum components (l, m), $r_n = r - R_n$ where R_n is the position of the nth scatterer, Y_L is a spherical harmonic, and j_l is the spherical Bessel function. The sum over L runs over all angular momentum components deemed important in the scattering process (having important phase shifts for instance). The outgoing scattered waves $\psi_{k,s}(r_n)$ emanating from the nth scatterer are obtained from Equation 6.2 by applying the T-matrix of Ar for each angular momentum. A reexpansion of the outgoing waves from each scatterer around each of the other scatterers generates an interdependence between all of the $B_{kL}^{(n)}$. The $N_S \times N_L$ coefficients $B_{kL}^{(n)}$, where N_L is the total number of retained angular momentum components, thus satisfy as many equations, which assure full self-consistency of the MS formulation. Caron and Sanche (2004) showed how periodicity in the structure, the screw symmetry for instance, could be used to reduce the number of coefficients. Once these coefficients are determined, one can calculate the total wave function anywhere outside or within the DNA segment by adding together the incident electron contribution and the ones from all scatterers. What Caron and Sanche then did is to calculate the total incoming wave function at the center R_C of the six-membered ring of PC. This takes the form

$$\psi_{k,i}^{PC}(\mathbf{r}) = e^{i\mathbf{k}\cdot\mathbf{r}} + \sum_{n' \notin PC} \psi_{k,s}(\mathbf{r}_{n'}) = 4\pi e^{i\mathbf{k}\cdot\mathbf{R}_C} \sum_L i^l [Y_L^*(\mathbf{k}) + C_{kL}] Y_L(\Omega_{\mathbf{r}_C}) j_l(k r_C) \quad (6.4)$$

in which $\mathbf{r}_C = \mathbf{r} - \mathbf{R}_C$. Following the discussion in Section 6.1 surrounding Equation 6.1, the one-center approximation to the T-matrix for DEA of the PC, were it a real molecule, would be proportional to

$$\sqrt{4\pi} V_{L_0}(R) \left[Y_L^*(\mathbf{k}) + C_{kL} \right]$$

in which the second term inside the brackets is a result of MS. This suggested defining a capture factor

$$\Gamma(L_0) = 4\pi \left| Y_L^*(\mathbf{k}) + C_{\mathbf{k}L} \right|^2, \tag{6.5}$$

and, in situations where $Y_L^*(\mathbf{k}) \neq 0$, a relative capture factor

$$\Gamma_{\text{rel}}(L_0) = \frac{\left| Y_L^*(\mathbf{k}) + C_{\mathbf{k}L} \right|^2}{\left| Y_L^*(\mathbf{k}) \right|^2}, \tag{6.6}$$

which provides a simple measure of the changes of the DEA cross section due to MS. Any value greater than one of $\Gamma_{\text{rel}}(L_0)$ implies an enhancement of the cross section relative to the one of an incident plane wave. It should also be noted that for $L_0 = (0, 0)$ and according to Equation 6.4, $\Gamma(L_0)$ is equal to the square of the incident wave function at the center of the PC ring.

Caron and Sanche started with calculations for a periodic spiral generated from the PMOLs of Figure 6.31 using the parameters of B-type DNA, that is, 10 base pairs in a turn of the helix having a pitch of 3.4 nm and thus a natural rise z_0 (distance between successive base pairs) equal to 0.34 nm. Figure 6.32 shows the results for the capture factor on the PC ring using an incident electron wave vector perpendicular to the spiral direction as in the experiments of Boudaïffa et al. (2000) and Martin et al. (2004). Note that $Y_{(L_0)}^*(\mathbf{k}) = 0$ for $l_0 + m_0 = $ odd when choosing the spiral axis as the z axis. Interestingly, there is a predicted enhancement of the capture probability by a factor of order 2 for $L_0 = (0,0)$, which reflects the near doubling of the square of the total wave function at the center of the PC ring due to MS in the energy range 12–16 eV. In order to assess the effect of the periodicity in base pair stacking on these results, the authors next modulated the rise for values from 0.8 z_0 to 1.2 z_0. Figure 6.33 shows the results for two values of resonance angular momentum. There is a clear shift of the capture factor modulations from high energy to low energy as

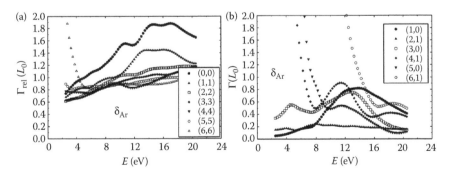

FIGURE 6.32 Capture factor at the center of the PC ring for the B-type DNA structure as a function of energy for various $l_0 + m_0 = $ even (a) and $l_0 + m_0 = $ odd (b) resonance channels as calculated by Caron and Sanche (2004) for scatterers with Ar phase shifts and electrons perpendicular to the helix axis. (Reprinted with permission from Caron, L. and L. Sanche. *Phys. Rev. A* 70(3), 032719-1-10, 2004. Copyright 2004 by the American Physical Society.)

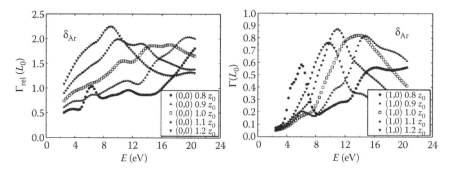

FIGURE 6.33 Capture factor for L_0 = (0,0), (1,0) at the center of the PC ring as a function of electron energy calculated by Carson and Sanche (2004) using Ar phase shifts and for various values of the interbase-pair distance and electrons perpendicular to the helix axis. z_0 is the crystallographic interbase-pair distance of B-type DNA. (Reprinted with permission from Caron, L. and L. Sanche. *Phys. Rev. A 70*(3), 032719-1-10, 2004. Copyright 2004 by the American Physical Society.)

the rise increases confirming that the modulations correlate with the inverse of the value of the rise that is, with the inverse of the wavelength associated with the natural "diffraction grating" defined by the base-pair stacking. Internal diffraction due mostly to MS is therefore important in DEA cross sections in DNA.

Caron and Sanche next studied the effect of scatterer strength and disorder on the capture factor modulations. They modified their PMOLs so that Ar was used for C atoms while Krypton was used as scatterer for O and N atoms. They generated four PMOLs assembled into two base-pair sets PC–PG and PA–PT. The capture factors for the regular PC–PG helix structure yielded capture factors at the center of the PC ring of similar size to the Ar only situation, but with modulations that are shifted in energy or showing new peaks. When randomizing the base-pair sequence within the unit cell of the helix having 10 base pairs, the only significant change in the capture factor is a general attenuation of the modulation amplitudes.

When using the parameters of A-type DNA which has 11 base pairs per twist of the helix, a smaller pitch of 2.8 nm, and a base-pair tilt (angle of the base-pair plane with the axis) of 20°, there is a spectacular increase of the capture factor at low energy as seen in Figure 6.34. The calculation provided the first evidence of the importance of these parameters on transient anion formation. Furthermore, Caron and Sanche (2003) predicted that l_0 should be smaller than 3 for energies less than 4 eV, thus providing theoretical evidence for the high efficiency of 0–4 eV electrons to break a single strand of DNA (Martin et al. 2004). It should be noted that since LEE experiments with DNA have been performed on dry films, in ultrahigh vacuum, the molecule adopts the A-form owing to nonstructural water loss (Swarts et al. 1992).

In their following paper, Caron and Sanche (2005) added a pseudobackbone (PB), again using Ar scatterers for the heavy atoms. The capture factors for odd resonance channels are relatively the same as without the backbone whereas those for even symmetry channels are twice as small. This may be due to a shadowing effect of the backbone. The capture factor on the pseudophosphate group was found to be

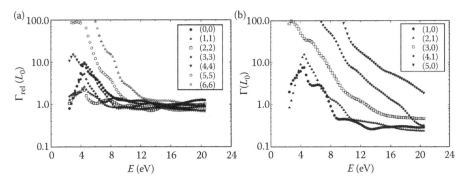

FIGURE 6.34 Capture factor at the center of the PC ring for the A-type DNA structure as a function of energy for various $l_0 + m_0 =$ even (a) and $l_0 + m_0 =$ odd (b) resonance channels as calculated by Caron and Sanche (2004) for scatterers with Ar phase shifts and electrons perpendicular to the helix axis. (Reprinted with permission from Caron, L. and L. Sanche. *Phys. Rev. A 70*(3), 032719-1-10, 2004. Copyright 2004 by the American Physical Society.)

quite large at perpendicular incidence. In order to study this more closely, calculations were also performed for a short segment of 11 base pairs, which was oriented either perpendicular or parallel to the incident electron direction. Results are shown in Figure 6.35. There is no significant difference due to orientation. At energies less than 4 eV, one would expect $l_0 \leq 2$ due to the small size of the phosphate group. Direct electron attachment to the phosphate group is predicted to be enhanced which is important for phosphodiester bond cleavage. Note that the trend in Figure 6.35 is

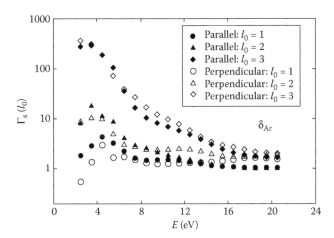

FIGURE 6.35 Surface-averaged partial capture factors at the center of the phosphate groups as function of electron energy using Ar phase shifts in Caron and Sanche (2005). Parallel and perpendicular segment orientations relative to the substrate are illustrated. Various entrance channels l_0 are shown. (Reprinted with permission from Caron, L. and L. Sanche. 2005. *Phys. Rev. A 72*(3), 032726-1-6, 2005. Copyright 2005 by the American Physical Society.)

reminiscent of the one in Figure 6.34 found for the A-form of DNA. What is common to the two systems? The structural parameters for the A-form of DNA point to closer proximity of the bases as probable cause of the large capture amplitudes. In the B-form with backbone, there is close proximity of the sugars to the bases and to the phosphate groups. The enhanced capture factors on the latter are thus possibly an intrabackbone effect emanating from the proximity of sugars and phosphates. As there are no important changes to the base capture factors, the sugar-backbone coupling seems weak. It is unfortunate that the capture factor on the sugars was not calculated. One should also remember that these are the predictions of such a toy model that may have predictive limits.

The last problem tackled by the toy model was on ss oligomers as described in Caron and Sanche (2006). This work was inspired by the experimental electron transmission experiments in films of self-assembled oligomers, both single stranded and double stranded, by Ray et al. (2005). One of the conclusions was that ds DNA does not capture electrons as efficiently as ss oligomers. The higher level of disorder in ss oligomers was thought to be the cause. Caron and Sanche constructed three ss decamers using a PB attached to PC PMOLS. Figure 6.36 shows the three geometries used: Regular, PC disorder, and PB disorder. The latter disorder interestingly mimics the "mushroom" appearance reported in Ray et al. (2006) Figure 6.37 summarizes the results for the capture factor at the center of the PC ring. A large diffraction peak is seen in the regular ss decamer (Figure 6.37a) which understandably disappears in the disordered structures. Both types of disorder lead to similar capture factors with large peaks around 4 eV, which here also result from the closer proximity between the PCs caused by disorder. This would not occur in ds decamers. The toy model thus confirms that capture is favored by the inherent disorder in ss DNA.

Surely the most complete and ambitious scattering study of a DNA segment to date is imputable to Orlando et al. (2008). These authors considered two hydrated ds dodecamers, B-DNA 5'-CCGGCGCCGG-3' and A-DNA 5'-CGCGAATTCGCG-3',

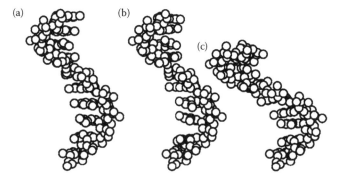

FIGURE 6.36 Side perspective of the three all-different PC decamers used in the study by Caron and Sanche (2006): (a) regular PB and helical placement of the PCs, (b) helical PB with PC orientation disorder, and (c) PB disorder resulting in a "mushroom" arrangement. (Reprinted with permission from Caron, L. and L. Sanche. *Phys. Rev. A 73*(6), 062707-1-7, 2006. Copyright 2006 by the American Physical Society.)

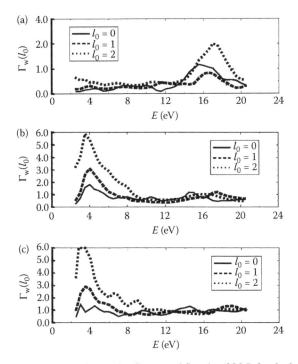

FIGURE 6.37 Partial capture factors by Caron and Sanche (2006) for the helically ordered ss decamer (a), the PC orientation disordered ss decamer (b), and the PB disordered ss decamer (c) as a function of energy. (Reprinted with permission from Caron, L. and L. Sanche. *Phys. Rev. A* 73(6), 062707-1-7, 2006. Copyright 2006 by the American Physical Society.)

on which they used their variant of an MS theory that was extended to first order only. Although Caron and Sanche (2003) found their implementation of MS to their toy decamer not to converge, Orlando et al. (2008) added a short inelastic mean free path of 10.6 Å which should make an expansion convergent. The first-order treatment is thus *a priori* justifiable. Each atomic scatterer was treated with core-excited Feshbach-type resonances of the complex (H_2O : DNA) phase shifts by the authors. This atomistic approach is somewhat evocative of the one used by Yalunin and Leble (2007) on uracil. Figure 6.38 illustrates the target and incident electron geometry. Figure 6.39 shows the electron intensity energy profiles on the sugar, phosphate, and water units. Surprisingly, only the water profiles, actually for those water molecules in the major groove, show any structure, in contrast to the findings of Caron and Sanche (2005) in which both the bases and, to a lesser extent, the phosphate group showed internal diffraction effects in their capture factors around 12 eV. It is unfortunate that Orlando et al. do not report any intensity profile for the bases. The authors then establish a correlation between the intensity peaks on the water molecules, the known DEA resonances of water, and the ss and ds strand break yield functions. The authors conclude that since diffraction intensity is localized on the water, core-excited Feshbach-type resonances of the complex (H_2O : DNA) are considered to contribute to the LEE-induced

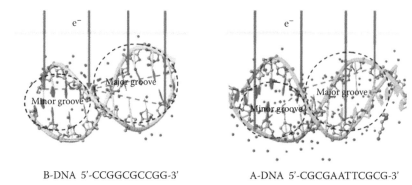

B-DNA 5'-CCGGCGCCGG-3'   A-DNA 5'-CGCGAATTCGCG-3'

FIGURE 6.38 (**See color insert**.) The two DNA targets used in the calculations of Orlando et al. (2008) are illustrated in Figure 1 of their paper. The detailed structural information was obtained from the protein data bank (PDB). The dots in and around target represent the structural water positions extracted from the database. For B-DNA 5'-CCGGCGCCGG-3' and A-DNA 5'-CGCGAATTCGCG-3', the minor and major grooves are indicated by the dashed-line circles. DNA strands are aligned with the symmetry axis parallel to the surface. Bold vertical lines describe the incident electrons whereas dashed lines are the first-order scattered components. (Reprinted with permission from Orlando, T. M. et al. 2008. *J. Chem. Phys.* *128*(19), 195102-1-7. Copyright 2008 American Institute of Physics.)

SSB and DSB probability. This is an interesting proposition since the existence of DNA–H_2O complexes has been observed by Ptasinska and Sanche (2007).

At about the same period in time as Orlando et al. were working on their model calculations, the work of Tonzani and Greene (2006a) motivated Caron et al. (2008) to move away from the toy model and use credible scattering information in the form of gas phase T-matrices for the bases. The inspiration for this came from the pioneering work of Fabrikant et al. (1997) and Fabrikant (2007) who used the R-matrix on molecules on the surface or in the bulk of thin films. The implementation by Caron et al. (2008) is quite different, however, being based on MS. It is not *a priori* obvious that gas phase T-matrices for molecules having an important electric dipole can be readily used in a multiple scattering theory when close proximity conditions prevail. This issue had to be explored first. Caron et al. (2007) undertook a study of the face-centered cubic phase of ice in which both conditions pertaining to molecules having a sizable dipole moment and close proximity prevail. The authors' findings were unanticipated. The correct energy-band structure of ice using MS and the water molecule T-matrix could only be properly reproduced if certain conditions were imposed:

1. The range of the dipole must be cutoff at a distance d where d is a characteristic length of ice which is used to tune the model and turned out to be of the order of the intermolecular distance.
2. An angular momentum cutoff $l \leq l_c$ has to be imposed on the T-matrix components in which l_c is a solution of

$$l_c(l_c + 1) = 2E d^2, \qquad (6.7)$$

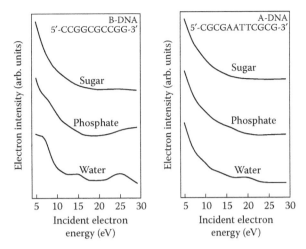

FIGURE 6.39 The calculated electron intensity resulting from elastic scattering within the two targets of Orlando et al. (2008) as a function of incident energy. Data for the sugar, phosphate, and water groups are shown. (Reprinted with permission from Orlando, T. M. et al. 2008. *J. Chem. Phys. 128*(19), 195102-1-7. Copyright 2008 American Institute of Physics.)

where both the electron energy E and d are in a.u. The reasoning is as follows. In order to interact with a molecule, electrons must have enough energy to get over the centrifugal barrier $l(l+1)/(2d^2)$ a.u. The condition $E \geq l(l+1)/(2d^2)$ leads to an upper angular momentum cutoff defined by Equation 6.7.

3. A two-point interpolation procedure is used for any scalar quantity that can only be calculated at integer values of the angular momentum cutoff (T-matrix elements are defined for integer values of l only), between the values floor (l_c) and ceiling (l_c).

These rules were also validated for various water dimer geometries (Bouchiha et al. 2008, Caprasecca et al. 2009) and work well for energies larger than 1 eV.

Caron et al. (2008) first built an idealized regularly sequenced B-form poly(A) •poly(T) decamer on which to try out the aforementioned rules with T-matrices calculated for the isolated bases. The authors found that they could observe the dominantly $L = 2, 3$ shape resonances of the gas phase A and T molecules (Tonzani and Greene 2006a) in the elastic cross section of the decamer below 3 eV (see Figure 6.40) if they used a characteristic length d of 11 a.u., which is not the axial-stacking distance (the rise) but rather a measure of the center to center distance in base pairs, the dimension of the bases themselves, and the size of the R-matrix sphere.

In their quest for signs of internal diffraction, ref136 defined a weighted capture factor

$$\Gamma_w(L_0) = \frac{\sum_n \gamma(l_0, \mathbf{R}_n)}{\sum_n}, \tag{6.8}$$

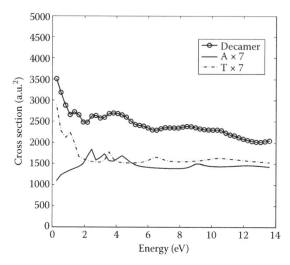

FIGURE 6.40 Interpolated total elastic cross section of Caron et al. (2008) for the poly(A) •poly(T) decamer compared with the single adenine and thymine R-matrix cross-section values as a function of incident electron energy and for normal electron incidence. (Reprinted with permission from Caron, L. et al. *Phys. Rev. A 78*(4), 042710-1-13, 2008. Copyright 2008 by the American Physical Society.)

where

$$\gamma(l_0, \mathbf{R_n}) = \frac{4\pi}{2l_0 + 1} \sum_{|m_0| \le l_0} \left| B^{(n)}_{\mathbf{k}l_0, m_0} \right|^2, \tag{6.9}$$

and $B^{(n)}_{\mathbf{k}l_0, m_0}$ is the angular momentum amplitude of the total MS wave function on the base at $\mathbf{R_n}$. For a plane wave (no MS), $B^{(n)}_{\mathbf{k}l_0, m_0} = Y^*_L(\mathbf{k}), \gamma(l_0, \mathbf{R_n}) = 1$ and $\Gamma_w(L_0) = 1$. Anything larger than 1 thus implies an enhancement of the average capture probability on the bases. Note that $\Gamma_w(0)$ measures the absolute square of the wave function averaged over all bases. Figure 6.41 shows $\Gamma_w(0)$ and the axially scattered current for normal incidence on the poly(A) •poly(T) decamer for two values of the rise. The boxed regions indicate a common energy range for a likely signature of internal diffraction. The encircled region in the axially scattered current points to a Bragg peak resulting from constructive interference of the scattered beams emanating from each base pair, in contrast to internal diffraction, which expresses itself through the amplitudes $B^{(n)}_{\mathbf{k}l_0, m_0}$. The authors next considered the A and B forms of a GCGAATTGGC decamer without a backbone.* They repeated the calculations of $\Gamma_w(0)$ and the axially scattered current for normal incidence and for two values of the rise on the B-form decamer. The results are shown in Figure 6.42. Internal diffraction is again seen at the same

---

* This sequence was extracted from the ideal 36 base-pair theoretical models of DNA in the atlas of macromolecules of the Protein Explorer at *http://www.umass.edu/microbio/chime/pe_beta/pe/atlas/ atlas.htm*

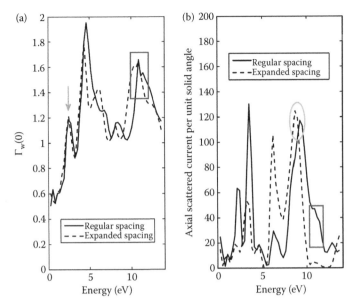

FIGURE 6.41 (**See color insert**.) (a) Interpolated value of the weighted partial capture factors for the $L = 0$ component, and (b) scattered current in a.u. in the $+ \hat{z}$ (axial) direction of the poly(A) •poly(T) decamer as a function of incident electron energy: For normal spacing and for an expanded spacing 1.05 times the regular one, from Caron et al. (2008). The encircled and boxed regions are discussed in the text. (Reprinted with permission from Caron, L. et al. *Phys. Rev. A 78*(4), 042710-1-13, 2008. Copyright 2008 by the American Physical Society.)

energy as for Figure 6.42. Internal diffraction is again seen at the same energy as for the poly(A) •poly(T) decamer (boxed regions). It is thus resistant to sequence disorder. The large peak previously identified as a Bragg peak has vanished, however, a victim of the sequence disorder. Figure 6.43 compares the average of the square of the wave function on the bases for both A- and B-forms. The energies were shifted by estimates of the polarization energy for each form. Interestingly, the peaks for the A-form (i.e., the form for DNA under ultrahigh vacuum conditions) at 1.3 and 2.8 eV (within the boxed regions) are in fair agreement with the experimental maxima in the yield function for SSB in Figure 6.4.

The puzzling presence of electron intensity structures on only the water molecules that was found by Orlando et al. cited in Caron et al. (2009) to try a new calculation including water molecules. The authors also considered it opportune to also include the effect of pair-mismatch defects. They focused on the inner part of the 1D80 dodecamer which consists of a GCGAATTGGC oligomer that has two base-pair mismatches: The third base pair from either end is G–G.[*] The backbone was removed and 23 innermost water molecules were retained. The unit is shown in Figure 6.44.

---

[*] The 1D80 dodecamer can be found in the Nucleic Acid Database at *http://ndbserver.rutgers.edu/* and the Protein Data Base at *http://www.pdb.org/pdb/home/home.do*

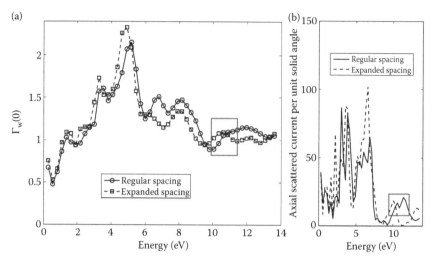

FIGURE 6.42 (**See color insert**.) (a) Interpolated value of the weighted partial capture factors for the $L = 0$ component, and (b) scattered current in a.u. in the $+ \hat{z}$ (axial) direction of the B-form GCGAATTGGC decamer as a function of incident electron energy: for normal spacing and for an expanded spacing 1.05 times the regular one, from Caron et al. (2008). The boxed regions are discussed in the text. (Reprinted with permission from Caron, L. et al. *Phys. Rev. A* 78(4), 042710-1-13, 2008. Copyright 2008 by the American Physical Society.)

Caron et al. (2009) first compared the results of their base-pair mismatched GCGAATTGGC decamer (without the water molecules) with those for the regular one in their previous paper (Caron et al. 2008).

This is shown in Figure 6.45. The structures are quite similar except for larger modulation amplitudes in the perfect decamer. The previously reached conclusion that sequence disorder does not fundamentally change the elastic scattering patterns thus also extends to structural disorder. The authors next compared results with or without the water molecules. This is presented in Figure 6.46. The scattering off of the bases clearly influences the contribution of the water molecules. The one exception is the 9 eV peak in the axial scattered current which seems to be base independent. At 2.5 eV, the square of the wave function shows a strong coupling between water and bases. It is interesting to note that around 7 eV, the square of the wave function on the water molecules bears a strong signature of base behavior, in the same region that the water intensity of Orlando et al. exhibits a maximum. Otherwise, the results of Caron et al. (2009) lead to conclusions quite different from those of Orlando et al. (2008). In an effort to reconcile these results, Caron et al. (2009) introduced an inelastic mean free path in their calculations which they varied from 60 to 30 a.u. Surprisingly, large resonances in the averaged square of the wave function appeared on the bases at energies below 5 eV (the smaller the mean free path, the larger the amplitude of the resonances). These resonances reveal the existence of localized electron states located mostly on the mismatched G–G pairs and the neighboring adenine bases. Such resonances might play a role

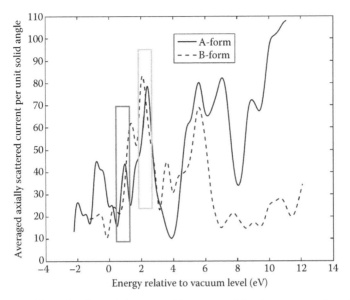

FIGURE 6.43 (**See color insert.**) Average of the square of the electron wave function on the bases over two incident and mutually orthogonal incident directions on the GCGAATTGGC A-form and B-form decamers as a function of incident electron energy corrected for the estimated polarization energy of each form from Caron et al. (2008, 2009). The two rectangular boxes are discussed in the text. (Reprinted with permission from Caron, L. et al. *Phys. Rev. A 78*(4), 042710-1-13, 2008. Copyright 2008 by the American Physical Society.)

FIGURE 6.44 (**See color insert.**) Side view of the decamer studied by Caron et al. (2009). The oxygen atom of the retained water molecules appear as isolated spheres. (Reprinted with permission from Caron, L. et al. *Phys. Rev. A 80*(1), 012705-1-6, 2009. Copyright 2009 by the American Physical Society.)

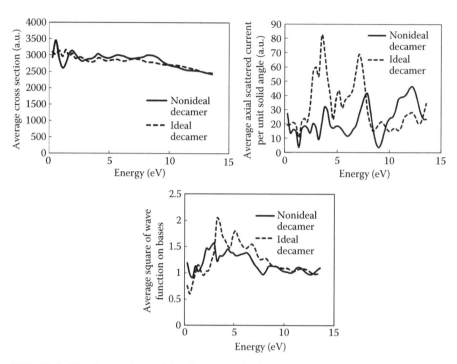

FIGURE 6.45 Comparison of simulation results for the ideal GCGAATTGGC decamer (Caron et al. 2008) and the defect one (Caron et al. 2009) with water molecules removed. (Reprinted with permission from Caron, L. et al. *Phys. Rev. A 80*(1), 012705-1-6, 2009. Copyright 2009 by the American Physical Society.)

in strand breaking because they could efficiently funnel LEEs to the DNA backbone.

6.3.1 Concluding Remarks on Large DNA Molecular Structures

Work on large DNA segments is still work in progress. The toy model of Caron and Sanche provided guidelines for more elaborate calculations. The approach of Orlando et al. has the advantage of simplicity. It would be interesting to pursue it a bit farther in perturbation or even to attempt a full self-consistent MS treatment. Comparison with Caron et al. (2008, 2009) would then be easier and probably quite enlightening. Perhaps the present disagreements between these two approaches could be smoothed out. The future lies in better T-matrices for the bases and inclusion of the backbone subunits, improving on methods based on single-center potentials, which presently yield shape resonances at too large an energy. There is also a need to incorporate core-excited resonances into the formalisms. The intriguing resonances found by Caron et al. (2009) should be more closely studied by alternate methods.

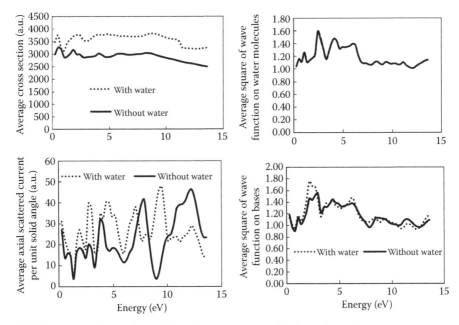

FIGURE 6.46 Comparison of simulation results for the defect GCGAATTGGC decamer (Caron et al. 2009) with and without the water molecules (bases only). (Reprinted with permission from Caron, L. et al. *Phys. Rev. A 80*(1), 012705-1-6, 2009. Copyright 2009 by the American Physical Society.)

6.4 APPENDIX: ABBREVIATIONS

A	Adenine
C	Cytosine or carbon atom
CASPT2	Complete Active Space with Second-Order Perturbation Theory
CASSCF	Complete Active Space Self-Consistent Field
CC	Close coupling
CPCM	Polarizable conductor model
CS	Cross section
d	Deoxyribose
dA	Deoxyadenosine
DBS	Dipole-bound state
dC	Deoxycytidine
DCS	Differential cross section
DEA	Dissociative electron attachment
DFT	Density functional theory
dG	Deoxyguanosine
DMP	Deoxyribose monophosphate

DNA	Deoxyribonucleic acid
DP	Diphosphate
ds	Double-stranded
DSB	Double-strand breaks
dT	Deoxythymidine
DZP	Double zeta with polarization basis set
EEL	Electron energy loss
eV	Electron volt
G	Guanine
H	Hydrogen atom
HF	Hartree–Fock
HOMO	Highest (fully)occupied molecular orbital
HTHF	Hydroxytetrahydrofuran
IAM	Independent atom model
ICS	Integral cross section
IEF-PCM	Integral equation formalism-polarized continuum model
LEE	Low-energy electron
MM	Molecular mechanics
MO	Molecular orbital
MP	Monophosphate
MPH	Monophosphate (protonated)
MTCS	Momentum-transfer cross section
N	Nitrogen atom
NO	Natural orbital
O	Oxygen atom
P	Phosphate or phosphorus atom
PB	Pseudobackbone
PCM	Polarizable continuum model
PES	Potential energy surface
PMOL	Pseudomolecule
QM	Quantum mechanics
RNA	Ribonucleic acid
SCAR	Screening-corrected additivity rule
SCF	Self-consistent field
SE	Static exchange
SEP	Static exchange plus polarization
SOMO	Singly occupied molecular orbital in radical anion
ss	Single-stranded
SSB	Single-strand break
T	Thymine
TDDFT	Time-dependent DFT
THF	Tetrahydrofuran
TNI	Transient negative ion
TS	Transition state
U	Uracil

ACKNOWLEDGMENT

This work was supported by the Canadian Institutes of Health Research and the Marie Curie international incoming fellowship program of the European Commission. The collaboration of Ms Francine Lussier in the elaboration of this article is gratefully acknowledged.

REFERENCES

Abdoul-Carime, H., P. Cloutier, and L. Sanche. 2001. Low-energy (5–40 eV) electron-stimulated desorption of anions from physisorbed DNA bases. *Radiat. Res. 155*(4), 625–633.

Abouaf, R. and H. Dunet. 2005. Structures in dissociative electron attachment cross-sections in thymine, uracil and halouracils. *Eur. Phys. J. D 35*(2), 405–410.

Aflatooni, K., G. A. Gallup, and P. D. Burrow. 1998. Electron attachment energies of the DNA bases. *J. Phys. Chem. A 102*(31), 6205–6207.

Aflatooni, K., A. M. Scheer, and P. D. Burrow. 2006. Total dissociative electron attachment cross sections for molecular constituents of DNA. *J. Chem. Phys. 125*(5), 054301.

Allan, M. 2007. Absolute angle-differential elastic and vibrational excitation cross sections for electron collisions with tetrahydrofuran. *J. Phys. B: At. Mol. Opt. Phys. 40*(17), 3531–3544.

Antic, D., L. Parenteau, M. Lepage, and L. Sanche. 1999. Low-energy electron damage to condensed-phase deoxyribose analogues investigated by electron stimulated desorption of H⁻ and electron energy loss spectroscopy. *J. Phys. Chem. B 103*(31), 6611–6619.

Anusiewicz, I., J. Berdys, M. Sobczyk, P. Skurski, and J. Simons. 2004. Effects of base π-stacking on damage to DNA by low-energy electrons. *J. Phys. Chem. A 108*(51), 11381–11387.

Baccarelli, I., F. A. Gianturco, A. Grandi, and N. Sanna. 2008b. Metastable anion fragmentations after resonant attachment: Deoxyribosic structures from quantum electron dynamics. *Int. J. Quantum Chem. 108*(11), 1878–1887. 3rd International Theoretical Biophysics Symposium, Cetraro, Italy, June 16–20, 2007.

Baccarelli, I., F. A. Gianturco, A. Grandi, N. Sanna, R. R. Lucchese, I. Bald, J. Kopyra, and E. Illenberger. 2007. Selective bond breaking in β-d-ribose by gas-phase electron attachment around 8 eV. *J. Am. Chem. Soc. 129*(19), 6269–6277. PMID: 17444644.

Baccarelli, I., N. Sanna, F. A. Gianturco, and F. Sebastianelli. 2008a. Molecular mechanisms for electron-induced damage to biocomponents: Cross sections and temporary anionic states for monosaccharides. *J. Phys. Conf. Ser. 115*(1), 012009.

Bader, G., G. Perluzzo, L. G. Caron, and L. Sanche. 1984. Structural-order effects in low-energy electron transmission spectra of condensed Ar, Kr, Xe, N₂, CO, and O₂. *Phys. Rev. B 30*(1), 78–84.

Bald, I., J. Kopyra, and E. Illenberger. 2006. Selective excision of C5 from D-ribose in the gas phase by low-energy electrons (0–1eV): Implications for the mechanism of DNA damage. *Angew. Chem., Int. Ed. 45*(29), 4851–4855.

Bao, X., J. Wang, J. Gu, and J. Leszczynski. 2006. DNA strand breaks induced by near-zero-electronvolt electron attachment to pyrimidine nucleotides. *Proc. Natl. Acad. Sci. U. S. A. 103*(15), 5658–5663.

Barrios, R., P. Skurski, and J. Simons. 2002. Mechanism for damage to DNA by low-energy electrons. *J. Phys. Chem. B 106*(33), 7991–7994.

Berdys, J., I. Anusiewicz, P. Skurski, and J. Simons. 2004a. Damage to model DNA fragments from very low-energy (≤1 eV) Electrons. *J. Am. Chem. Soc. 126*(20), 6441–6447. PMID: 15149241.

Berdys, J., I. Anusiewicz, P. Skurski, and J. Simons. 2004b. Theoretical study of damage to DNA by 0.2–1.5 eV electrons attached to cytosine. *J. Phys. Chem. A 108*(15), 2999–3005.

Berdys, J., P. Skurski, and J. Simons. 2004c. Damage to model DNA fragments by 0.25–1.0 eV electrons attached to a thymine π^* orbital. *J. Phys. Chem. B 108*(18), 5800–5805.

Bernas, A., C. Ferradini, and J.-P. Jay-Gerin. 1997. On the electronic structure of liquid water: Facts and reflections. *Chem. Phys. 222*(2–3), 151–160.

Blanco, F. and G. García. 2007. Calculated cross sections for electron elastic and inelastic scattering from DNA and RNA bases. *Phys. Lett. A 360*(6), 707–712.

Bouchiha, D. 2007. *Étude théorique des collisions d'électrons de basse énergie avec des molécules d'intérêt biologique*. Ph. D. thesis, Université de Sherbrooke.

Bouchiha, D., L. G. Caron, J. D. Gorfinkiel, and L. Sanche. 2008. Multiple scattering approach to elastic low-energy electron collisions with the water dimer. *J. Phys. B: At. Mol. Opt. Phys. 41*(4), 045204.

Bouchiha, D., J. Gorfinkiel, L. Caron, and L. Sanche. 2006. Low-energy electron collisions with tetrahydrofuran. *J. Phys. B: At. Mol. Opt. Phys. 39*(4), 975–986.

Boudaïffa, B., P. Cloutier, D. Hunting, M. Huels, and L. Sanche. 2000. Resonant formation of DNA strand breaks by low-energy (3 to 20 eV) electrons. *Science (Washington, DC, U. S.) 287*(5458), 1658–1660.

Bremmer, L., M. Curtis, and I. Walker. 1991. Electronic states of some simple ethers studied by vacuum ultraviolet-absorption and near-threshold electron energy-loss spectroscopy. *J. Chem. Soc. Faraday T. 87*(8), 1049–1055.

Breton, S.-P., M. Michaud, C. Jäggle, P. Swiderek, and L. Sanche. 2004. Damage induced by low-energy electrons in solid films of tetrahydrofuran. *J. Chem. Phys. 121*(22), 11240–11249.

Burke, P. G. and K. A. Berrington. 1993. *Atomic and Molecular Processes: An R-Matrix Approach*. IOP Publishing, Bristol.

Burrow, P. D. 2005. Comment on "Radiation damage of biosystems mediated by secondary electrons: Resonant precursors for uracil molecules" [J. Chem. Phys. [120], 7446 (2004)]. *J. Chem. Phys. 122*(8), 087105.

Burrow, P. D., G. A. Gallup, and A. Modelli. 2008. Are there π^* shape resonances in electron scattering from phosphate groups? *J. Phys. Chem. A 112*(17), 4106–4113. PMID: 18366196.

Caprasecca, S., J. D. Gorfinkiel, D. Bouchiha, and L. G. Caron. 2009. Multiple scattering approach to elastic electron collisions with molecular clusters. *J. Phys. B: At. Mol. Opt. Phys. 42*(9), 095205.

Car, R. and M. Parrinello. 1985. Unified Approach for molecular dynamics and density-functional theory. *Phys. Rev. Lett. 55*(22), 2471–2474.

Caron, L., D. Bouchiha, J. D. Gorfinkiel, and L. Sanche 2007. Adapting gas-phase electron scattering R-matrix calculations to a condensed-matter environment. *Phys. Rev. A 76*(3), 032716.

Caron, L. and L. Sanche. 2004. Diffraction in resonant electron scattering from helical macromolecules: *A*- and *B*-type DNA. *Phys. Rev. A 70*(3), 032719-1-10.

Caron, L. and L. Sanche. 2005. Diffraction in resonant electron scattering from helical macromolecules: Effects of the DNA backbone. *Phys. Rev. A 72*(3), 032726-1-6.

Caron, L. and L. Sanche. 2006. Temporary electron localization and scattering in disordered single strands of DNA. *Phys. Rev. A 73*(6), 062707-1-7.

Caron, L., L. Sanche, S. Tonzani, and C. H. Greene. 2008. Diffraction in low-energy electron scattering from DNA: Bridging gas-phase and solid-state theory. *Phys. Rev. A 78*(4), 042710-1-13.

Caron, L., L. Sanche, S. Tonzani, and C. H. Greene. 2009. Low-energy electron scattering from DNA including structural water and base-pair irregularities. *Phys. Rev. A 80*(1), 012705-1-6.

Caron, L. G. and L. Sanche. 2003. Low-energy electron diffraction and resonances in DNA and other helical macromolecules. *Phys. Rev. Lett. 91*(11), 113201-1-4.

Cecchini, S., S. Girouard, M. A. Huels, L. Sanche, and D. J. Hunting. 2005. Interstrand cross-links: A new type of γ-ray damage in bromodeoxyuridine-substituted DNA. *Biochemistry 44*(6), 1932–1940. PMID: 15697218.

Colyer, C. J., V. Vizcaino, J. P. Sullivan, M. J. Brunger, and S. J. Buckman. 2007. Absolute elastic cross-sections for low-energy electron scattering from tetrahydrofuran. *New J. Phys. 9*(2), 41.

Crewe, A. V., M. Isaacson, and D. Johnson. 1971. Electron energy loss spectra of the nucleic acid bases. *Nature (London, U. K.) 231*(5300), 262–263.

Dabkowska, I., J. Rak, and M. Gutowski. 2005. DNA strand breaks induced by concerted interaction of H radicals and low-energy electrons—A computational study on the nucleotide of cytosine. *Eur. Phys. J. D 35*(2), 429–435.

Dampc, M., A. R. Milosavljević, I. Linert, B. P. Marinković, and M. Zubek. 2007. Differential cross sections for low-energy elastic electron scattering from tetrahydrofuran in the angular range 20°–180°. *Phys. Rev. A 75*(4), 042710.

Davidson, E. and D. Feller. 1986. Basis set selection for molecular calculations. *Chem. Rev. (Washington, DC, U. S.) 86*(4), 681–696.

Denifl, S., S. Ptasińska, M. Cingel, S. Matejcik, P. Scheier, and T. D. Märk. 2003. Electron attachment to the DNA bases thymine and cytosine. *Chem. Phys. Lett. 377*(1–2), 74–80.

Denifl, S., S. Ptasińska, G. Hanel, B. Gstir, M. Probst, P. Scheier, and T. D. Märk. 2004a. Electron attachment to gas-phase uracil. *J. Chem. Phys. 120*(14), 6557–6565.

Denifl, S., S. Ptasińska, M. Probst, J. Hrušák, P. Scheier, and T. D. Märk. 2004b. Electron attachment to the gas-phase DNA bases cytosine and thymine. *J. Phys. Chem. A 108*(31), 6562–6569.

Dillon, M., H. Tanaka, and D. Spence. 1989. The electronic-spectrum of adenine by electron-impact methods. *Radiat. Res. 117*(1), 1–7.

Dora, A., J. Tennyson, L. Bryjko, and T. van Mourik 2009. *R*-matrix calculation of low-energy electron collisions with uracil. *J. Chem. Phys. 130*(16), 164307.

Fabrikant, I. I. 2007. Dissociative electron attachment on surfaces and in bulk media. *Phys. Rev. A 76*(1), 012902.

Fabrikant, I. I., K. Nagesha, R. Wilde, and L. Sanche. 1997. Dissociative electron attachment to CH_3Cl embedded into solid krypton. *Phys. Rev. B 56*(10), R5725–R5727.

Gianturco, F. and R. Lucchese. 2004. Radiation damage of biosystems mediated by secondary electrons: Resonant precursors for uracil molecules. *J. Chem. Phys. 120*(16), 7446–7455.

Gianturco, F. A., F. Sebastianelli, R. R. Lucchese, I. Baccarelli, and N. Sanna. 2008. Ring-breaking electron attachment to uracil: Following bond dissociations via evolving resonances. *J. Chem. Phys. 128*(17), 174302.

Gianturco, F. A., F. Sebastianelli, R. R. Lucchese, I. Baccarelli, and N. Sanna. 2009. Erratum: "Ring-breaking electron attachment to uracil: Following bond dissociations via evolving resonances" [J. Chem. Phys. 128, 174302 (2008)]. *J. Chem. Phys. 131*(24), 249901.

Gillan, C. J., J. Tennyson, and P. G. Burke. 1995. *Computational Methods for Electron-Molecule Collisions* (1st ed.)., Chapter 10, pp. 239–254. New York, Plenum Press: Springer.

Grandi, A., F. A. Gianturco, and N. Sanna. 2004. *H*⁻ desorption from uracil via metastable electron capture. *Phys. Rev. Lett. 93*(4), 048103.

Gu, J., J. Wang, and J. Leszczynski. 2006. Electron attachment-induced DNA single strand breaks: $C_{3'}$–$O_{3'}$ σ-bond breaking of pyrimidine nucleotides predominates. *J. Am. Chem. Soc. 128*(29), 9322–9323. PMID: 16848454.

Gu, J., J. Wang, J. Rak, and J. Leszczynski. 2007. Findings on the electron-attachment-induced abasic site in a DNA double helix. *Angew. Chem., Int. Ed. 46*(19), 3479–3481.

Gu, J., Y. Xie, and H. F. Schaefer. 2005. Glycosidic bond cleavage of pyrimidine nucleosides by low-energy electrons: A theoretical rationale. *J. Am. Chem. Soc. 127*(3), 1053–1057. PMID: 15656644.

Hanel, G., B. Gstir, S. Denifl, P. Scheier, M. Probst, B. Farizon, M. Farizon, E. Illenberger, and T. D. Märk. 2003. Electron attachment to uracil: Effective destruction at subexcitation energies. *Phys. Rev. Lett. 90*(18), 188104.

Hickel, B. and K. Sehested. 1985. Activation energy for the reaction $H + OH^- \rightarrow e_{aq}^-$ Kinetic determination of the enthalpy and entropy of solvation of the hydrated electron. *J. Phys. Chem. 89*(24), 5271–5274.

Huels, M. A., I. Hahndorf, E. Illenberger, and L. Sanche. 1998. Resonant dissociation of DNA bases by subionization electrons. *J. Chem. Phys. 108*(4), 1309–1312.

Isaacson, M. 1972. Interaction of 25 keV electrons with the nucleic acid bases, adenine, thymine, and uracil. I. Outer shell excitation. *J. Chem. Phys. 56*(5), 1803–1812.

Jortner, J. and R. M. Noyes. 1966. Some thermodynamic properties of the hydrated electron. *J. Phys. Chem. 70*(3), 770–774.

Kobyłecka, M., J. Gu, J. Rak, and J. Leszczynski. 2008b. Barrier-free proton transfer in the valence anion of 2′-deoxyadenosine-5′-monophosphate. II. A computational study. *J. Chem. Phys. 128*(4), 044315.

Kobyłecka, M., J. Leszczynski, and J. Rak. 2008a. Valence anion of thymine in the DNA π-stack. *J. Am. Chem. Soc. 130*(46), 15683–15687. PMID: 18954049.

Kobyłecka, M., J. Leszczynski, and J. Rak. 2009. Stability of the valence anion of cytosine is governed by nucleobases sequence in the double stranded DNA π-stack: A computational study. *J. Chem. Phys. 131*(8), 085103.

Koch, W. and M. C. Holthausen. 2001. *A Chemist's Guide to Density Functional Theory* (2nd ed.). Weinheim, Germany: Wiley-VCH.

König, C., J. Kopyra, I. Bald, and E. Illenberger. 2006. Dissociative electron attachment to phosphoric acid esters: The direct mechanism for single strand breaks in DNA. *Phys. Rev. Lett. 97*(1), 018105.

Kumar, A. and M. D. Sevilla. 2007. Low-energy electron attachment to 5′-thymidine monophosphate: Modeling single strand breaks through dissociative electron attachment. *J. Phys. Chem. B 111*(19), 5464–5474. PMID: 17429994.

Kumar, A. and M. D. Sevilla. 2008a. *Radiation Induced Molecular Phenomena in Nucleic Acids* (1st ed.)., Chapter XX—Radiation effects on DNA: Theoretical investigations of electron, hole and excitation pathways to DNA damage, pp. 577–617. *A Comprehensive Theoretical and Experimental Analysis (Challenges and Advances in Computational Chemistry and Physics)*. Vol. 5, Amsterdam: Springer

Kumar, A. and M. D. Sevilla. 2008b. The role of πσ* excited states in electron-induced DNA strand break formation: A time-dependent density functional theory study. *J. Am. Chem. Soc. 130*(7), 2130–2131.

Kumar, A. and M. D. Sevilla. 2009. Role of excited states in Low-Energy Electron (LEE) induced strand breaks in DNA model systems: Influence of aqueous environment. *Chem. Phys. Chem. 10*(Special Issue 9–10), 1426–1430.

Kumar, A. and M. D. Sevilla. 2010. Low Energy Electron (LEE) induced DNA damage: Theoretical approaches to modeling experiment. In M. Shukla and J. Leszczynski (Eds.), *Vademecum of Computational Chemistry Volume 2C: Applications—Biomolecules*. Singapore: World Scientific.

Lepage, M., S. Letarte, M. Michaud, F. Motte-Tollet, M. Hubin-Franskin, D. Roy, and L. Sanche. 1998. Electron spectroscopy of resonance-enhanced vibrational excitations of gaseous and solid tetrahydrofuran. *J. Chem. Phys. 109*(14), 5980–5986.

Levesque, P. L., M. Michaud, W. Cho, and L. Sanche. 2005. Absolute electronic excitation cross sections for low-energy electron (5–12 eV) scattering from condensed thymine. *J. Chem. Phys. 122*(22), 224704.

Li, X., L. Sanche, and M. Sevilla. 2006. Base release in nucleosides induced by low-energy electrons: A DFT study. *Radiat. Res. 165*(6), 721–729.

Li, X., L. Sanche, and M. D. Sevilla. 2002. Dehalogenation of 5-halouracils after low energy electron attachment: A density functional theory investigation. *J. Phys. Chem. A 106*(46), 11248–11253.

Li, X., L. Sanche, and M. D. Sevilla. 2004a. Low energy electron interactions with uracil: The energetics predicted by theory. *J. Phys. Chem. B 108*(17), 5472–5476.

Li, X., M. D. Sevilla, and L. Sanche. 2003a. Density functional theory studies of electron interaction with DNA: Can zero eV electrons induce strand breaks? *J. Am. Chem. Soc. 125*(45), 13668–13669. PMID: 14599198.

Li, X., M. D. Sevilla, and L. Sanche. 2003b. DFT investigation of dehalogenation of adenine-halouracil base pairs upon low-energy electron attachment. *J. Am. Chem. Soc. 125*(29), 8916–8920. PMID: 12862488.

Li, X., M. D. Sevilla, and L. Sanche. 2004b. Hydrogen atom loss in pyrimidine DNA bases induced by low-energy electrons: Energetics predicted by theory. *J. Phys. Chem. B 108*(49), 19013–19019.

Loos, P.-F., E. Dumont, A. D. Laurent, and X. Assfeld. 2009. Important effects of neighbouring nucleotides on electron induced DNA single-strand breaks. *Chem. Phys. Lett. 475*(1–3), 120–123.

Marques, M. A. L. and K. U. Gross. 2003. Time-dependent density functional theory. In C. Fiolhais, F. Nogueira, and M. A. L. Marques (Eds.), *A Primer in Density Functional Theory*. Springer-Verlag, Berlin Heidelberg.

Martin, F., P. D. Burrow, Z. Cai, P. Cloutier, D. Hunting, and L. Sanche. 2004. DNA strand breaks induced by 0–4 ev electrons: The role of shape resonances. *Phys. Rev. Lett. 93*(6), 068101.

Milosavljević, A. R., F. Blanco, J. B. Maljković, D. Šević, G. García, and B. P. Marinković. 2008a. Absolute cross sections for elastic electron scattering from 3-hydroxytetrahydro-furan. *New J. Phys. 10*(10), 103005.

Milosavljević, A. R., D. Šević, and B. P. Marinković. 2008b. Electron interaction with deoxy-ribose analogue molecules in gaseous phase. *J. Phys. Conf. Ser. 101*(1), 012014.

Możejko, P., E. Ptasińska-Denga, A. Domaracka, and C. Szmytkowski. 2006. Absolute total cross-section measurements for electron collisions with tetrahydrofuran. *Phys. Rev. A 74*(1), 012708.

Możejko, P. and L. Sanche. 2005. Cross sections for electron scattering from selected components of DNA and RNA. *Radiat. Phys. Chem. 73*(2), 77–84.

Nguyen, M. T., R. Zhang, P.-C. Nam, and A. Ceulemans. 2004. Singlet-triplet energy gaps of gas-phase RNA and DNA bases. A quantum chemical study. *J. Phys. Chem. A 108*(31), 6554–6561.

O'Malley, T. F. and H. S. Taylor. 1968. Angular dependence of scattering products in electron-molecule resonant excitation and in dissociative attachment. *Phys. Rev. 176*(1), 207–221.

Orlando, T. M., D. Oh, Y. Chen, and A. B. Aleksandrov. 2008. Low-energy electron diffraction and induced damage in hydrated DNA. *J. Chem. Phys. 128*(19), 195102-1-7.

Pan, X., P. Cloutier, D. Hunting, and L. Sanche. 2003. Dissociative electron attachment to DNA. *Phys. Rev. Lett. 90*(20), 208102.

Pan, X. and L. Sanche. 2006. Dissociative electron attachment to DNA basic constituents: The phosphate group. *Chem. Phys. Lett. 421*(4–6), 404–408.

Panajotović, R., M. Michaud, and L. Sanche. 2007. Cross sections for low-energy electron scattering from adenine in the condensed phase. *Phys. Chem. Chem. Phys. 9*(1), 138–148.

du Penhoat, M. A. H., M. A. Huels, P. Cloutier, J. P. Jay-Gerin, and L. Sanche. 2001. Electron stimulated desorption of H-from thin films of thymine and uracil. *J. Chem. Phys. 114*(13), 5755–5764.

Pimblott, S. M. and J. A. LaVerne. 2007. Production of low-energy electrons by ionizing radiation. *Radiat. Phys. Chem. 76*(8–9), 1244–1247. Proceedings of the 11th Tihany Symposium on Radiation Chemistry.

Ptasińska, S., S. Denifl, B. Mróz, M. Probst, V. Grill, E. Illenberger, P. Scheier, and T. D. Märk. 2005. Bond selective dissociative electron attachment to thymine. *J. Chem. Phys. 123*(12), 124302.

Ptasińska, S., S. Denifl, P. Scheier, and T. D. Märk. 2004. Inelastic electron interaction (attachment/ionization) with deoxyribose. *J. Chem. Phys. 120*(18), 8505–8511.

Ptasińska, S. and L. Sanche. 2007. Dissociative electron attachment to hydrated single DNA strands. *Phys. Rev. E 75*(3), 031915.

Ray, S., S. Daube, and R. Naaman. 2005. On the capturing of low-energy electrons by DNA. *Proc. Natl. Acad. Sci. U. S. A. 102*(1), 15–19.

Ray, S. G., S. S. Daube, G. Leitus, Z. Vager, and R. Naaman. 2006. Chirality-induced spin-selective properties of self-assembled monolayers of DNA on gold. *Phys. Rev. Lett. 96*(3), 036101.

Rienstra-Kiracofe, J. C., G. S. Tschumper, H. F. Schaefer, S. Nandi, and G. B. Ellison. 2002. Atomic and molecular electron affinities: Photoelectron experiments and theoretical computations. *Chem. Rev. (Washington, DC, U. S.) 102*(1), 231–282.

Sanche, L. 2008. *Radiation Induced Molecular Phenomena in Nucleic Acids* (1 ed.)., Chapter XIX—Low energy electron damage to DNA, pp. 531–575. *A Comprehensive Theoretical and Experimental Analysis (Challenges and Advances in Computational Chemistry and Physics)*. Vol. 5, Amsterdam: Springer.

Sanche, L. 2009. Radical and radical ion reactivity in nucleic acid chemistry, Chapter 9—*Low-Energy Electron Interaction with DNA: Bond Dissociation and Formation of Transient Anions, Radicals, and Radical Anions*, pp. 239–293. Hoboken, NJ, USA: John Wiley & Sons, Inc.

Scheer, A., C. Silvernail, J. Belot, K. Aflatooni, G. Gallup, and P. Burrow. 2005. Dissociative electron attachment to uracil deuterated at the N_1 and N_3 positions. *Chem. Phys. Lett. 411*(1–3), 46–50.

Scheer, A. M., K. Aflatooni, G. A. Gallup, and P. D. Burrow. 2004. Bond breaking and temporary anion states in uracil and halouracils: Implications for the DNA bases. *Phys. Rev. Lett. 92*(6), 068102.

Schwarz, H. A. 1991. Enthalpy and entropy of formation of the hydrated electron. *J. Phys. Chem. 95*(17), 6697–6701.

Schyman, P. and A. Laaksonen. 2008. On the effect of low-energy electron induced DNA strand break in aqueous solution: A theoretical study indicating guanine as a weak link in DNA. *J. Am. Chem. Soc. 130*(37), 12254–12255. PMID: 18715005.

Schyman, P., A. Laaksonen, and H. W. Hugosson. 2008. Phosphodiester bond rupture in 5′ and 3′ cytosine monophosphate in aqueous environment and the effect of low-energy electron attachment: A Car-Parrinello QM/MM molecular dynamics study. *Chem. Phys. Lett. 462*(4–6), 289–294.

Shukla, M. and J. Leszczynski. 2004. TDDFT investigation on nucleic acid bases: Comparison with experiments and standard approach. *J. Comput. Chem. 25*(5), 768–778.

Sobolewski, A. and W. Domcke. 2002. On the mechanism of nonradiative decay of DNA bases: *ab initio* and TDDFT results for the excited states of 9H-adenine. *Eur. Phys. J. D 20*, 369–374. 10.1140/epjd/e2002-00164-5.

Springborg, M. 2000. *Methods of Electronic-Structure Calculations, From Molecules to Solids*. Chichester, England: John Wiley & Sons, Inc.

Stibbe, D. T. and J. Tennyson. 1996. Time-delay matrix analysis of resonances in electron scattering: e^-–H_2 and H_2^+. *J. Phys. B: At. Mol. Opt. Phys. 29*(18), 4267–4283.

Sulzer, P., S. Ptasinska, F. Zappa, B. Mielewska, A. R. Milosavljević, P. Scheier, T. D. Märk, et al. 2006. Dissociative electron attachment to furan, tetrahydrofuran, and fructose. *J. Chem. Phys. 125*(4), 044304.

Svensson, M., S. Humbel, R. D. J. Froese, T. Matsubara, S. Sieber, and K. Morokuma. 1996. ONIOM: A multilayered integrated MO + MM method for geometry optimizations and single point energy predictions. A test for Diels–Alder reactions and Pt(P(t-Bu)$_3$)$_2$ + H$_2$ oxidative addition. *J. Phys. Chem.* *100*(50), 19357–19363.

Swarts, S., M. Sevilla, D. Becker, C. Tokar, and K. Wheeler. 1992. Radiation-induced DNA damage as a function of hydration: I. Release of Unaltered Bases. *Radiat. Res.* *129*(3), 333–344.

Takayanagi, T., T. Asakura, and H. Motegi. 2009. Theoretical study on the mechanism of low-energy dissociative electron attachment for uracil. *J. Phys. Chem. A* *113*(16), 4795–4801.

Théodore, M., M. Sobczyk, and J. Simons. 2006. Cleavage of thymine N$_3$–H bonds by low-energy electrons attached to base π* orbitals. *Chem. Phys.* *329*(1–3), 139–147. Electron Correlation and Multimode Dynamics in Molecules—(in honour of Lorenz S. Cederbaum).

Tomasi, J., B. Mennucci, and R. Cammi. 2005. Quantum mechanical continuum solvation models. *Chem. Rev. (Washington, DC, U. S.)* *105*(8), 2999–3094.

Tonzani, S. and C. H. Greene. 2005. Electron–molecule scattering calculations in a 3D finite element *R*-matrix approach. *J. Chem. Phys.* *122*(1), 014111.

Tonzani, S. and C. H. Greene. 2006a. Low-energy electron scattering from DNA and RNA bases: Shape resonances and radiation damage. *J. Chem. Phys.* *124*(5), 054312.

Tonzani, S. and C. H. Greene. 2006b. Radiation damage to DNA: Electron scattering from the backbone subunits. *J. Chem. Phys.* *125*(9), 094504.

Trevisan, C. S., A. E. Orel, and T. N. Rescigno. 2006. Elastic scattering of low-energy electrons by tetrahydrofuran. *J. Phys. B: At. Mol. Opt. Phys.* *39*(12), L255–L260.

Tsolakidis, A. and E. Kaxiras. 2005. A TDDFT study of the optical response of DNA bases, base pairs, and their tautomers in the gas phase. *J. Phys. Chem. A* *109*(10), 2373–2380.

Vizcaino, V., J. Roberts, J. P. Sullivan, M. J. Brunger, S. J. Buckman, C. Winstead, and V. McKoy. 2008. Elastic electron scattering from 3-hydroxytetrahydrofuran: experimental and theoretical studies. *New J. Phys.* *10*(5), 053002.

Winstead, C. and V. McKoy. 2006a. Interaction of low-energy electrons with the purine bases, nucleosides, and nucleotides of DNA. *J. Chem. Phys.* *125*(24), 244302.

Winstead, C. and V. McKoy. 2006b. Low-energy electron collisions with gas-phase uracil. *J. Chem. Phys.* *125*(17), 174304.

Winstead, C. and V. McKoy. 2006c. Low-energy electron scattering by deoxyribose and related molecules. *J. Chem. Phys.* *125*(7), 074302.

Winstead, C. and V. McKoy. 2008a. Comment on "Ring-breaking electron attachment to uracil: Following bond dissociations via evolving resonances" [J. Chem. Phys. 128, 174302 2008]. *J. Chem. Phys.* *129*(7), 077101.

Winstead, C. and V. McKoy. 2008b. Resonant interactions of slow electrons with DNA constituents. *Radiat. Phys. Chem.* *77*(10–12), 1258–1264. The International Symposium on Charged Particle and Photon Interaction with Matter—ASR 2007.

Winstead, C., V. McKoy, and S. d'Almeida Sanchez. 2007. Interaction of low-energy electrons with the pyrimidine bases and nucleosides of DNA. *J. Chem. Phys.* *127*(8), 085105.

Xie, H. and Z. Cao. 2007. Electron attachment to the DNA bases adenine and guanine and dehydrogenation of their anionic derivatives: Density functional study. *Int. J. Quantum Chem.* *107*(5), 1261–1269.

Xie, H., R. Wu, F. Xia, and Z. Cao. 2008. Effects of electron attachment on C$_{5'}$–O$_{5'}$ and C$_{1'}$–N$_1$ bond cleavages of pyrimidine nucleotides: A theoretical study. *J. Comput. Chem.* *29*(12), 2025–2032.

Yalunin, S. and S. B. Leble. 2007. Multiple-scattering and electron-uracil collisions at low energies. *Eur. Phys. J. Special Topics* *144*, 115–122. 10.1140/epjst/e2007-00115-x.

Zecca, A., C. Perazzolli, and M. J. Brunger. 2005. Positron and electron scattering from tetrahydrofuran. *J. Phys. B: At. Mol. Opt. Phys.* *38*(13), 2079–2086.

Zhang, R. B. and L. A. Eriksson. 2006. The role of nucleobase carboradical and carbanion on DNA lesions: A theoretical study. *J. Phys. Chem. B 110*(46), 23583–23589. PMID: 17107214.

Zhang, R. B., F. X. Gao, and L. A. Eriksson. 2007. Radical-induced damage in 3′-dTMP insights into a mechanism for DNA strand cleavage. *J. Chem. Theory Comput. 3*(3), 803–810.

Zhang, R. B., K. Zhang, and L. A. Eriksson. 2008. Theoretical studies of damage to 3′-uridine monophosphate induced by electron attachment. *Chem. Eur. J. 14*(9), 2850–2856.

Zheng, Y., P. Cloutier, D. J. Hunting, L. Sanche, and J. R. Wagner. 2005. Chemical basis of DNA sugar-phosphate cleavage by low-energy electrons. *J. Am. Chem. Soc. 127*(47), 16592–16598. PMID: 16305248.

Zheng, Y., P. Cloutier, D. J. Hunting, J. R. Wagner, and L. Sanche. 2006a. Phosphodiester and N-glycosidic bond cleavage in DNA induced by 4–15 eV electrons. *J. Chem. Phys. 124*(6), 064710.

Zheng, Y., J. R. Wagner, and L. Sanche. 2006b. DNA damage induced by low-energy electrons: Electron transfer and diffraction. *Phys. Rev. Lett. 96*, 208101.

7 Low-Energy Electron Scattering at Surfaces

Anne Lafosse and Roger Azria

CONTENTS

7.1 INTRODUCTION

This chapter deals with the experimental aspects of low-energy electron scattering at surfaces. A monochromatic low-energy electron beam (1–20 eV) impinges onto a surface along a well-defined direction. Part of the incident beam is backscattered away from the substrate toward vacuum, either directly or after multiple scattering. The emerging beam is polychromatic and covers a broad range of outgoing angles. To perform spectroscopic measurements, the backscattered electrons are analyzed in energy and angle. General principles of low-energy electron scattering will be addressed, with special emphasis on the involved vibrational excitation mechanisms. Then the associated spectroscopic technique, namely the high-resolution electron energy loss spectroscopy (HREELS) will be described and the different types of recorded spectra listed. Many reviews were dedicated to this technique, each of them focusing on a particular class of substrates or presenting the insights that a given scientific community gains at using HREEL spectroscopy (Ibach and Mills 1982, Mills 1985, Lucas et al. 1986, Thiry et al. 1986, 1987, Egdell et al. 1987, Jones 1992, Ibach et al. 1992, Palmer and Rous 1992, Ibach 1994, Rocca 1995, Richardson 1997, Rizzi 1997, Grégoire et al. 1999, Roy et al. 2000, Conrad and Kordesch 2009). Most often this technique is used to probe the vibrational pattern of surfaces or interfaces, including adsorbate vibrations as well as lattice vibrations, that is, phonons. In addition, the electronic structure of the probed sample does influence strongly the low-energy electron scattering. Electron elastic reflectivity gives insight into the sample conduction-band density-of-states (DOS), which modulates the probability of inelastic scattering for electrons coming in the close vicinity of the substrate. To illustrate the importance of a structured reflectivity on HREELS analysis and to emphasize its possible use as an analysis tool, hydrogenated or deuterated synthetic polycrystalline diamond films will be the systems considered in this chapter.

HREEL spectroscopy is extensively applied in surface science to characterize prepared substrates due to its intrinsic surface sensitivity, in contrast to optical vibrational techniques such as Raman or IR spectroscopy, and its ability to probe hydrogen bonding, in contrast to x-ray photoelectron spectroscopy (XPS). On flat single crystals, of metals or oxide thin layers, adsorbates were sought for and identified by their vibrational signature-recording specular and off-specular energy loss spectra. Adsorption sites and bonding configurations were determined by taking advantage of the selection rules governing the electron inelastic scattering on conductive surfaces. For example, hydrogen (Müssig et al. 1994, Oura et al. 1999, Eggeling et al. 1999a), oxygen (Vattuone et al. 1994, Bădescu et al. 2002), hydroxyl (Bedürftig et al. 1999), and carbon monoxide adsorption (Froitzheim et al. 1977, Wang et al. 2001a) on metals or thin oxide layers (Wang and Wöll 2009) have been extensively studied. Prepared phases and substrates were also characterized by their phonons, including dispersion band mapping (Mills 1985, Nienhaus and Mönch 1995), and low-energy plasmons (Rowe et al. 1990, Rocca 1995, Roy et al. 2000). Also insulators and insulator layers were probed by HREELS (Thiry et al. 1985, 1986, Liehr et al. 1986).

From the chemistry point of view, many studies on elementary processes and reaction intermediates involved in substrate oxidation (Völkening et al. 1999) and in heterogeneous catalysis have been conducted by HREELS. For example, the catalytic oxidation of carbon monoxide CO into carbon dioxide CO_2 (Wang et al. 2001b), the decomposition of CO_2 (Freund and Roberts 1996), and reaction pathways leading from molecular nitrogen and molecular hydrogen to ammonia were investigated on different monocrystalline phases of metals (Bassignana et al. 1986, Wang et al. 2002).

Also, orientations and conformations of macromolecules can be studied on flat metal surfaces. For example, in the context of the molecular switches, the photoinduced ring-opening reaction of nitrospiropyran was observed to lead to an impressive and clear transformation of the vibrational energy loss spectra (Piantek et al. 2009). HREELS is also often used to follow chemical composition modifications of molecular ices, that is, films of molecules physisorbed onto a substrate cooled down to cryogenic temperatures, as a consequence of irradiation-induced synthesis and decomposition reactions (Sanche 1990, Bass and Sanche 1998, Göötz et al. 2001, Swiderek and Burean 2007, Swiderek et al. 2007, Lafosse et al. 2009, Arumainayagam et al. 2010). Finally, HREEL spectroscopy has been applied to characterize numbers of molecular films, ordered or disordered, like self-assembled monolayers (SAMs) of thiolated hydrocarbons (Vilar et al. 1998, Kato et al. 2002, Magnée et al. 2003, Duwez 2004, Kudelski 2005) as well as of thiolated DNA strands (Vilar et al. 2008, Vilar and do Rego 2009), Langmuir–Blodgett films (Wandass and Gardella 1987), and polymer layers (Pireaux et al. 1986, 1991, Apai and McKenna 1991, Akavoor et al. 1996).

Facing the variety of substrates and interfaces, and the variety of physical or chemical characteristics that were probed, and for clarity reasons, the examples selected to illustrate this chapter will be focused on hydrogenated synthetic polycrystalline diamond films. These films provide large, robust, and inert platforms to develop electronic devices, electrodes for electrochemical applications, and chemical sensors (Ushizawa et al. 2002, Yang et al. 2002, Knickerbocker et al. 2003, Härtl et al. 2004, Lasseter-Clare et al. 2005, Hamers et al. 2005, Manivannan et al. 2005, Shin et al. 2006). The chemical inertness (Laikhtman et al. 2004) and the electronic properties of these substrates (negative electron affinity, high surface conductivity) (Krainsky et al. 1996, Cui et al. 1998, Ristein et al. 2001, Ristein 2006) are strongly dependent on the degree of hydrogen coverage and can thus be tuned at will by the adaptation of the degree of hydrogenation. Furthermore, hydrogen is a key component of the gas mixture usually used for diamond nucleation and growth by chemical vapor deposition (CVD) methods (Michaelson et al. 2007a). So that chemical and physical characterization of the uppermost surface atomic layers of diamond films represents a great challenge for a better understanding of CVD growth as well as for many application purposes of diamond. Accordingly, hydrogen content and possible hydrogen-bonding configurations on diamond surfaces have been extensively studied (Küppers 1995, Kawarada 1996) using different vibrational spectroscopy techniques. On one hand, optical spectroscopies like sum frequency generation (SFG) spectroscopy (Chin et al. 1995, Takaba et al. 2001), infrared (IR) spectroscopy (Dischler et al. 1993, Chang et al. 1995, Cheng et al. 1997a, McNamara et al. 2004), and Raman spectroscopy (Ushizawa et al. 1999) were used. On the other hand, a number of HREELS studies have been reported on monocrystalline (Aizawa et al. 1993, Lee and Apai 1993, Thoms and Butler 1994, 1995, Kinsky et al. 2002, Pehrsson et al. 2002), polycrystalline (Sun et al. 1993, Thoms et al. 1994, Aizawa et al. 1995, Michaelson et al. 2006, 2007b), and in particular nanocrystalline (Michaelson and Hoffman 2006, Hoffman et al. 2006, Haensel et al. 2009, Michaelson et al. 2010), hydrogenated and deuterated diamond surfaces. These studies were mainly aimed at resolving as much as possible the different peaks which appear in the energy loss spectra of these materials, and at identifying them as diamond lattice related or sp^m-hybridized CH_x ($m = 1-3, x = 1-3$) hydrogen terminations related. However, loss overlapping and mode mixing (Alfonso et al. 1995, Sandfort et al. 1995, 1996, Smirnov and Rasseev 2000) resulting in broad features in energy loss spectra prevent a complete and definitive attribution of the observed vibrational modes. The involved vibrational excitation mechanisms were investigated through off-specular measurements, leading to angular-scattering behaviors of selected loss features, and more rarely through delicate measurements of excitation functions (Thoms and Butler 1994).

In addition to the large absolute band gap separating the conduction band from the valence band, the diamond electronic structure possesses a second absolute band gap above vacuum level splitting the conduction band. Hydrogenated diamond substrates maintain the diamond bulk structure up to the surface and consequently their electronic band structure exhibits a second absolute band gap at about 13 eV above the vacuum level (Painter et al. 1971, Morar et al. 1985, 1986a,b, Comelli et al. 1988, Weng et al. 1989, Huang and Ching 1993, Sokolov et al. 2003). Electrons impinging from the vacuum at an energy resonant with the second band gap cannot propagate into the sample since no electronic states are available in this energy region, which leads to an enhancement of the electron-backscattering probability. This strong structuration of the sample DOS influences the elastic- and inelastic-scattering behavior of low-energy electrons, and thus modulates the measured electron reflectivity curves and vibrational excitation functions. Hydride-terminated polycrystalline diamond films can be considered as model systems for a comprehensive study on low-energy electron scattering on semiconductor substrate presenting an absolute band gap above vacuum level. The observed behavior for electron-scattering results from the close interplay between the three following characteristics: (i) the density of states of the substrate, (ii) the vibrational excitation mechanisms (dipolar and/or impact scattering including resonant scattering), and (iii) the surface/lattice character of the excited vibrational modes.

7.2 ELECTRON SCATTERING

Electron scattering onto a surface is schematized in Figure 7.1. An incident beam of electrons of well-defined energy $E_0 \pm \Delta E_0$ is directed toward the substrate surface along a well-defined direction characterized by the incident angle θ_i, generally measured from the surface normal. Therefore, the incident electrons have a well-defined wave-vector $\mathbf{k}_i$, which can be decomposed into two contributions, a normal $\mathbf{k}_{i\perp}$ and a parallel $\mathbf{k}_{i\parallel}$ one. The scattered electrons travel away from the substrate with the kinetic energy $E_f \pm \Delta E_f$, along a direction characterized by the angle θ_f, and with the wave-vector $\mathbf{k}_f$.

Two conservation laws govern electron scattering at surfaces.

- Energy conservation: The energy E_{loss}, lost by the electron on scattering at the surface, is equal to the difference between its incident energy E_0 and its final energy E_f.

$$E_0 = E_f + E_{loss} \quad \text{or}$$
$$E_{loss} = |\Delta E| = E_0 - E_f.$$

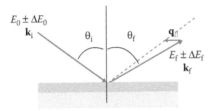

FIGURE 7.1 **(See color insert.)** Schematic plot of the electron-scattering geometry in the plane of incidence. $\mathbf{k}_i$ and $\mathbf{k}_f$ are the initial and final electron wave-vectors, respectively. The angles θ are defined with respect to the normal to the surface. $\mathbf{q}_\parallel$ is the surface parallel excitation wave-vector.

- The momentum conservation law expressed in the plane of the substrate is of particular interest when dealing with the excitation of phonons (surface as well as bulk ones), whose (two-dimensional) wave-vector $\mathbf{q}_\parallel$ lies within the Brillouin zone appropriate to the surface geometry. The wave-vector components parallel to the surface are conserved to within a (two-dimensional) reciprocal lattice vector $\mathbf{G}_\parallel$ (Mills 1985, Nienhaus and Mönch 1995), which leads to the following relationship between the two initial and final parallel components of the electron wave-vector $\mathbf{k}_\parallel$:

$$\mathbf{k}_{i\parallel} = \mathbf{k}_{f\parallel} + \mathbf{q}_\parallel + \mathbf{G}_\parallel.$$

- As the excited mode energy is generally much smaller than the absolute electron energy values, E_0 and consequently E_f, the absolute value of the parallel wave-vector transfer is determined by the simplified relationship:

$$q_\parallel = \mathbf{k}_i |\sin\theta_i - \sin\theta_f|.$$

Energy loss spectroscopy is performed over a broad range of incident electron energy. The considered energy loss range(s) determine the probed physical and/or chemical characteristics of the sample. Extended electron energy loss fine structures (EXELFS) spectroscopy concerns high incident electron energy and high energy losses ($E_{loss} > 50$ eV). Electron energy loss near-edge structures (ELNES) spectroscopy involves high-energy incident electrons ($E_0 \sim 120$ keV) and energy losses in the range of 0.5–50 eV. The incoming electrons induce core and inner-shell ionizations, providing information about the element composition, dielectric properties, and interatomic distances in the sample. When using a transmission electron microscope, the energy loss spectroscopy analysis, then called Energy Filtering Transmission Electron Microscopy (EFTEM), can be performed with a remarkable bidimensional spatial resolution, reaching 10 nm, since such high-energy beams can be extremely well focused on the sample (Hofer et al. 1995, Rizzi 1997). Electron Energy Loss Spectroscopy (EELS) involves lower incident energies, typically in the range of a few tens of eV up to a few hundreds of eV for energy losses extending from 100 meV up to 50 eV. Mostly, these are the electronic properties of the system which are probed. Finally, high-resolution electron energy loss spectroscopy (HREELS) deals with the lower incident electron energies, most often comprised between 1 and 25 eV, for energy losses smaller than 1000 meV. The millimeter-scale dimension of such low-energy electron beams implies that HREELS is a global technique, probing vibrational modes of adsorbates and of the substrate lattice (phonons), as well as low-energy plasmons (Rocca 1995).

In low-energy electron scattering on surfaces and interfaces, the density of targets (typically 10^{15} atoms/cm², or 10^{23} atoms/cm₃) is much greater than in the molecular beams used in gas-phase experiments. Thus multiple scattering participates in electron scattering. A single dipolar inelastic-scattering event is schematized in Figure 7.2 and compared to other scenarios, also referred to as dipolar inelastic scattering, but involving supplementary elastic-scattering events from the substrate. Elastic scattering at a crystal surface results in electron diffraction and is exploited through Low-Energy Electron Diffraction (LEED). The scattered electron beam analyzed in HREELS, also called near-specular peak, can be regarded as the zero diffraction order beam of the incident electron beam (McRae 1979, Rocca and Moresco 1994). The quality of the substrates in terms of roughness and long-range periodicity will directly influence the relative intensities of the elastic and inelastic contributions to the scattered beam, and the energy resolution worsening happening on scattering ($\Delta E_f > \Delta E_0$). Thus, spectra recorded for flat monocrystalline substrates will display a preserved energy resolution and a strong contrast between the elastic and inelastic contributions, whilst for polycrystalline rough surfaces these two contributions will differ by only one order of magnitude or even less.

When probing a substrate with an electron beam, some electrons are transmitted through the sample, some of them can be trapped, and the rest is backscattered

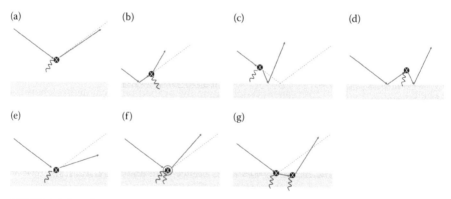

FIGURE 7.2 (See color insert.) Schematic representation of seven possible scattering scenarios for an incoming electron at a substrate surface. Crosses denote the points where inelastic-scattering events take place. Schemes (a)–(d) all correspond to dipolar scattering and result in a single energy loss event taking place at large distance. Schemes (b)–(d) picture the involvement of elastic scattering away from the substrate. (Adapted from Thiry, P. A. et al. 1987. *Phys. Scripta* 35(3), 368–379.) Scheme (e) pictures one possible scenario of impact scattering. Further scenarios involving elastic scattering can be derived from it. Schemes (f) and (g) picture multiple energy loss events. In scheme (f), the multiple energy loss proceeds through a single-scattering event, involving the temporary trapping of the electron (resonant scattering), and in scheme (g) it proceeds through successive inelastic impact-scattering events.

elastically or inelastically away from the surface into the vacuum. Thus HREELS analysis is only possible for samples that are partially conductive. Otherwise an efficient charging of the sample would take place, especially at low energy. Even for a modest (negative) charging, incoming electrons of moderate incident energy would no longer be able to reach the sample surface and would be backscattered by the resulting long-range Coulomb repulsion. The conductive properties of a substrate are directly governed by its electronic band structure. Thus, the scattering behavior of the electrons impinging from the vacuum onto the substrate is partly governed by its density-of-states above vacuum level E_{vac}. If the incident electrons achieve the sample at an energy resonant with a low DOS energy range (e.g., a band gap), they will not be able to propagate into the sample, and thus the scattered part of the beam will be strongly dominant with respect to the transmitted part (Palmer and Rous 1992, Rocca and Moresco 1994). The absolute electron energy, sample DOS, and scattering geometry strongly influence the probing depth, in particular through the inelastic mean free path of electrons in the considered interface. Although HREELS is well known for its surface sensitivity, the excited vibrational modes can involve atomic motions, either mostly in the outermost atomic layer (such modes will be referred to as surface modes in the following) or over several atomic layers below the surface (lattice modes).

7.2.1 EXCITATION MECHANISMS

Electron-induced vibrational excitation of a substrate proceeds through two interaction mechanisms: namely dipolar and impact excitations. They differ by the electron—substrate distance at which they take place. While the dipolar scattering is a long-range Coulombic interaction (taking place at about ~10–100 Å), the impact scattering takes place at much shorter distance (~1 Å), since these are the very atomic potentials of the substrate which are probed. Both mechanisms contribute simultaneously, although their relative contributions depend on the scattering conditions, in particular on the incident electron energy and on the considered scattering geometry. The chief characteristics of both mechanisms are addressed here, with particular attention paid to their behavior as a function of the incident electron energy and the involvement of multiple losses. The coexistence of the two vibrational excitation mechanisms renders any adsorbate quantitative analysis using HREEL spectroscopy delicate, although possible in some cases (Bass and Sanche 1998, Burean and Swiderek 2008). The complexity of the scattering process in general and, in particular, the coupling of electronic and vibrational degrees of freedom explain the lack of complete theoretical developments describing electron scattering at surfaces. For a detailed introduction into the theoretical aspects, see, for example, the review by Palmer (Palmer and Rous 1992).

7.2.1.1 Dipolar Excitation Mechanism

On dipolar scattering, the excitation of vibrational modes proceeds via the interaction of the incoming electrons with the dynamic dipoles of the substrate. Since the electrons are moving, the sample sees a time-varying field. Analogous to the vibrational excitation mechanism involved in infrared absorption spectroscopy, only infrared-active modes will contribute to dipolar scattering, meaning that only changes during the vibrations can be probed by HREELS. Another consequence is that fundamental losses will be dominantly excited, by opposition to multiple losses. Due to the long-range nature of the Coulomb potential, the interaction between the electron and the substrate can be pictured by a global-reflecting potential. The interaction occurs from a relatively large distance (~10–100 Å), and during a rather long time. Loss events involving small momentum transfers are favored ($\Delta k \ll k$). This implies on one hand that the scattering takes place in the forward direction, as schematized in Figure 7.2a, and that the scattered electrons are predominantly found in a narrow lobe around the specular direction. On the other hand, the excitation of surface modes characterized by small wave-vectors (long wavelengths) is favored. The probability for vibrational excitation by dipolar scattering decreases for increasing vibrational mode energy $E_{loss} \sim \hbar\omega$ and decreases continuously with increasing incident electron energy E_0 (Thiry et al. 1987).

When working on samples conductive enough, a supplementary selection rule restricts the efficient vibrational excitation to dynamic dipoles perpendicular to the surface. The substrate electrons can follow instantaneously the variation of the dipole moment. This charge image effect leads to the reinforcement of the dynamic dipoles perpendicular to the substrate surface, and to the screening of the parallel ones (Thiry et al. 1987, Kudelski 2005).

7.2.1.2 Impact and Resonant Excitation Mechanisms

On impact scattering, the vibrational excitation is operative via short-range (~1 Å) interaction mechanisms relying on the direct scattering of the incoming electrons by the substrate atomic potentials (Figure 7.2e). Through this mechanism, the impinging electrons can experience drastic changes of momentum. Thus, the specular direction has no particular significance, and the lobe of scattered electrons expands over a wide range of angles. Large momentum transfers take place and therefore short wavelength surface excitations are involved. On interaction at such short range, several inelastic (as well as elastic)-scattering events can take place before the electron is backscattered away from the surface, which results in an enhanced probability of multiple loss excitation (see Figure 7.2g). In principle all modes are allowed, although impact scattering has its own selection rules for ordered substrates (see Thiry et al. 1987 for further discussion). The probability of vibration excitation via impact scattering increases with increasing the incident electron energy E_0 (keeping it below 30 eV).

A particular case of impact scattering is resonant scattering (Thiry et al. 1987, Palmer and Rous 1992). Resonant scattering for chemisorbed systems involves the temporary capture of the incident electron, either into an empty orbital of the adsorbate forming a negative ion resonance (Schulz 1973, Gadzuk 1983a,b) or by the crystal surface giving surface state or surface resonance (SR) (Baró et al. 1979, Conrad et al. 1986, Millo et al. 1989, Petaccia et al. 1999). Therefore, resonant scattering only takes place in few energy windows, when the incoming electrons have energy resonant with the energy of the temporarily populated state. Then the vibrational excitation probability is not only enhanced for the fundamental loss, but also for its overtones, corresponding to the excitation of several quanta of vibration through a single inelastic-scattering event, as schematized in Figure 7.2f.

To get an insight into the vibrational excitation mechanisms involved in low-energy electron scattering, characteristic signatures will be sought for in the different types of spectra that can be recorded (discussed next), in particular in extended energy loss spectra, off-specular energy loss spectra, and vibrational excitation functions.

7.2.2 HREEL Spectroscopy

Surface and interface sensitive vibrational spectroscopies involve different probing particles, characterized by their respective penetration depths (Lucas et al. 1986). At one extreme, photon spectroscopies [Reflection-Absorption IR Spectroscopy (RAIRS), Multiple Internal Reflection Spectroscopy (MIR, alternatively called Attenuated Total Reflection ATR), Raman, and Diffuse Reflectance IR spectroscopy (DRIFTS) (Chesters 1986, Raval 1995, Kudelski 2009)] have large penetration depths on an atomic scale, by opposition to atomic inelastic-scattering spectroscopy, for which the penetration is zero, to SFG spectroscopy, which is intrinsically specific to the interfaces where the centro-symmetry is broken (Shen 1984, Tadjeddine and Peremans 1996, Tadjeddine 2000), and finally to Surface-Enhanced Raman Spectroscopy (SERS) concerning molecules adsorbed on specific surfaces (e.g., Ag, Au, Cu) inducing a very large increase in the intensity of Raman spectrum (Kudelski 2009). HREELS enjoys an intermediate position, the electron penetration depth is smaller than 1 nm (Lucas et al. 1986).

Now restricting the discussion to the most common surface vibrational IR spectroscopy and HREELS, the spectral range routinely probed by IR ($600\text{--}3600$ cm^{-1}, i.e., 75–450 meV) is limited when going toward low-energy vibrational modes, as opposed to HREELS, where the whole range is in principle covered down to 0 meV. The limitation comes from the broadness of the elastic peak ΔE_{FWHM} (mentioned later), that is, from the effectively achieved resolution. For example, in the favorable case of OH adsorption on Pt(111) (Bedürftig et al. 1999), the probed spectral range was 10–500 meV (80–4000 cm^{-1}), allowing the observation of hindered translational modes of adsorbates with respect to the substrate. However, HREEL spectroscopy suffers a modest energy resolution. In the most favorable cases, that is, flat well-ordered conducting substrates, the achieved resolution is about 2 meV (16 cm^{-1}) (Tautz and Schaefer 1998, Bedürftig et al. 1999), which remains modest in comparison with standard resolutions achieved by IR spectroscopy (1–4 cm^{-1} typically, i.e., 0.1–0.5 meV). HREEL spectroscopy can be conveniently performed only on sufficiently conducting substrates, and disorder or roughness leads to direct deterioration of the energy resolution. Also free carrier quasielastic contributions destroy the achievable resolution in the case of doped semiconductors (Chen et al. 1994, Richardson 1997). To drive low-energy electron beams through the whole spectrometer, including through the scattering chamber, and to preserve their energy, can only be done at the cost of two strong practical conditions, counting as drawbacks when opposed to infrared spectroscopy. It requires a system (i) put under ultrahigh vacuum and (ii) properly shielded from any magnetic field (μ-metal shielding, amagnetic materials, and careful current delivery).

In contrast to the aforementioned global vibrational spectroscopy techniques, inelastic electron tunnelling spectroscopy performed under an STM tip (STM-IETS) leads to bidimensional vibrational mapping of prepared substrates and of supported molecular objects, with a resolution reaching molecular dimensions (Stipe et al. 1999,

Lauhon and Ho 2000, Kim et al. 2002, Sainoo et al. 2005). The tunnelling-active modes are observed as peaks in the second derivative of the tunnelling current d^2I/dV^2. The involved excitation mechanisms are under discussion in the dedicated literature. The energy resolution of the obtained vibrational spectra is rather modest and most of the time the studies are restricted to the stretching mode region (higher-energy vibrational modes).

Surface densities of unoccupied states, above vacuum level, are generally probed experimentally using near-edge x-ray absorption fine structure (NEXAFS) spectroscopy and ELNES, techniques for which incident photons or fast electrons induce transitions from atomic core levels to unoccupied electronic states of the material. HREELS allows probing the surface DOS above vacuum level, keeping the electronic core structure intact.

7.2.2.1 HREEL Spectrometer: Instrument Description

Electron energy loss spectrometers are always designed along the same sequence: An electron source (cathode), extraction and guiding lenses, a monochromating system, focalization and guiding lenses, a collision chamber ensuring a well-defined potential, again extraction and guiding lenses, an analyzing system, and a detector. The nature of the monochromating and analyzing systems can vary from one design to the other (Thiry et al. 1987, Ibach et al. 1992, Palmer and Rous 1992, Richardson 1997, LeGore et al. 2002). We describe here briefly the HREEL spectrometer, model IB500 by OMICRON (Omicron 2000), schematized in Figure 7.3, specially designed to measure energy loss spectra as well as excitation functions. Housed in UHV μ-metal shielded setup, the HREEL spectrometer consists of two successive electrostatic monochromators and a single rotatable electrostatic analyzer. The quasi-monochromatic

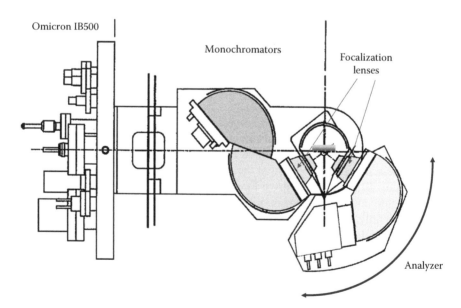

FIGURE 7.3 (**See color insert**.) Schematic representation of the Omicron IB500 HREEL spectrometer. (Adapted from Omicron. 2000. *IB 500 User's Guide, version 1.0.*)

electron beam delivered at the exit of the second monochromator is accelerated at the nominal energy and concomitantly focused by a set of adaptable lenses. The scattered electron beam is collected by a symmetric set of lenses and fed into the analyzer. The effective impact energy of the electrons onto the sample is obtained by correcting the nominal instrumental energy from the difference in work function between the mono-chromator slit and the sample (Palmer and Rous 1992, Rocca and Moresco 1994, Omicron 2000). This correction can reach about 1 eV. The purely instrumental resolution is measured when the instrument is set in the straight-through position, meaning that the analyzer is set online with the monochromator. Then one can reasonably consider that $\Delta E_0 \sim \Delta E_f \sim \Delta E_{FWHM}$. Some operative characteristics are listed in Table 7.1.

7.2.2.2 Different Types of Spectra for Electron Scattering at Surfaces

Electron energy loss spectra $I = I(E_{loss})$ are recorded at fixed incident electron energy E_0. The number of detected electrons is measured as a function of the energy E_{loss} they have lost on scattering on the sample. The probed range is usually 0–500 meV, extended up to 800 meV when multiple losses are considered. The overall resolution is estimated as the Full Width at Half Maximum of the elastic peak ΔE_{FWHM}, and it varies from <2 meV, for flat single crystals, up to ~15 meV for polycrystalline rough samples or complex molecular films. Energy loss spectra can be recorded at various incident electron energies E_0 (typically 1–25 eV), incident angles θ_i (generally 45–60°), and scattering angles θ_f. These spectra are recorded either in the specular geometry $\theta_i = \theta_f$, or in the off-specular geometry, $\theta_i \neq \theta_f$. The scattering geometries are strongly con-strained by the instrumentally required space for the rotating monochromator or ana-lyzer. Off-specular measurements are most often performed by changing the scattering angle θ_f (the electron detection direction). Angular profiles $I_{XXmeV} = I(\theta_f)$ of selected losses are obtained by recording energy loss spectra at various-scattering angles but constant incident energy, and by following the intensity of the loss features $E_{loss} = XX$ meV of interest (Thoms and Butler 1994). Note that in some experimental setup, this is the incident angle which is varied. Substrate phonon dispersion curves $\hbar\omega = \hbar\omega\,(q_\parallel)$ are determined by following the energy position of losses in different (off-specular) scattering geometries. The incident angle θ_i and/or the scattering angle θ_f are adjusted in combination to the incident electron energy E_0 (consequently k_i) to cover the whole Brillouin zone range (Mills 1985, Frederick et al. 1991, Nienhaus and Mönch 1995). Vibrational mode dispersion curves were measured by HREELS

TABLE 7.1

Main Operative Characteristics of the Omicron IB500 HREEL Spectrometer

Filament current (LaB$_6$)	1.55 A
Delivered current	~0.1 nA
Beam dimensions	~2 mm
Instrumental resolution (ΔE_0)	<1 meV
Incident angle	55°
Scattering angle	54–90°

for hydrogenated/deuterated monocrystalline phases of diamond C(111) and C(100), showing clear dispersion for the diamond phonon modes spanning from ~20 meV up to 90 meV across the Brillouin zone, by opposition to the hydrogen termination modes which are nondispersive (Aizawa et al. 1995, Kinsky et al. 2002). To enhance the resolution of energy loss spectra, different approaches were proposed. They rely on data recovery using maximum likelihood in alliance with Bayesian probability theory (Frederick et al. 1996, Richardson 1997) or deconvolution with the elastic peak shape (Apai et al. 1995). Fourier deconvolution was also proposed to remove phonon–phonon and phonon–adsorbate combination losses from HREEL spectra of adsorbate-covered oxide crystals (Petrie and Vohs 1991).

Quasi-elastic (elastic reflectivity) and inelastic (vibrational) excitation functions $I_{XXmeV} = I(E_0)$ are recorded at fixed energy loss $E_{loss} = XX$ meV and scattering geometry, but at varying incident electron energy. The elastic reflectivity curve is measured by following the elastic peak absolute intensity variation as a function of the incident electron energy E_0. Inelastic excitation functions are recorded by following the variation of the number of detected electrons, having lost the considered amount of energy E_{loss} on surface scattering (peak count rate given in number of counts per second), as a function of the incident electron energy E_0. Most often series of energy loss spectra are recorded at incident electron energy increasing/decreasing with a step of 0.5–1 eV (Thoms and Butler 1994, Apai et al. 1995). More rarely, when the instrument allows it, these excitation functions are acquired quasi-continuously (Azria et al. 2011). The influence of the instrument transmission function on the recorded elastic reflectivity and vibrational excitation functions (VEF) is widely discussed in the literature (Palmer and Rous 1992, Thoms and Butler 1994, Eggeling et al. 1999b). The HREEL spectrometer transmission function is not easily determined, so that the variations and energy positions of maxima and minima observed for the excitation functions are reliable, but not the absolute values.

7.3 VIBRATIONAL PATTERN ANALYSIS THROUGH ENERGY LOSS SPECTRA

Any HREEL study of a sample starts with the measurement of energy loss spectra. They give an overview of the low-energy electron-scattering behavior. All spectra presented below were acquired at room temperature and in the specular geometry along an incident direction of 55° with respect to the surface normal.

All hydrogenated synthetic diamond samples were provided by the group of Prof. Alon Hoffman, of the Technion, Israel Institute of Technology (Haifa). These synthetic polycrystalline diamond films were grown *ex situ* on p-type-doped silicon substrates by Hot Filament Chemical Vapor Deposition (HF-CVD) and direct current Glow Discharge Chemical Vapor Deposition (dc GD CVD) using conditions described elsewhere (Michaelson et al. 2007a). HF CVD polycrystalline diamond films were grown from a hydrogen-rich gas mixture (CH_4/H_2 ratio was 1/99). The micron thick films consist of 300–400 nm crystallites. The as-grown films are characterized by high diamond quality (>98% sp^3) and well-defined fully hydrogenated diamond facets. The samples could be further hydrogenated, either *ex situ*, by hydrogen MicroWave (MW) plasma treatment, or *in situ*, by exposure to activated hydrogen generated by flowing H_2 over a heated W filament (Lafosse et al. 2006). Hydrogen-free (or bare) microcrystalline diamond films were prepared *in situ* by several annealing cycles to 1000°C. dc GD CVD films were grown from hydrogen-rich plasma (CH_4/H_2 ratio was 9/91) (Michaelson and Hoffman 2006, Hoffman et al. 2006). The nanodiamond crystallites (final size of ~5 nm) are embedded in hydrogenated amorphous carbon matrix and are covered by an upper layer consisting of hydrogenated amorphous carbon. This latter layer is removed *ex situ* by a further hydrogen MW plasma treatment. To obtain deuterated dc GD CVD samples, fully deuterated compounds (CD_4 and D_2) were used for film growth as well as for the final MW plasma treatment. Before performing measurements, the diamond samples were annealed to 350–450°C in UHV in order to desorb all species possibly physisorbed on their surface, like water or hydrocarbons. Hydrogenated and deuterated polycrystalline diamond films are characterized by surface conductivities high enough to allow convenient HREELS investigation.

Considering the polycrystalline nature of the films, different bonding configurations of hydrogen are expected to occur due to bonding to carbon atoms positioned in different surface planes, grain boundaries, and surface defects. In general, the hydrogen-bonding configurations to these surfaces can be noted by $sp^m\text{-}CH_x$, where m and x may obtain values of 1, 2, or 3 depending on the carbon atom hybridization (local bonding configuration).

7.3.1 ENERGY LOSS SPECTRA OF HYDROGENATED MICROCRYSTALLINE AND NANOCRYSTALLINE DIAMOND FILMS

The energy loss spectra recorded at 5 eV incident electron energy for an as-grown hydrogenated microcrystalline film and an *ex situ* hydrogenated nanocrystalline diamond film are compared in Figure 7.4. They appear to be quite similar, although the achievable resolution for nanocrystalline films is moderate. Such spectra are fairly well known and understood (Sun et al. 1993, Thoms et al. 1994, Küppers 1995, Aizawa et al. 1995, Lafosse et al. 2005, 2006, Michaelson et al. 2006, 2007b, Hoffman et al. 2008).

The stretching modes of the hydride groups $v(CH_x)$, which are decoupled from the lattice modes (Sandfort et al. 1995, 1996, Smirnov and Raseev 2000), are surface modes giving rise to an unresolved energy loss band in the region 340–420 meV. The broad loss structure observed in the region 125–190 meV is attributed to the CH_x-bending modes and lattice modes (phonons) (Thoms et al. 1994, Aizawa et al. 1995). Below 160–165 meV, the bending modes are quasi-resonant in energy with lattice modes, which leads in fact to mixed modes (Lee and Apai 1993, Alfonso et al. 1995, Sandfort et al. 1995, 1996, Frauenheim et al. 1996, Smirnov and Raseev 2000). The observed unresolved band will be referred to as a hydrogen-bending lattice band in the following. The relative contributions, of the hydrogen-termination-bending vibration and of the lattice vibration, to mixed modes vary. Some modes possess a strong phonon character and involve the motion of several atomic layers of the films (lattice modes), while other modes involve mainly the motion of the hydrogen terminations of the outermost layer (surface modes).

Multiple losses, associated with the fundamental band around 150 meV, are observed at 300, 450, and 600 meV. The additional peak at 510 meV is attributed to

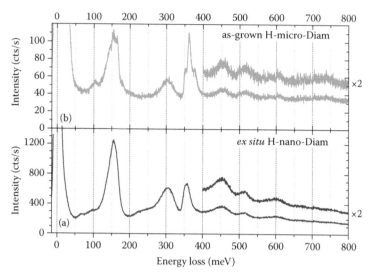

FIGURE 7.4 **(See color insert.)** Energy loss spectra measured at $E_0 = 5$ eV for (a) an *ex situ* MW-hydrogenated nanocrystalline (5 nm) dc GD CVD diamond film with a typical resolution of $\Delta E_{FWHM} \sim 12$ meV; and (b) an as-grown-hydrogenated microcrystalline (300 nm) HF CVD diamond film with a typical resolution of $\Delta E_{FWHM} \sim 5$ meV.

the combination of the hydrogen-bending lattice band (150 meV) with the hydrogen-stretching band (Waclawski et al. 1982, Okuyama et al. 2003), and the shoulder at 720 meV to a multiple loss of the fundamental CH_x-stretching band.

In order to clarify the nature of the observed energy losses, as diamond lattice or hydride termination-related ones, Michaelson and coworkers prepared a series of isotopically substituted (H/D and $^{12}C/^{13}C$) microcrystalline diamond films (Michaelson et al. 2006, 2007b). The isotopic exchanges were performed by using appropriate gas mixtures during the HF CVD growth ($^{12}CH_4 + H_2$, $CD_4 + D_2$, or $^{13}CH_4 + H_2$) and the further *ex situ* MW-H or -D plasma treatment. By comparing the energy loss spectra recorded over a range extending up to 800 meV, distinct series of vibrational losses were observed to shift, depending on the performed isotopic substitution. The set of multiple losses observed at 300, 450, and 600 meV could be doubtless assigned to multiple losses associated with pure diamond lattice modes $v(C-C)$, which contribute to the fundamental feature located at 150 meV. These multiple losses are absent from the energy loss spectra of amorphous carbon films (Biener et al. 1993, 1994, Horn et al. 1995), but are observed on the well-defined (100), (111) diamond surfaces (Sun et al. 1993, Thoms and Butler 1995, Pehrsson et al. 2002). However, in the case of a hydrogenated sample, the hydrogen-bending lattice band observed at $\sim$150 meV is ascribed to two overlapping contributions, associated with the vibrational excitation of diamond lattice $v(C-C)$ modes and of hydride termination-bending $\delta(CH_x)$ modes. Finally, the 510 meV mode was demonstrated to result from two multiple loss contributions, the hydrogen-stretching modes (around 360 meV) combined with both, the diamond lattice and the bending $\delta(CH_x)$ vibrations, the two latter vibrations contributing to the spectrum at about 150 meV.

The attribution of the unresolved contributions to the hydrogen-bending lattice band cannot be done solely on the basis of energy loss spectra analysis, even when considering isotope-substituted diamond films. The comparison of associated vibrational excitation functions is a powerful tool to go further in the analysis and attribution of the involved vibrational modes.

7.3.2 COMPOSITION ANALYSIS OF THE SP^M-CH$_x$ TERMINATIONS OF MICROCRYSTALLINE DIAMOND–HYDROGEN MW PLASMA TREATMENT AGAINST AMORPHIZATION INDUCED BY ION BOMBARDMENT

The amorphous and graphitic content of synthetic diamond films influences, in particular, their conductivity properties. Thus it is of prime interest to check the deposited films for the presence and amount of such matrices. For polycrystalline films, such non-diamond areas are found at grain boundaries, in voids, and in defective regions and are strongly dependent on the diamond growth mechanisms and crystallite sizes (Lee and Apai 1993, Dischler et al. 1993, Küppers 1995, Stiegler et al. 1999, Reichart et al. 2004, McNamara et al. 2004, Michaelson et al. 2007c).

The shape of the stretching band $\nu(CH_x)$ located in the region 340–420 meV and the existence of resolved peaks depend on the relative contributions of the participating-stretching modes, related to the actual hydride composition of the film. In the case of as-grown microcrystalline-hydrogenated diamond films (see Figures 7.4b and 7.5a), the peaks observed at 350, 362, and 376 meV cannot be attributed to some major-stretching modes of sp^m-CH$_x$ species, because of the polycrystalline nature of the samples and because a resolution sufficient to separate the different contributions cannot be achieved using HREEL spectroscopy.

However, the sp^m hybridization of the involved carbons can be determined to some extent according to the stretching energy domains. sp^3- (-CH$_x$), sp^2- (=CH$_x$), and sp-hybridized (≡CH) species are expected to give rise to hydride-stretching losses in the regions ~345–372, ~372–384, and ~384–415 meV, respectively (Ibach and Mills 1982, Küppers 1995, Ushizawa et al. 1999). Accordingly, the two dominant low-energy-stretching mode features of the as-grown-hydrogenated microcrystalline film are ascribed to sp^3-hybridized CH$_x$ groups, while the weaker contribution at higher energy loss is indicative of the vibrational excitation of sp^2-hybridized species. Such sp^2-CH$_x$ groups have already been mentioned for both hydrogenated diamond mono-crystalline phases C(100) and C(111) (Lee and Apai 1993, Ushizawa et al. 1999) and polycrystalline films, where they are located at grain boundaries, in voids, and in defective regions, but not at the surface of well-defined diamond crystallites. Therefore, the sp^2-CH$_x$ surface density is much smaller than the sp^3-CH$_x$ surface density.

The effect of an *ex situ* hydrogen MW plasma treatment on the stretching band $\nu(sp^m$-CH$_x)$ for an as-grown hydrogenated microcrystalline diamond film appears clearly in Figure 7.5a. As expected, the $\nu(sp^2$-CH$_x)$ contribution at about 376 meV almost vanishes, while the relative intensities of the contributions $\nu(sp^3$-CH$_x)$ at 350 and 362 meV change slightly. Exposure of diamond films to a hydrogen MW-plasma is known to etch nonuniformly the diamond facets having different orientations (Cheng et al. 1997b), to etch preferentially amorphous carbon phases at the grain boundaries and/or surface defects and to result in hydrogen saturation of the surface

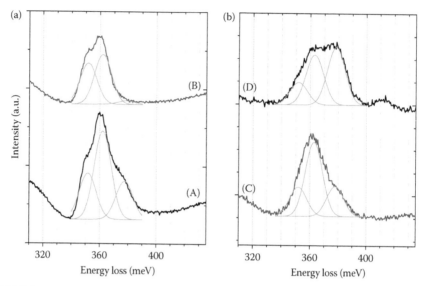

FIGURE 7.5 (**See color insert.**) (a): Hydrogen termination-stretching band ν(sp^m-CH$_x$) of as-grown hydrogenated HF CVD microcrystalline diamond films after annealing to 350°C (A), and after subsequent *ex situ* MW-hydrogenation and annealing to 350°C (B) (Adapted from Lafosse, A. et al. 2008. *Diam. Relat. Mater. 17*(6), 949–953.) (b) Hydrogen termination-stretching band ν(sp^m-CH$_x$) of *in situ*-hydrogenated microcrystalline diamond films annealed to 600°C (C) and after further Ar$^+$ ion irradiation ($E = 1$ keV, ~10^{15} cm^{-2}) (D) (Adapted from Bertin, M. et al. 2007. *Phys. Lett. 90*(6), 061918.) The represented Gaussians are only introduced to guide the eye, and are indicative of the different main sp^m-CH$_x$ contributions. All spectra were recorded at $E_0 = 4$ eV and with a resolution ΔE_{FWHM} ~ 9–14 meV.

(Bobrov et al. 2003, Michaelson and Hoffman 2006, Michaelson et al. 2007a,c). In accordance, no contribution of sp^2-CH$_x$ groups to hydride-stretching band is observed around 376 meV in the energy loss spectrum measured for the *ex situ*-hydrogenated nanocrystalline diamond shown in Figure 7.4a.

To characterize the stability of hydrogenated synthetic diamond films under different irradiation sources, *in situ*-hydrogenated microcrystalline diamond was exposed to an Ar$^+$ ion beam operated at 1 keV kinetic energy (dose ~10^{15} ions/cm^2). The induced deterioration and amorphization are followed in Figure 7.5b (spectra (C), (D)). The *in situ* exposure to activated hydrogen of a bare microcrystalline diamond surface results in the anchoring of hydride groups, sp^3- as well as sp^2-hybridized, as evidenced by the contributions observed at about 350 and 362 meV on one hand, and at ~376 meV on the other hand. This *in situ* hydrogenation procedure appears to be less efficient than the *ex situ* hydrogen MW plasma treatment in etching the amorphous parts and defects of the films. The dramatic changes observed in the HREEL spectrum upon ion processing consist mainly in the striking increase of the contribution of the ν(sp^2-CH$_x$) modes at ~380 meV energy loss and the appearance of a new contribution to the hydride-stretching band at ~410 meV. The appearance of a new loss at 90 meV was concomitantly observed, which is attributed to the vibrational excitation of hydrogen-free surface carbon dimers (not shown here, see Bertin et al.

2007). Thus ion bombardment creates defects in the now, at least partially, disordered near surface region. They consist in the formation of hydrogen-free carbon dimers and an increased content of hydrogenated amorphous phase, as clearly demonstrated by the vibrational signatures of sp^2- (=CH_x) and even sp-hybridized (≡CH) hydride groups (Biener et al. 1993).

7.4 QUASI-ELASTIC ELECTRON REFLECTIVITY: GLOBAL PROBE OF THE DENSITY-OF-STATES ABOVE VACUUM LEVEL

The scattering behavior of the low-energy electrons impinging from the vacuum onto the substrate is partly governed by its density-of-states above vacuum level E_{vac}. The measurement of quasi-elastic reflectivity curves, by following the intensity variation of the elastic peak as a function of the incident electron energy, gives insight into the sample global conduction-band DOS. Thereby, inelastic scattering for electrons coming in the close vicinity of the substrate can be predicted as significantly modulated if the sample DOS is structured enough.

The specular elastically backscattered electron curves (reflectivity curves) measured for incident electron energies in the 3–18 eV range for three hydride-terminated HF-CVD microcrystalline diamond films, which were hydrogenated/deuterated by different methods, were observed to have very similar behaviors (not shown here, see Lafosse et al. 2003, Azria et al. 2011). They are characterized by an intense broad peak at about 13.5 eV incident electron energy, a marked structure around 8.0 eV, and a minor one around 10.6 eV. The reflectivity curve of the hydrogen-free microcrystalline diamond presents completely different features, it mainly decreases continuously without intense structure, and presents two weak features at 11.5 and 8.5 eV incident electron energy. For clarity reasons, the elastic reflectivity curve for the *in situ*-hydrogenated microcrystalline diamond surface alone is compared to the reflectivity curve of the hydrogen-free microcrystalline diamond in the upper panel of Figure 7.6. These curves are plotted on an energy scale having the sample Valence Band Maximum (VBM) for origin, and after having taken the HREELS energy correction into account. The vacuum-level positions E_{vac} were estimated at 3.6 eV above the VBM (indicated by an arrow) and 5.3 eV for the hydrogenated and bare diamond samples, respectively. Note that the energetic location of E_{vac} depends on the diamond phase and on the hydrogen coverage (Cui et al. 1998, Maier et al. 2001), which introduces an uncertainty related to the quality of the surface saturation achieved by the *in situ* hydrogenation.

First, the resemblance between the *ex situ/in situ*-hydrogenated/deuterated microcrystalline diamond reflectivities demonstrates that the observed behavior depends neither on the hydrogenation procedure nor on the involved isotope. Second, this behavior differs strikingly from what is observed for the bare microcrystalline diamond. Third, as demonstrated by NEXAFS spectroscopy (Morar et al. 1985, 1986a,b, Comelli et al. 1988, Hoffman et al. 1999), the full hydrogenation of the diamond results in the saturation of the dangling bonds by hydrogen atoms and ensures the preservation of the bulk electronic band structure up to the surface. The bare diamond surface undergoes dimer reconstructions, which introduce supplementary DOS contributions at energies located in the bulk band gaps (Winn et al. 1997, Hoffman et al. 1999). According to these three remarks, causes of electronic nature have to be considered

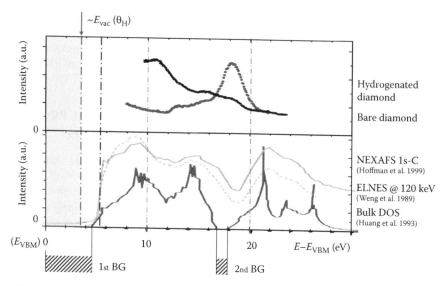

FIGURE 7.6 (**See color insert.**) Upper panel: Elastic reflectivity curves measured for the *in situ* hydrogenated microcrystalline diamond surface (in blue) and bare microcrystalline diamond surface (in black). Lower panel: Comparison of the bulk diamond calculated DOS (Huang and Ching 1993) with the diamond C(1s) ELNES (Weng et al. 1989) (dashed line) and NEXAFS (Hoffman et al. 1999) (full line) spectra. The five spectra are drawn using a common energy scale, whose origin was chosen as the VBM of the substrate ($E_{VBM} = 0$ eV). The relative intensities of the spectra have been arbitrarily fixed. (Adapted from Lafosse, A. et al. 2005. *Surf. Sci.* 587(1–2), 134–141.)

to explain the observations, and more precisely, the electronic band structure of the bulk diamond above vacuum level.

Bulk diamond has a fairly well-known band structure, whose conduction band is split into two parts by a second absolute band gap with a center estimated at about 17.2–18.5 eV above the VBM and a width of 0.5–4.5 eV, depending on the considered calculations (Painter et al. 1971, Weng et al. 1989, Huang and Ching 1993, Sokolov et al. 2003). Conduction band DOS calculations extending up to 30 eV above the VBM require special attention and have been performed by Huang and Ching (1993) and Weng et al. (1989). The density of unoccupied states can be probed experimentally using near-edge spectroscopic techniques like NEXAFS and ELNES, for which incident photons or fast electrons induce transitions from core states C(1s) to unoccupied electronic states of the material. In the lower panel of Figure 7.6, NEXAFS and ELNES spectra are plotted together with the DOS calculated by Huang et al. using the same energy scale than in the upper panel ($E_{VBM} = 0$ eV). The NEXAFS spectrum (full line curve) was recorded by measuring the partial yield of secondary electrons having a kinetic energy of 8 eV over the range 280–340 eV of photon energy (Hoffman et al. 1999). The ELNES spectrum (dashed line curve) measured by Weng and coworkers is an energy loss spectrum recorded with an electron microscope operated at 120 keV equipped with an energy loss spectrometer (Weng et al. 1989).

For clarity, only one NEXAFS spectrum and one calculated DOS are given for comparison with the measured elastic reflectivity of the hydrogenated substrate. For example, the NEXAFS spectra (measured by Morar et al. 1985, 1986a,b, Comelli et al. 1988, Ma et al. 1992, Nithianandam 1992) and the DOS calculated by Weng et al. (1989) compare also very well with the curves presented in Figure 7.6. NEXAFS and ELNES curves reproduce quite well the variations of the calculated conduction band DOS. One recognizes between 0 and 5.5 eV the first band gap, which is much larger than the second absolute band gap observed experimentally as 2 eV and shifted toward higher energy (around 19 eV). The local minimum at about 12 eV above the VBM is also confirmed in core-level excitation experiments.

Maxima observed in the elastic reflectivity curve of hydrogenated microcrystalline diamond correspond to minima in the ELNES, NEXAFS, and DOS curves. In particular, the dominant feature peaking at 18.2 eV above the VBM (and involving incident electrons having 13.5 eV kinetic energy) is related to the existence of the diamond second band gap. This band gap shows up in spite of the substrate polycrystalline nature because of its absolute character. The probability for an electron to be backscattered increases dramatically when the density-of-states available at the resonant energy is low, since the probability for the electron to propagate into bulk diamond decreases. The feature of the elastic reflectivity peaking at about 13 eV above the VBM (associated with 8.3 eV electron energy) is tentatively attributed to the broad local DOS minimum. The shoulder located at about 15.3 eV above the VBM (associated with 10.6 eV electron energy) is not understood at the moment.

For hydrogen-free microcrystalline surface, the reflectivity curve probes the electronic structure of the reconstructed diamond surface, for which new states located in the second diamond band gap are created upon reconstruction, explaining the absence of the diamond characteristic maximum in the reflectivity curve.

For hydrogenated as well as deuterated nanocrystalline diamond, the electron elastic reflectivity curves measured were comparable to the reflectivity curves for hydride-terminated microcrystalline diamond commented earlier. In particular, the strong enhancement related to the second absolute band gap of diamond was also observed at ~13 eV electron energy. This demonstrated that the diamond-like character in the hydride-terminated nanocrystallites is preserved up to the surface (Hoffman et al. 2008, Amiaud et al. 2011).

When probing the sample electron elastic reflectivity using HREEL spectroscopy, the whole area reached by the incident electron beam is contributing to the backscattered-detected signal. (No chemical entities are selected.) Thus, it is the global reflectivity of the surface which is probed.

7.5 VIBRATIONAL EXCITATION FUNCTIONS: VIBRATIONAL MODE CHARACTERIZATION

The characteristics of low-energy electron inelastic scattering at substrate surfaces are dictated by a close interplay between (i) the density-of-states of the environment in which the probed species are embedded (or local DOS), (ii) the vibrational excitation mechanisms (dipolar and/or impact scattering including resonant scattering), and (iii) the surface/lattice character of the excited vibrational modes. Excitation function measurement provides a tool to attribute losses observed in the energy loss spectra and to classify them according to their surface/lattice character. The situation is different for inelastic excitation functions than for elastic reflectivity, since VEF are recorded for given energy losses. Thereby, given species are selected and the excitation functions are related to the local DOS, that is, the DOS of the environment in which the species are embedded.

7.5.1 INTERPLAY BETWEEN EXCITATION MECHANISMS, LOCAL DOS EFFECT, AND THE SURFACE/LATTICE CHARACTER OF THE EXCITED VIBRATIONAL MODES

Owing to the vibrational mode frequency shift associated with the isotope substitution, deuterated nanocrystalline diamond is a model system for which pure hydride termination related, namely surface bending $\delta(CD_x)$ and stretching $v(CD_x)$ modes, and pure phonon $v(CC)$-related modes can be resolved and isolated in its energy loss spectrum (not shown here, see Amiaud et al. 2011). The global diamond-like character of this type of sample gives rise to a strong enhancement of the electron reflectivity at about 13 eV electron energy, which is used as a tool to analyze the surface/lattice character of chosen vibrational modes.

The VEF for the bending $\delta(CD_x)$ observed at 107 meV, the phonon $v(CC)$ at 152 meV, and stretching $v(sp^3\text{-}CD_x)$ at 272 meV modes are compared in Figure 7.7 in the case of deuterated nanodiamond (Aizawa et al. 1995, Kinsky et al. 2002, Pehrsson et al. 2002). These curves are shown after proper background subtraction (Lafosse et al. 2005, Hoffman et al. 2008). The behaviors as a function of the electron incident energy are strikingly different. The VEF for $\delta(CD_x)$ is mimicking the shape of the sample global elastic reflectivity, which is very similar to the elastic reflectivity curve shown in Figure 7.6 for hydrogenated microcrystalline diamond. It presents a strong maximum around 13 eV and a smaller contribution at 8 eV. The VEF for $v(CD_x)$ shows a reduced contribution of the 13 eV feature and an enhanced contribution of the 8 eV feature. The VEF for $v(CC)$ shows two maxima around 8 and 11.5 eV and two weak shoulders around 5 and 14.5 eV.

The VEF for the bending mode $\delta(CD_x)$ shows in general an increasing trend, indicating that this mode is dominantly excited by impact scattering. Vibrational excitation is then induced by electrons approaching the substrate close enough to probe the atomic potentials, in other words to probe the substrate density of states. The strong maximum around 13 eV in the VEF is therefore a direct consequence of the enhanced electron reflectivity. Even the shoulder at about 8 eV can be related to the local maximum observed in the electron reflectivity.

The VEF for the loss observed at about 300 meV, also shown in Figure 7.7, has a completely different shape than the $\delta(CD_x)$ one. The excitation probability around 13 eV is almost completely cancelled out. This behavior is a direct consequence of a multiple-scattering process associated with the excitation of pure diamond lattice-related vibrational modes. One can reasonably assume that the associated atomic motions involve several layers (Alfonso et al. 1995, Sandfort et al. 1995, 1996, Smirnov and Raseev 2000). The incoming electrons have to undergo two successive inelastic-scattering events (each of them resulting in a loss of about 150 meV)

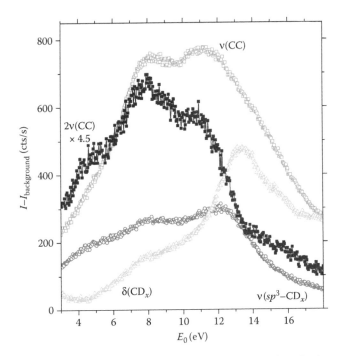

FIGURE 7.7 (**See color insert**.) Selected vibrational excitation functions for deuterated nano-crystalline dc GD CVD diamond, after subtraction of the background: v(CC) at 152 meV (red), δ(CD$_x$) at 107 meV (green), v(sp^3-CD$_x$) at 272 meV (violet), and 2 v(CC) at 300 meV (blue, intensity multiplied by a factor 4.5). (Adapted from Amiaud, L. et al. 2011. *Phys. Chem. Chem. Phys. 13(24)*, 11495–11502.)

before being detected, which renders this type of events highly improbable in the energy range associated with the second absolute band gap. This is in complete accordance with the isotope substitution experiments on hydrogenated/deuterated microcrystalline diamond discussed in Section 7.3.1 (Michaelson et al. 2006, 2007b), which demonstrated that the loss at ~300 meV was exclusively related to lattice v(CC) modes. Interestingly three maxima at 5, 8, and 11.5 eV are clearly resolved in the VEF attributed to 2v(CC). Their presence is an indication that excitation mainly proceeds via resonant scattering. Resonances were also observed in the literature for H$^-$ formation from hydrogenated microcrystalline diamond (Hoffman et al. 2001) and for gas phase hydrocarbons (Allan and Andric 1996) in a similar energy range. Note that evidences for the resonance at 4–5 eV were already observed in the case of hydrogenated micro- and nanocrystalline diamond in the energy loss spectra shown in Figure 7.4. The observed series of multiple losses at 300, 450, and 600 meV are ascribed to the excitation of overtones associated with the fundamental mode at 150 meV.

The VEF of the fundamental mode at 152 meV differs strongly in shape from the VEF of the loss at 300 meV discussed in the previous paragraph. It shows no evidence for a dominant impact or dipolar-scattering mechanism. Since the isotope shift accompanying the deuteration of nanodiamond shifts down in energy the contribution of the deuterium termination bending modes, the loss at 150 meV is attributed to a dominant contribution of diamond phonon modes v(CC). The vibration excitation probability of this multilayer mode is as expected low in the energy range of the second absolute band gap around 13 eV, since the incoming electrons do not have available states to propagate into the sample. The resonances identified at 8 and 11 eV give clearly two peaks in the VEF, and the third low-energy resonance can be associated to the weak shoulder observed at 5 eV. However, the bending modes $\delta(CH_x)$ of the residual hydrogen terminations present at the deuterated nanocrystalline diamond surface are contributing to the energy loss at 152 meV. This leads to the reduced extinction of the vibrational excitation at about 13 eV.

The VEF for the $v(sp^3\text{-}CD_x)$ mode again shows a different shape than the other VEFs shown in Figure 7.7. A global decreasing dependence is observed, meaning that the excitation mechanism dominantly involved is dipolar scattering. Nevertheless, a strong contribution of resonant excitation via the same resonances as discussed above is identified at about 4.5 and 8 eV. However, the maximum of the high energy contribution is shifted to 12.2 eV instead of 11.5 eV as observed for the phonon mode, and instead of 13 eV as observed for the bending mode. This shift results from the contribution of the resonant excitation at 11.5 eV overlapping with the residual contribution of the strongly enhanced reflectivity around 13 eV.

VEF measurement provides a powerful tool to understand the origin of losses observed in energy loss spectra. The involved excitation mechanisms (dipolar and/or impact scattering including resonant scattering), the DOS-related electron reflectivity variations, and the surface/lattice character of the excited vibrational modes have to be taken into account to understand fully the vibrational excitation induced by low-energy electron scattering.

7.5.2 LOCAL PROBE OF THE DENSITY-OF-STATES: ENVIRONMENT IN WHICH THE EXCITED SPECIES ARE EMBEDDED

Stretching mode $v(sp^m\text{-}CH_x)$ VEF measured for as-grown-hydrogenated microcrystalline diamond films illustrate and emphasize to which extent the local DOS (by opposition to global DOS) governs the observed VEF trends (Lafosse et al. 2006).

The evolution, as a function of the incident electron energy, of the hydrogen termination stretching band of the energy loss spectra is shown in Figure 7.8a. The three resolved peaks at 350, 362, and 376 meV are ascribed to the vibrational excitation of $sp^3\text{-}CH_x$, $sp^3\text{-}CH_x$, and $sp^2\text{-}CH_x$ groups, respectively (as discussed above in Section 7.3.2). Globally, the contribution of the $sp^2\text{-}CH_x$ groups decreases in intensity when the incident electron energy increases, while the contributions of the $sp^3\text{-}CH_x$ groups increase. These trends can be directly recognized when considering the VEF associated with the losses at 350, 362, and 380 meV shown in Figure 7.8b. The third function was recorded at 380 meV (and not at the maximum of the $sp^2\text{-}CH_x$ related peak) to avoid any possible contribution of the intense $sp^3\text{-}CH_x$ loss at 362 meV. Whereas the two functions associated with the excitation of $sp^3\text{-}CH_x$ groups behave similarly, it is not the case for the VEF corresponding to sp^2-hybridized groups. The trend

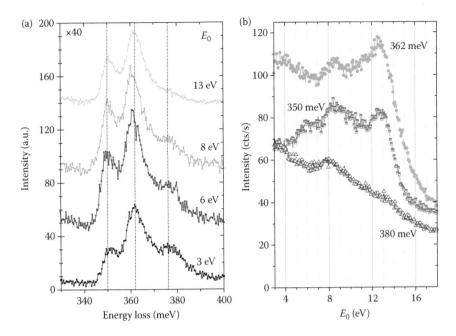

FIGURE 7.8 (**See color insert.**) (a) Hydrogen termination-stretching band $\nu(sp^m\text{-}CH_x)$ of as-grown-hydrogenated HF CVD microcrystalline diamond films recorded at increasing incident electron energy. Figure (b) Specular vibrational excitation functions recorded for the energy losses 362, 350, and 380 meV, respectively, attributed to $sp^3\text{-}CH_x$, $sp^3\text{-}CH_x$, and $sp^2\text{-}CH_x$ groups. (Adapted from Lafosse, A. et al. 2006. *Phys. Rev. B 73*(19), 195308.)

obtained for the two $\nu(sp^3\text{-}CH_x)$ VEF is quite comparable to the behavior discussed in the previous section for the VEF associated with the $\nu(sp^3\text{-}CD_x)$ mode, contributing at 272 meV when working on deuterated nanocrystalline diamond surfaces. Here also, the two main structures at ~8.4 and 12.8 eV, superimposed on a decreasing background attributed to dipole scattering, result from contributions of resonant-scattering mechanisms and of the electron reflectivity enhancement related to the diamond second absolute band gap.

In contrast the excitation function measured for the loss 380 meV, belonging to the $sp^2\text{-}CH_x$-stretching mode energy region, decreases quasi monotonically with a weak peak at 8 eV, attributed to a negative-ion-resonance vibrational excitation. Such a mechanism has been observed around 7.5 eV in the excitation of CH stretching modes for unsaturated hydrocarbons in the gas phase (Walker et al. 1978). As expected, no peak is observed around 13 eV in the $\nu(sp^2\text{-}CH_x)$ VEF. The probed unsaturated sp^2-hybridized CH_x groups, selected by their stretching energy loss, cannot be embedded in a diamond-like environment, for which each carbon atom is sp^3-hybridized. The $sp^2\text{-}CH_x$ groups are rather associated with hydrogen adsorption in grain boundaries and/or highly defective regions, and thus DOS governing the VEF is dictated by the DOS of the local environment in which the probed species are embedded.

7.6 CONCLUSION

Experimental aspects of low-energy electron scattering at surfaces have been reviewed. A monochromatic low-energy electron beam (1–20 eV) impinges onto a surface along a well-defined direction. Part of the incident beam is backscattered away from the substrate toward vacuum, either directly or after multiple scattering. The emerging beam is polychromatic and covers a broad range of outgoing angles. To perform spectroscopic measurements, the backscattered electrons are analyzed in energy and angle. The associated spectroscopic technique, namely the HREELS, is extensively applied in surface science to characterize surfaces and interfaces, due to its intrinsic surface sensitivity and its ability to probe hydrogen bonding.

Most of the studies mainly aim at resolving as much as possible the different peaks which appear in the energy loss spectra, and at attributing them to vibrational modes. However, loss overlapping and mode mixing resulting in broad features prevent a complete and definitive attribution of the observed vibrational modes. The comparison of associated vibrational excitation functions allows going further in the analysis of the involved vibrational modes.

To illustrate the effect of a structured reflectivity on HREELS analysis and to emphasize its possible use as a tool, hydrogenated/deuterated synthetic polycrystalline diamond films were chosen as model systems. Their electronic density-of-states is strongly structured above vacuum level, which influences the elastic- and inelastic-scattering behavior of low-energy electrons, and modulates the measured electron elastic reflectivity curves and vibrational excitation functions. The observed behavior for electron scattering results from the close interplay between the three following characteristics: (i) the density of states of the matrix in which the probed species are embedded, (ii) the vibrational excitation mechanisms (dipolar and/or impact scattering including resonant scattering), and (iii) the surface/lattice character of the excited vibrational modes.

ACKNOWLEDGMENT

We gratefully thank Prof. Alon Hoffman (Technion, Haifa, Israel) for the diamond film elaborations and for very fruitful scientific discussions. This work has been partly performed within the frame of the ECCL COST Action CM0601 (Electron Controlled Chemical Lithography). The acquisition of the HREELS equipment at the LCAM-ISMO was financially supported by Conseil Général de l'Essonne.

REFERENCES

Aizawa, T., T. Ando, M. Kamo, and Y. Sato. 1993. High-resolution electron-energy-loss spectroscopic study of epitaxially grown diamond (111) and (100) surfaces. *Phys. Rev. B* 48(24), 18348–18351.

Aizawa, T., T. Ando, K. Yamamoto, M. Kamo, and Y. Sato. 1995. Surface vibrational studies of CVD diamond. *Diam. Relat. Mater.* 4(5–6), 600–606.

Akavoor, P., W. Menezes, L. L. Kesmodel, G. Apai, and W. P. McKenna. 1996. High-resolution electron energy loss spectroscopy and infrared spectroscopy of polymer surfaces: High-resolution and orientation effects of polytetrafluoroethylene films. *J. Vac. Sci. Technol. 14*, 95–103.

Alfonso, D., D. Drabold, and S. Ulloa. 1995. Phonon modes of diamond (100) surfaces from ab initio calculations. *Phys. Rev. B 51*(3), 1989–1992.

Allan, M. and L. Andric. 1996. σ* resonances in electron impact-induced vibrational excitation of *n*-propane, cyclopropane, ethylene oxide, cyclopentane, and cyclohexane. *J. Chem. Phys. 105*(9), 3559–3568.

Amiaud, L., I. Martin, A. R. Milosavljević, Sh. Michaelson, A. Hoffman, R. Azria, and A. Lafosse. 2011. Low-energy electron scattering on deuterated nanocrystalline diamond films—A model system for understanding the interplay between density-of-states, excitation mechanisms and surface versus lattice contributions. *Phys. Chem. Chem. Phys. 13*(24), 11495–11502.

Apai, G. and W. P. McKenna. 1991. Surface analysis of polycarbonate thin films by High-Resolution Electron Energy Loss Spectroscopy. Negative ion resonances and surface vibrations. *Langmuir 7*(10), 2266–2272.

Apai, G., W. P. McKenna, P. Akavoor, and L. L. Kesmodel. 1995. HREELS detection of resonance electron scattering in oriented polytetrafluoroethylene films. *J. Electron Spectrosc. 73*(3), R5–R10.

Arumainayagam, C. R., H.-L. Lee, R. B. Nelson, D. R. Haines, and R. P. Gunawardane. 2010. Low-energy electron-induced reactions in condensed matter. *Surf. Sci. Rep. 65*(1), 1–44.

Azria, R., A. Lafosse, L. Amiaud, Sh. Michaelson, and A. Hoffman. 2011. Hydrogenated polycrystalline diamond films: Elastic and inelastic electron reflectivity. *Prog. Surf. Sci. 86*(5–8), 94–114.

Bădescu, S. C., P. Salo, T. Ala-Nissila, S. C. Ying, K. Jacobi, Y. Wang, K. Bedürftig, and G. Ertl. 2002. Energetics and vibrational states for hydrogen on Pt(111). *Phys. Rev. Lett. 88*(13), 136101.

Baró, A. M., H. Ibach, and H. Bruchmann. 1979. Vibrational modes of hydrogen adsorbed on Pt(111): Adsorption site and excitation mechanism. *Surf. Sci. 88*(2–3), 384–398.

Bass, A. D. and L. Sanche. 1998. Absolute and effective cross-sections for low-energy electron-scattering processes within condensed matter. *Radiat. Environ. Biophys. 37*, 243–257.

Bassignana, I., K. Wagemann, J. Küppers, and G. Ertl. 1986. Adsorption and thermal decomposition of ammonia on a Ni(110) surface: Isolation and identification of adsorbed NH_2 and NH. *Surf. Sci. 175*(1), 22–44.

Bedürftig, K., S. Volkening, Y. Wang, J. Wintterlin, K. Jacobi, and G. Ertl. 1999. Vibrational and structural properties of OH adsorbed on Pt(111). *J. Chem. Phys. 111*(24), 11147–11154.

Bertin, M., A. Lafosse, R. Azria, S. Michaelson, O. Ternyak, and A. Hoffman. 2007. Vibrational study of hydrogen bonding to ion irradiated diamond surfaces. *Appl. Phys. Lett. 90*(6), 061918.

Biener, J., A. Schenk, B. Winter, C. Lutterloh, U. Schubert, and J. Küppers. 1993. Spectroscopic identification of C–H species in C:H films using HREELS. *Surf. Sci. Lett. 291*(1–2), L725–L729.

Biener, J., A. Schenk, B. Winter, C. Lutterloh, U. A. Schubert, and J. Küppers. 1994. Hydrogenation of amorphous C: H surfaces by thermal H (D) atoms. *Surf. Sci. 307-309*(Part 1), 228–234.

Bobrov, K., A. Mayne, G. Comtet, G. Dujardin, L. Hellner, and A. Hoffman. 2003. Atomic-scale visualization and surface electronic structure of the hydrogenated diamond *C*(100)–(2 × 1):*H* surface. *Phys. Rev. B 68*(19), 195416.

Burean, E. and P. Swiderek. 2008. Electron-induced reactions in condensed acetaldehyde: Identification of products and energy-dependent cross sections. *J. Phys. Chem. C 112*(49), 19456–19464.

Chang, H.-C., J.-C. Lin, J.-Y. Wu, and K.-H. Chen. 1995. Infrared spectroscopy and vibrational relaxation of CH_x and CD_x stretches on synthetic diamond nanocrystal surfaces. *J. Phys. Chem.* 99(28), 11081–11088.

Chen, P., J. Rowe, and J. Yates Jr. 1994. Electron-energy-loss investigation of hole-plasmon excitation due to thermal indiffusion boron doping of Si (111) surfaces. *Phys. Rev. B* 50(24), 18134–18141.

Cheng, C.-L., H.-C. Chang, J.-C. Lin, K.-J. Song, and J.-K. Wang. 1997b. Direct observation of hydrogen etching anisotropy on diamond single crystal surfaces. *Phys. Rev. Lett.* 78(19), 3713–3716.

Cheng, C.-L., J.-C. Lin, and H.-C. Chang. 1997a. The absolute absorption strength and vibrational coupling of CH stretching on diamond C(111). *J. Chem. Phys.* 106(17), 7411–7421.

Chesters, M. 1986. Infrared spectroscopy of molecules on metal single-crystal surfaces. *J. Electron Spectrosc.* 38, 123–140.

Chin, R. P., J. Y. Huang, Y. R. Shen, T. J. Chuang, and H. Seki. 1995. Interaction of atomic hydrogen with the diamond C(111) surface studied by infrared-visible sum-frequency-generation spectroscopy. *Phys. Rev. B* 52(8), 5985–5995.

Comelli, G., J. Stöhr, C. J. Robinson, and W. Jark. 1988. Structural studies of argon-sputtered amorphous carbon films by means of extended X-ray-absorption fine structure. *Phys. Rev. B* 38(11), 7511–7519.

Conrad, H. and M. E. Kordesch. 2009. High resolution electron energy loss spectroscopy. In: John Lindon (ed.), *Encyclopedia of Spectroscopy and Spectrometry,* 2nd ed., Elsevier Ltd. (ISBN: 978-0-12-374413-5). pp. 865–876.

Conrad, H., M. E. Kordesch, R. Scala, and W. Stenzel. 1986. Surface resonances on Pd(111)/H observed with HREELS. *J. Electron Spectrosc.* 38, 289–298.

Cui, J. B., J. Ristein, and L. Ley. 1998. Electron affinity of the bare and hydrogen covered single crystal diamond (111) surface. *Phys. Rev. Lett.* 81(2), 429–432.

Dischler, B., C. Wild, W. Müller-Sebert, and P. Koidl. 1993. Hydrogen in polycrystalline diamond: An infrared analysis. *Physica B: Condensed Matter* 185(1–4), 217–221.

Duwez, A. S. 2004. Exploiting electron spectroscopies to probe the structure and organization of self-assembled monolayers: A review. *J. Electron Spectrosc.* 134, 97–138.

Egdell, R., W. Flavell, Z. Gray-Grychowski, R. Stradling, B. Joyce, and J. Neave. 1987. Surface and interface phonon and plasmon excitations in iii-v semiconductor materials. *J. Electron Spectrosc.* 45, 177–187.

Eggeling, J., G. R. Bell, and T. S. Jones. 1999a. Vibrational excitation mechanisms in electron energy loss spectroscopy studies of hydrogen adsorbed on Si(100) and Ge(100). *J. Phys. Chem. B* 103(44), 9683–9689.

Eggeling, J., E. A. H. Griffiths, I. R. Gould, and T. S. Jones. 1999b. Negative ion resonances and vibrational mode selectivity in inelastic electron scattering studies of hydrogen and diethylsilane adsorbed on Si(100). *Chem. Phys. Lett.* 313(5–6), 805–811.

Frauenheim, T., T. Köhler, M. Sternberg, D. Porezag, and M. R. Pederson. 1996. Vibrational and electronic signatures of diamond surfaces. *Thin Solid Films* 272(2), 314–330.

Frederick, B., G. Apai, and T. Rhodin. 1991. Surface phonons in thin aluminum oxide films: Thickness, beam-energy, and symmetry-mixing effects. *Phys. Rev. B* 44(4), 1880–1890.

Frederick, B., B. Frederick, and N. Richardson. 1996. Multiple scattering contributions and defining the background for resolution enhancement in HREELS. *Surf. Sci.* 368(1–3), 82–95.

Freund, H. J. and M. W. Roberts. 1996. Surface chemistry of carbon dioxide. *Surf. Sci. Rep.* 25(8), 225–273.

Froitzheim, H., H. Ibach, and S. Lehwald. 1977. Surface vibrations of CO on W(100). *Surf. Sci.* 63, 56–66.

Gadzuk, J. W. 1983a. Shape resonances, overtones, and electron energy loss spectroscopy of gas phase and physisorbed diatomic molecules. *J. Chem. Phys. 79*, 3982–3987.

Gadzuk, J. W. 1983b. Vibrational excitation in molecule–surface collisions due to temporary negative molecular ion formation. *J. Chem. Phys. 79*(12), 6341–6348.

Göötz, B., O. Kröhl, and P. Swiderek. 2001. High-resolution electron-energy-loss spectroscopy of molecular multilayer film growth and properties. *J. Electron Spectrosc. 114–116*, 569–574.

Grégoire, C., L. M. Yu, F. Bodino, M. Tronc, and J. J. Pireaux. 1999. Separation of dipole and impact scatterings in high resolution electron energy loss spectroscopy: Experiment from model organic material. *J. Electron Spectrosc. 98–99*, 67–82.

Haensel, T., J. Uhlig, R. J. Koch, S. I. U. Ahmed, J. A. Garrido, D. Steinmüller-Nethl, M. Stutzmann, and J. A. Schaefer. 2009. Influence of hydrogen on nanocrystalline diamond surfaces investigated with HREELS and XPS. *Phys. Status Solidi A 206*(9), 2022–2027.

Hamers, R. J., J. E. Butler, T. Lasseter, B. M. Nichols, J. N. Russell, Jr., K.-Y. Tse, and W. Yang. 2005. Molecular and biomolecular monolayers on diamond as an interface to biology. *Diam. Relat. Mater. 14*(3–7), 661–668.

Härtl, A., E. Schmich, J. Garrido, J. Hernando, S. Catharino, S. Walter, P. Feulner, A. Kromka, D. Steinmüller, and M. Stutzmann. 2004. Protein-modified nanocrystalline diamond thin films for biosensor applications. *Nat. Mater. 3*(10), 736–742.

Hofer, F., P. Warbichler, and W. Grogger. 1995. Imaging of nanometer-sized precipitates in solids by electron spectroscopic imaging. *Ultramicroscopy 59*(1–4), 15–31.

Hoffman, A., I. Gouzman, and S. Michaelson. 2006. Formation mechanism of nano-diamond films from energetic species: From experiment to theory. *Thin Solid Films 515*(1), 14–26.

Hoffman, A., A. Lafosse, S. Michaelson, M. Bertin, and R. Azria. 2008. Nano size effects in the high resolution electron energy loss spectra and excitation function of hydrogenated diamond films. *Surf. Sci. 602*(18), 3026–3032.

Hoffman, A., A. Laikhtman, S. Ustaze, M. H. Hamou, M. N. Hedhili, J.-P. Guillotin, Y. Le Coat, D. T. Billy, R. Azria, and M. Tronc. 2001. Dissociative electron attachment and dipolar dissociation of H⁻ electron stimulated desorption from hydrogenated diamond films. *Phys. Rev. B 63*(4), 045401.

Hoffman, A., M. Petravic, G. Comtet, A. Heurtel, L. Hellner, and G. Dujardin. 1999. Photon-stimulated desorption of H⁺ and H⁻ ions from diamond surfaces: Evidence for direct and indirect processes. *Phys. Rev. B 59*(4), 3203–3209.

Horn, A., J. Biener, A. Schenk, C. Lutterloh, and J. Küppers. 1995. H/D exchange reaction at graphitic CH groups by thermal H(D) atoms. *Surf. Sci. 331–333*(Part 1), 178–182.

Huang, M. and W. Ching. 1993. Calculation of optical excitations in cubic semiconductors. I. Electronic structure and linear response. *Phys. Rev. B 47*(15), 9449–9463.

Ibach, H. 1994. Electron energy loss spectroscopy: The vibration spectroscopy of surfaces. *Surf. Sci. 299–300*, 116–128.

Ibach, H., M. Balden, D. Bruchmann, and S. Lehwald. 1992. Electron energy loss spectroscopy: Recent advances in technology and application. *Surf. Sci. 269–270*, 94–102.

Ibach, H. and D. L. Mills. 1982. *Electron Energy Loss Spectroscopy and Surface Vibrations.* New York, NY: Academy Press.

Jones, T. 1992. Applications of electron energy loss spectroscopy in surface science. *Vacuum 43*(3), 177–183.

Kato, H. S., J. Noh, M. Hara, and M. Kawai. 2002. An HREELS study of alkanethiol self-assembled monolayers on Au(111). *J. Phys. Chem. B 106*(37), 9655–9658.

Kawarada, H. 1996). Hydrogen-terminated diamond surfaces and interfaces. *Surf. Sci. Rep. 26*(7), 205–206.

Kim, Y., T. Komeda, and M. Kawai. 2002. Single-molecule reaction and characterization by vibrational excitation. *Phys. Rev. Lett. 89*(12), 126104.

Kinsky, J., R. Graupner, M. Stammler, and L. Ley. 2002. Surface vibrations on clean, deuterated, and hydrogenated single crystal diamond(100) surfaces studied by high-resolution electron energy loss spectroscopy. *Diam. Relat. Mater. 11*(3–6), 365–370.

Knickerbocker, T., T. Strother, M. P. Schwartz, J. N. Russell, J. Butler, L. M. Smith, and R. J. Hamers. 2003. DNA-modified diamond surfaces. *Langmuir 19*(6), 1938–1942.

Krainsky, I. L., V. M. Asnin, G. T. Mearini, and J. A. Dayton. 1996. Negative-electron-affinity effect on the surface of chemical-vapor-deposited diamond polycrystalline films. *Phys. Rev. B 53*(12), R7650–R7653.

Kudelski, A. 2005. Characterization of thiolate-based mono- and bilayers by vibrational spectroscopy: A review. *Vib. Spectrosc. 39*(2), 200–213.

Kudelski, A. 2009. Raman spectroscopy of surfaces. *Surf. Sci. 603*(10–12), 1328–1334.

Küppers, J. 1995. The hydrogen surface chemistry of carbon as a plasma facing material. *Surf. Sci. Rep. 22*(7–8), 249–321.

Lafosse, A., M. Bertin, and R. Azria. 2009. Electron driven processes in ices: Surface functionalization and synthesis reactions. *Prog. Surf. Sci. 84*(5–6), 177–198.

Lafosse, A., M. Bertin, S. Michaelson, R. Azria, R. Akhvlediani, and A. Hoffman. 2008. Surface defects induced by *in-situ* annealing of hydrogenated polycrystalline diamond studied by high resolution electron energy loss spectroscopy. *Diam. Relat. Mater. 17*(6), 949–953.

Lafosse, A., D. Cáceres, M. Bertin, A. Hoffman, and R. Azria. 2005. Role of electronic band structure and resonances on electron-scattering. The case of the hydrogenated polycrystalline diamond. *Surf. Sci. 587*(1–2), 134–141.

Lafosse, A., A. Hoffman, M. Bertin, D. Teillet-Billy, and R. Azria. 2006. Density-of-states effect on surface and lattice vibrational modes in hydrogenated polycrystalline diamond. *Phys. Rev. B 73*(19), 195308.

Lafosse, A., D. Teillet Billy, J.-P. Guillotin, Y. Le Coat, R. Azria, A. Laikhtman, and A. Hoffman. 2003. Role of electronic band structure and resonances on electron reflectivity and vibrational excitation functions: The case of hydrogenated diamond. *Phys. Rev. B 68*(23), 235421.

Laikhtman, A., A. Lafosse, Y. Le Coat, R. Azria, and A. Hoffman. 2004. Interaction of water vapor with bare and hydrogenated diamond film surfaces. *Surf. Sci. 551*(1–2), 99–105.

Lasseter-Clare, T., B. H. Clare, B. M. Nichols, N. L. Abbott, and R. J. Hamers. 2005. Functional monolayers for improved resistance to protein adsorption: Oligo(ethylene glycol)-modified silicon and diamond surfaces. *Langmuir 21*(14), 6344–6355.

Lauhon, L. J. and W. Ho. 2000. Single-molecule chemistry and vibrational spectroscopy: Pyridine and benzene on Cu(001). *J. Phys. Chem. A 104*(11), 2463–2467.

Lee, S.-T. and G. Apai. 1993. Surface phonons and CH vibrational modes of diamond (100) and (111) surfaces. *Phys. Rev. B 48*(4), 2684–2693.

LeGore, L. J., R. H. Jackson, Z. Yang, P. Kleban, L. K. DeNoyer, and B. G. Frederick. 2002. Advantages of maximum likelihood methods for PRBS modulated TOF electron spectrometry. *Surf. Sci. 502–503*, 232–239.

Liehr, M., P. A. Thiry, J. J. Pireaux, and R. Caudano. 1986. Characterization of insulators by high-resolution electron-energy-loss spectroscopy: Application of a surface-potential stabilization technique. *Phys. Rev. B 33*(8), 5682–5697.

Lucas, A. A., J.-P. Vigneron, P. Lambin, P. A. Thiry, J.-J. Pireaux, and R. Caudano. 1986. Recent advances in electron energy loss spectroscopy of surface and interface vibrations of layered materials. *Phys. Scripta* (T13), 150–154.

Ma, Y., N. Wassdahl, P. Skytt, J. Guo, J. Nordgren, P. D. Johnson, J.-E. Rubensson, T. Boske, W. Eberhardt, and S. D. Kevan. 1992. Soft-X-ray resonant inelastic scattering at the C K edge of diamond. *Phys. Rev. Lett. 69*(17), 2598–2601.

Magnée, R., M. Maazouz, C. Doneux, F. Bodino, P. Rudolf, D. Teillet-Billy, and J.-J. Pireaux. 2003. Resonant electron scattering of 11-mercaptoundecanoic acid self-assembled monolayer adsorbed on Au (111). *J. Phys. Chem. B 107*(19), 4567–4572.

Maier, F., J. Ristein, and L. Ley. 2001. Electron affinity of plasma-hydrogenated and chemically oxidized diamond (100) surfaces. *Phys. Rev. B 64*(16), 165411.

Manivannan, A., L. Ramakrishnan, M. Seehra, E. Granite, J. Butler, D. Tryk, and A. Fujishima. 2005. Mercury detection at boron doped diamond electrodes using a rotating disk technique. *J. Electroanal. Chem. 577*(2), 287–293.

McNamara, K., B. Williams, K. Gleason, and B. Scruggs. 2004. Identification of defects and impurities in chemical-vapor-deposited diamond through infrared spectroscopy. *J. Appl. Phys. 76*(4), 2466–2472.

McRae, E. G. 1979. Electronic surface resonances of crystals. *Rev. Mod. Phys. 51*(3), 541–568.

Michaelson, S. and A. Hoffman. 2006. Hydrogen bonding, content and thermal stability in nano-diamond films. *Diam. Relat. Mater. 15*(4–8), 486–497.

Michaelson, S., A. Hoffman, and Y. Lifshitz. 2006. Determination of vibrational modes in electron energy loss spectroscopy of polycrystalline diamond surfaces by isotopic exchange. *Appl. Phys. Lett. 89*, 223112.

Michaelson, S., Y. Lifshitz, and A. Hoffman. 2007b. High resolution electron energy loss spectroscopy of hydrogenated polycrystalline diamond: Assignment of peaks through modifications induced by isotopic exchange. *Diam. Relat. Mater. 16*(4–7), 855–860.

Michaelson, S., A. Stacey, R. Akhvlediani, S. Prawer, and A. Hoffman. 2010. High resolution electron energy loss spectroscopy surface studies of hydrogenated detonation nano-diamond spray-deposited films. *Surf. Sci. 604*(15–16), 1326–1330.

Michaelson, S., O. Ternyak, R. Akhvlediani, A. Hoffman, A. Lafosse, R. Azria, O. A. Williams, and D. M. Gruen. 2007a. Hydrogen concentration and bonding configuration in polycrystalline diamond films: From micro-to nanometric grain size. *J. Appl. Phys. 102*(11), 113516.

Michaelson, S., O. Ternyak, A. Hoffman, O. A. Williams, and D. M. Gruen. 2007c. Hydrogen bonding at grain surfaces and boundaries of nanodiamond films detected by high resolution electron energy loss spectroscopy. *Appl. Phys. Lett. 91*(10), 103104.

Millo, O., Y. Goldstein, A. Many, and J. I. Gersten. 1989. Resonant enhancement of the electron-energy-loss cross section due to excitation of surface plasmons. *Phys. Rev. B 39*(14), 9937–9946.

Mills, D. 1985. Studies of surface phonons by inelastic electron scattering: A brief review. *Surf. Sci. 158*(1–3), 411–421.

Morar, J. F., F. J. Himpsel, G. Hollinger, and J. L. Jordan. 1985. Observation of a C-1s core exciton in diamond. *Phys. Rev. Lett. 54*(17), 1960–1963.

Morar, J. F., F. J. Himpsel, G. Hollinger, J. L. Jordan, G. Hughes, and F. R. Mcfeely. 1986a. C 1s excitation studies of diamond(111). I. Surface core levels. *Phys. Rev. B 33*, 1340–1345.

Morar, J. F., F. J. Himpsel, G. Hollinger, J. L. Jordan, G. Hughes, and F. R. Mcfeely. 1986b. C 1s excitation studies of diamond(111). II. Unoccupied surface states. *Phys. Rev. B 33*, 1346–1349.

Müssig, H.-J., W. Stenzel, Y. Song, and H. Conrad. 1994. Hydrogen adsorption on Rh(110) studied by high resolution electron energy loss spectroscopy. *Surf. Sci. 311*(3), 295–307.

Nienhaus, H. and W. Mönch. 1995). Surface phonons in InP(110). *Surf. Sci. 328*(3), L561–L565.

Nithianandam, J. 1992. Synchrotron X-ray photoemission and reflectance study of the dipole forbidden diamond core exciton. *Phys. Rev. Lett. 69*(21), 3108–3111.

Okuyama, H., S. Thachepan, T. Aruga, T. Ando, and M. Nishijima. 2003. Overtones of the C-H stretch vibrations on C(001)(2×1)-H. *Chem. Phys. Lett. 381*(5–6), 535–540.

Omicron. 2000. *IB 500 User's Guide, version 1.0.*

Oura, K., V. G. Lifshits, A. A. Saranin, A. V. Zotov, and M. Katayama. 1999. Hydrogen interaction with clean and modified silicon surfaces. *Surf. Sci. Rep. 35*(1–2), 1–69.

Painter, G., D. Ellis, and A. Lubinsky. 1971. Ab initio calculation of the electronic structure and optical properties of diamond using the discrete variational method. *Phys. Rev. B* *4*(10), 3610–3622.

Palmer, R. E. and P. J. Rous. 1992. Resonances in electron scattering by molecules on surfaces. *Rev. Mod. Phys.* *64*(2), 383–440.

Pehrsson, P. E., T. W. Mercer, and J. A. Chaney. 2002. Thermal oxidation of the hydrogenated diamond (100) surface. *Surf. Sci.* *497*(1–3), 13–28.

Petaccia, L., L. Grill, M. Zangrando, and S. Modesti. 1999. Dispersion and intrinsic width of image resonances measured by resonant inelastic electron scattering: The α phase of Pb/Ge(111). *Phys. Rev. Lett.* *82*(2), 386–389.

Petrie, W. and J. Vohs. 1991. An HREELS investigation of the adsorption and reaction of formic acid on the (0001)-Zn surface of ZnO. *Surf. Sci.* *245*(3), 315–323.

Piantek, M., G. Schulze, M. Koch, K. J. Franke, F. Leyssner, A. Kruger, C. Navio, J. Miguel, M. Bernien, M. Wolf, W. Kuch, P. Tegeder, and J. I. Pascual. 2009. Reversing the thermal stability of a molecular switch on a gold surface: Ring-opening reaction of nitrospiropyran. *J. Am. Chem. Soc.* *131*(35), 12729–12735.

Pireaux, J. J., C. Gregoire, R. Caudano, M. R. Vilar, R. Brinkhuis, and A. J. Schouten. 1991. Electron-induced vibrational spectroscopy. A new and unique tool to unravel the molecular structure of polymer surfaces. *Langmuir* *7*(11), 2433–2437.

Pireaux, J. J., P. A. Thiry, R. Caudano, and P. Pfluger. 1986. Surface analysis of polyethylene and hexatriacontane by high resolution electron energy loss spectroscopy. *J. Chem. Phys.* *84*(11), 6452–6457.

Raval, R. 1995. Probing the nature of molecular chemisorption using RAIRS. *Surf. Sci.* *331-333*(Part 1), 1–10.

Reichart, P., G. Datzmann, A. Hauptner, R. Hertenberger, C. Wild, and G. Dollinger. 2004. Three-dimensional hydrogen microscopy in diamond. *Science* *306*(5701), 1537–1540.

Richardson, N. V. 1997. High resolution electron energy loss spectroscopy. *Curr. Opin. Solid St. M.* *2*(5), 517–524.

Ristein, J. 2006. Surface science of diamond: Familiar and amazing. *Surf. Sci.* *600*(18), 3677–3689.

Ristein, J., F. Maier, M. Riedel, M. Stammer, and L. Ley. 2001. Diamond surface conductivity experiments and photoelectron spectroscopy. *Diam. Relat. Mater.* *10*(3–7), 416–422.

Rizzi, A. 1997. Electron energy loss spectroscopy and its application in material science. *Fresen. J. Anal. Chem.* *358*, 15–24.

Rocca, M. 1995. Low-energy EELS investigation of surface electronic excitations on metals. *Surf. Sci. Rep.* *22*(1–2), 1–71.

Rocca, M. and F. Moresco. 1994. LEED fine structures and trapping phenomena in inelastic scattering of electrons off Ag(001) and Ag(110). *Phys. Rev. B* *50*(24), 18621–18628.

Rowe, J., R. Malic, E. Chaban, R. Headrick, and L. Feldman. 1990. Energy-dependent vibrational spectra of the Si(111)-B surface. *J. Electron Spectrosc.* *54-55*, 1115–1122.

Roy, D., M. Portail, and J. Layet. 2000. High resolution electron energy loss spectroscopy applied to a graphite surface modified by ion bombardment. *Surf. Rev. Lett.* *7*, 463–473.

Sainoo, Y., Y. Kim, T. Okawa, T. Komeda, H. Shigekawa, and M. Kawai. 2005. Excitation of molecular vibrational modes with inelastic scanning tunneling microscopy processes: Examination through action spectra of *cis*-2-butene on Pd(110). *Phys. Rev. Lett.* *95*(24), 246102.

Sanche, L. 1990. Low-energy electron scattering from molecules on surfaces. *J. Phys. B* *23*(10), 1597–1624.

Sandfort, B., A. Mazur, and J. Pollmann. 1995. Surface phonons of hydrogen-terminated semiconductor surfaces. II. The H:C(111)-(1x1) system. *Phys. Rev. B* *51*(11), 7150–7156.

Sandfort, B., A. Mazur, and J. Pollmann. 1996. Surface phonons of hydrogen-terminated semi-conductor surfaces. III. Diamond (001) monohydride and dihydride. *Phys. Rev. B* *54*(12), 8605–8615.

Schulz, G. J. 1973. Resonances in electron impact on diatomic molecules. *Rev. Mod. Phys.* *45*(3), 423–486.

Shen, Y. R. 1984. *The Principles of Non-linear Optics*, Chapter 25. New York, NY: Wiley.

Shin, D., N. Tokuda, B. Rezek, and C. E. Nebel. 2006. Periodically arranged benzene-linker molecules on boron-doped single-crystalline diamond films for DNA sensing. *Electrochem. Commun.* *8*(5), 844–850.

Smirnov, K. and G. Raseev. 2000. A tight-binding molecular dynamics study of vibrational spectra of H-covered diamond (100) surfaces. *Surf. Sci.* *459*(1–2), 124–134.

Sokolov, A., E. Kurmaev, S. Leitch, A. Moewes, J. Kortus, L. Finkelstein, N. Skorikov, C. Xiao, and A. Hirose. 2003. Band dispersion of MgB_2, graphite and diamond from resonant inelastic scattering. *J. Phys. Condens. Mat.* *15*, 2081–2089.

Stiegler, J., J. Michler, and E. Blank. 1999. An investigation of structural defects in diamond films grown at low substrate temperatures. *Diam. Relat. Mater.* *8*(2–5), 651–656.

Stipe, B., M. Rezaei, and W. Ho. 1999. Localization of inelastic tunneling and the determination of atomic-scale structure with chemical specificity. *Phys. Rev. Lett.* *82*(8), 1724–1727.

Sun, B., X. Zhang, and Z. Lin. 1993. Growth mechanism and the order of appearance of diamond (111) and (100) facets. *Phys. Rev. B* *47*(15), 9816–9824.

Swiderek, P. and E. Burean. 2007. Analysis of band broadening in vibrational high-resolution electron-energy-loss spectra of condensed methane. *J. Chem. Phys.* *127*(21), 214506.

Swiderek, P., C. Jäggle, D. Bankmann, and E. Burean. 2007. Fate of reactive intermediates formed in acetaldehyde under exposure to low-energy electrons. *J. Phys. Chem. C* *111*(1), 303–311.

Tadjeddine, A. 2000. Spectroscopic investigation of surfaces and interfaces by using infrared–visible sum and difference frequency generation. *Surf. Rev. Lett.* *7*(4), 423–436.

Tadjeddine, A. and A. Peremans. 1996. Vibrational spectroscopy of the electrochemical interface by visible infrared sum-frequency generation. *Surf. Sci.* *368*(1–3), 377–383.

Takaba, H., K. Kusafuka, M. Nishitani-Gamo, Y. Sato, T. Ando, J. Kubota, A. Wada, and C. Hirose. 2001. Vibrational sum-frequency observation of synthetic diamonds. *Diam. Relat. Mater.* *10*(9–10), 1643–1646.

Tautz, F. S. and J. A. Schaefer. 1998. Ultimate resolution electron energy loss spectroscopy at H/Si(100) surfaces. *J. Appl. Phys.* *84*(12), 6636–6643.

Thiry, P. A., M. Liehr, J. J. Pireaux, and R. Caudano. 1986. High resolution electron energy loss spectroscopy of insulators. *J. Electron Spectrosc.* *39*, 69–78.

Thiry, P. A., M. Liehr, J. J. Pireaux, and R. Caudano. 1987. Electron interaction mechanisms in high resolution electron energy loss spectroscopy. *Phys. Scripta* *35*(3), 368–379.

Thiry, P. A., M. Liehr, J. J. Pireaux, R. Sporken, R. Caudano, J. P. Vigneron, and A. A. Lucas. 1985. Vibrational study of the SiO_2/Si interface by high resolution electron energy loss spectroscopy. *J. Vac. Sci. Technol.* *3*(4), 1118–1121.

Thoms, B. and J. Butler. 1994. HREELS scattering mechanism from diamond surfaces. *Phys. Rev. B* *50*(23), 17450–17455.

Thoms, B. and J. Butler. 1995. HREELS and LEED of H/C (100): The 2×1 monohydride dimer row reconstruction. *Surf. Sci.* *328*(3), 291–301.

Thoms, B. D., P. E. Pehrsson, and J. E. Butler. 1994. A vibrational study of the adsorption and desorption of hydrogen on polycrystalline diamond. *J. Appl. Phys.* *75*(3), 1804–1810.

Ushizawa, K., M. N.-Gamo, Y. Kikuchi, I. Sakaguchi, Y. Sato, and T. Ando. 1999. Surface-enhanced Raman spectroscopic study of hydrogen and deuterium chemisorption on diamond (111) and (100) surfaces. *Phys. Rev. B* *60*(8), R5165–R5168.

Ushizawa, K., Y. Sato, T. Mitsumori, T. Machinami, T. Ueda, and T. Ando. 2002. Covalent immobilization of DNA on diamond and its verification by diffuse reflectance infrared spectroscopy. *Chem. Phys. Lett. 351*(1–2), 105–108.

Vattuone, L., M. Rocca, P. Restelli, M. Pupo, C. Boragno, and U. Valbusa. 1994. Low-temperature dissocation of O_2 on Ag(110): Surface disorder and reconstruction. *Phys. Rev. B 49*(7), 5113–5116.

Vilar, M. R., A. M. Botelho do Rego, A. M. Ferraria, Y. Jugnet, C. Nogues, D. Peled, and R. Naaman. 2008. Interaction of self-assembled monolayers of DNA with electrons: HREELS and XPS studies. *J. Phys. Chem. B 112*(23), 6957–6964.

Vilar, M. R., Y. Bouali, N. Kitakatsu, P. Lang, R. Michalitsch, F. Garnier, and P. Dubot. 1998. HREELS characterization of surfaces and interfaces in self-assembled molecular mono-layers. *Thin Solid Films 327–329*, 236–240.

Vilar, M. R. and A. B. do Rego. 2009. High-resolution electron energy loss spectroscopy as a probe of SAMs. *J. Electron Spectrosc. 172*(1–3), 27–35.

Völkening, S., K. Bedürftig, K. Jacobi, J. Wintterlin, and G. Ertl. 1999. Dual-path mechanism for catalytic oxidation of hydrogen on platinum surfaces. *Phys. Rev. Lett. 83*(13), 2672–2675.

Waclawski, B. J., D. T. Pierce, N. Swanson, and R. J. Celotta. 1982. Direct verification of hydrogen termination of the semiconducting diamond(111) surface. *J. Vac. Sci. Technol. 21*(2), 368–370.

Walker, I. C., A. Stamatovic, and S. F. Wong. 1978. Vibrational excitation of ethylene by electron impact: 1–11 eV. *J. Chem. Phys. 69*(12), 5532–5537.

Wandass, J. H. and J. A. Gardella. 1987. Vibrational spectroscopic analysis of Langmuir–Blodgett multilayers by HREELS: Sampling depth and scattering mechanisms. *Langmuir 3*(2), 183–188.

Wang, J., C. Y. Fan, K. Jacobi, and G. Ertl. 2001a. Adsorption and reaction of CO on RuO_2 (110) surfaces. *Surf. Sci. 481*(1–3), 113–118.

Wang, J., Y. Wang, and K. Jacobi. 2001b. Dissociation of CO on the Ru(11$\bar{2}$0) surface. *Surf. Sci. 488*(1–2), 83–89.

Wang, Y., A. Lafosse, and K. Jacobi. 2002. Stepwise dehydrogenation of NH_3 at the Ru(11$\bar{2}$0) surface. *Surf. Sci. 507-510*, 773–777.

Wang, Y. and C. Wöll. 2009. Chemical reactions on metal oxide surfaces investigated by vibrational spectroscopy. *Surf. Sci. 603*(10–12), 1589–1599.

Weng, X., P. Rez, and O. Sankey. 1989. Pseudo-atomic-orbital band theory applied to electron-energy-loss near-edge structures. *Phys. Rev. B 40*(8), 5694–5704.

Winn, M. D., M. Rassinger, and J. Hafner. 1997. Atomic and electronic structure of the diamond (100) surface: Reconstructions and rearrangements at high hydrogen coverage. *Phys. Rev. B 55*(8), 5364–5375.

Yang, W., O. Auciello, J. Butler, W. Cai, J. Carlisle, J. Gerbi, D. Gruen, et al. 2002. DNA-modified nanocrystalline diamond thin-films as stable, biologically active substrates. *Nat. Mater. 1*(4), 253–257.

8 Vibrational Excitations of Polyatomic Molecules

Petr Čársky and Roman Čurík

CONTENTS

8.1 INTRODUCTION

Vibrationally inelastic electron scattering calculations are much more difficult than those for elastic scattering. This applies particularly to polyatomic targets. In 2004, Itikawa characterizes the situation as that "a lot of problems are still to be solved" and "compared to the large number of theoretical studies of vibrational excitation of diatomic molecules, the number of theoretical works for polyatomic molecules is very limited" (Itikawa 2004). Whereas with very small molecules a sophisticated and rigorous approach can be employed (see, e.g., the application to the water molecule and CO_2 (Haxton et al. 2004, Rescigno et al. 2002)), for larger molecules some approximations must be assumed. As in the theory of photon vibrational spectra, the only manageable approach to polyatomics seems to be that based on Born–Oppenheimer approximation and harmonic force field. Obviously such simplifications are open to uncertainties. Primarily it is the nonadiabaticity at low electron energies, for which rotations and electronic–vibrational coupling cannot be disregarded. Also the use of harmonic force field has its limitations. It disregards coupling between vibrational modes, its use is limited to $1 \leftarrow 0$ transitions, not speaking about the effects of anharmonicity of potential curves along the vibrational coordinates. Still we considered it expedient to explore performance of such a simple theoretical model. Experimentally it is hard to determine cross sections for vibrational excitations with accuracy better than 30% (Kato et al. 2009) and if the problems noted need not be considered, the calculations of this type may be beneficial. Also some observations for larger molecules, particularly resonances, have been explained only by phenomenological models (Wong and Schulz 2003). Our objective was to develop a theory and software that would make rigorous *ab initio* calculations on inelastically vibrational electron scattering feasible on commonly used desktop computers for larger molecules than previously possible.

8.2 BASIC THEORY

The essence of our theoretical model is a two-channel approach expressed by a pair of coupled Lippmann–Schwinger equations:

$$T_{10} = U_{10} + U_{10}G_0T_{00} + U_{11}G_1T_{10} \tag{8.1}$$

$$T_{00} = U_{00} + U_{00}G_0T_{00} + U_{01}G_1T_{10}. \tag{8.2}$$

The subscripts 0 and 1, respectively, stand for the ground and vibrationally excited states of the molecule. In actual calculations, the Lippmann–Schwinger equation is solved in a chosen basis set and therefore hereafter the symbols T, U, and G will have a meaning of matrices. Hence, the transition moments (amplitudes) for elastic and vibrationally inelastic electron scattering are given as

$$f_{00} = -\frac{1}{4\pi}T_{00}, \tag{8.3}$$

$$f_{10} = -\frac{1}{4\pi}T_{10}, \tag{8.4}$$

where T_{00} and T_{10} are following matrix elements:

$$T_{00} = \langle \varphi_0^{out} \mid T \mid \varphi_0^{in} \rangle, \tag{8.5}$$

$$T_{10} = \langle \varphi_1^{out} \mid T \mid \varphi_0^{in} \rangle, \tag{8.6}$$

for functions φ representing the final states of the $N+1$ electron system before and after the electron collision. In the next section, we show how to make this general scheme computationally feasible.

8.3 DISCRETE MOMENTUM REPRESENTATION METHOD

Computationally we find it profitable to cast the system of Equations 8.5 and 8.6 into a super-matrix form

$$\begin{pmatrix} \mathbf{T}_{00} & \mathbf{T}_{01} \\ \mathbf{T}_{10} & \mathbf{T}_{11} \end{pmatrix} = \begin{pmatrix} \mathbf{U}_{00} & \mathbf{U}_{01} \\ \mathbf{U}_{10} & \mathbf{U}_{11} \end{pmatrix} + \begin{pmatrix} \mathbf{U}_{00} & \mathbf{U}_{01} \\ \mathbf{U}_{10} & \mathbf{U}_{11} \end{pmatrix} \cdot \begin{pmatrix} \mathbf{G}_0 & 0 \\ 0 & \mathbf{G}_1 \end{pmatrix} \cdot \begin{pmatrix} \mathbf{T}_{00} & \mathbf{T}_{01} \\ \mathbf{T}_{10} & \mathbf{T}_{11} \end{pmatrix}. \tag{8.7}$$

Then, this Lippmann–Schwinger equation is solved by the matrix inversion as is usual. Explicit integration of the UGT term is bypassed by means of numerical discretization

$$\int dE \int d\Omega\, f(E,\Omega) = \sum_p^{N_{rad}} w_p \sum_i^{N_{ang}} w_i f(E_p, \Omega_i).$$ (8.8)

The p and i indices run through abscissae of the radial and angular numerical quadratures, respectively. We therefore call this approach the Discrete Momentum Representation (DMR) method (Čurík and Čársky 2003, Čársky and Čurík 2006). Our aim was to develop a computational method for large polyatomic molecules, and as already mentioned in Section 8.1, the only manageable approach for this purpose is the harmonic approximation. Hence, the exact $N+1$ wave functions in Equations 8.5 and 8.6 for the initial and final states are represented as

$$\varphi_0^{in} = \Phi_0 \chi_0 \mathbf{k}_0^{in},$$ (8.9)

$$\varphi_1^{out} = \Phi_0 \chi_1 \mathbf{k}_1^{out}.$$ (8.10)

Here Φ_0 is the Slater determinant of the ground state of the molecule obtained by Hartree–Fock calculation. By this, we introduced the static-exchange approximation. Accounting for polarization–correlation effects will be shown in Section 8.7. Roots of the numerical quadrature (Equation 8.8) determine the plane-wave basis, in which the U, G, and T matrix elements are to be evaluated. Hence, the matrix elements of the interaction potential U become (Čurík and Čársky 2003)

$$\left[U_{00}\right]_{pi,qj} = \langle \chi_0 \mathbf{k}_{pi} \,|\, U \,|\, \chi_0 \mathbf{k}_{qj} \rangle$$ (8.11)

$$\left[U_{01}\right]_{pi,qj} = \langle \chi_1 \mathbf{k}_{pi} \,|\, U \,|\, \chi_0 \mathbf{k}_{qj} \rangle = \left[U_{10}\right]_{pi,qj},$$ (8.12)

$$\left[U_{11}\right]_{pi,qj} = \langle \chi_1 \mathbf{k}_{pi} \,|\, U \,|\, \chi_1 \mathbf{k}_{qj} \rangle.$$ (8.13)

For convenience, U is defined as a double of the interaction potential ($U = 2V$). For numerical stability and computational efficiency, the radial integration region $(0, \infty)$ is restricted (Čurík and Čársky 2003) to (k_{min}, k_{max}) by using the substitution

$$k = k_0 \frac{a+x}{a-x},$$ (8.14)

where the variable x is from the region $(-1,1)$. Choice of the parameter a determines the limiting values of k

$$k_{min} = k_0 \frac{a-1}{a+1},$$ (8.15)

$$k_{max} = k_0 \frac{a+1}{a-1}.$$ (8.16)

The optimum choice of a will be discussed in the next section. Once a is fixed, the matrix elements of Green's function become

$$G_l = -i\delta_{ij}\delta_{pq}\frac{\pi k_0 w_j}{2} \qquad l=0,1 \quad \text{for radial points } k_p = k_0 \text{ or } k_p = k_1,$$

$$G_l = \delta_{ij}\delta_{pq}\frac{2aw_p w_j k_p^2 k_0}{(a-x_p)^2(k_l^2-k_p^2)} l=0,1 \quad \text{for radial points } k_p \neq k_0 \text{ or } k_p = k_1. \tag{8.17}$$

In the static-exchange approximation, the U_{00} operator looks like a Hartree–Fock operator without the kinetic energy operator (see, e.g., Lane 1980) and hence for construction of its matrix elements we need to evaluate nuclear attraction, Coulomb, and exchange integrals in the plane-wave basis. In harmonic approximation, the matrix elements of the U_{10} operator are obtained as derivatives of the U_{00} matrix elements with respect to the vibrational coordinate of the ith vibrational mode (Nishimura and Itikawa 1995, 1996):

$$\langle \chi_1(\xi_i)\mathbf{k}_1 \mid U(\mathbf{R}_\alpha) \mid \chi_0(\xi_i)\mathbf{k}_0 \rangle = \frac{1}{\sqrt{2}}\frac{\partial}{\partial \xi_i}\langle \chi_0(\xi_i)\mathbf{k}_1 \mid U(\mathbf{R}_\alpha) \mid \chi_0(\xi_i)\mathbf{k}_0 \rangle. \tag{8.18}$$

Hence, in addition to nuclear attraction, Coulomb, and exchange integrals, we also need to evaluate their derivatives with respect to atomic coordinates of the molecule. Evaluation of Coulomb and exchange integrals is a bottleneck of scattering calculations not in the DMR approach alone (see, e.g., a review on variational Schwinger method (Winstead and Mckoy 1995)). Therefore, we paid great attention to make their evaluation efficient. In Sections 8.5 and 8.6, we described in some detail how this can be done without compromising the rigor of calculations and accuracy of integrals and their derivatives. Memory limitations with modern computers are less topical than the computer time limitations. Still we considered it useful to reduce the memory requirements by super-matrices appearing in Equation 8.7. If the size of the plane-wave basis is N, the size of super-matrices is $2N \times 2N$. In the two-channel approach, represented by Equation 8.7, the cross sections for elastic electron scattering differ slightly for different vibrational modes. However, the effect of vibrational channels on the elastic scattering is small, which is in accordance with the general experience that cross sections for elastic scattering can be calculated routinely with a single-channel approach:

$$\mathbf{T}_{00} = \mathbf{U}_{00} + \mathbf{U}_{00}\mathbf{G}_0\mathbf{T}_{00}. \tag{8.19}$$

This suggests that the $\mathbf{T}_{00}$ matrix so obtained can be inserted in Equation 8.2 by defining the $\mathbf{A}_{11}$ and $\mathbf{B}_{10}$ matrices as (Čársky and Čurík 2008)

$$\mathbf{A}_{11} = (\mathbf{1}-\mathbf{U}_{11}\mathbf{G}_1)^{-1}, \tag{8.20}$$

$$\mathbf{B}_{10} = \mathbf{U}_{10}(\mathbf{1}+\mathbf{G}_0\mathbf{T}_{00}), \tag{8.21}$$

and the required matrix of amplitudes for each vibrational mode can be obtained as

$$\mathbf{T}_{10} = \mathbf{A}_{11}\mathbf{B}_{10}. \tag{8.22}$$

The size of all matrices in Equations 8.19 through 8.22 is only $N \times N$. The test calculations on integral and differential cross section showed (Čársky and Čurík 2008) that the error introduced by this approximation was marginal.

8.4 NUMERICAL QUADRATURE

A standard evaluation of the UGT and UGU terms in *ab initio* calculations of electron scattering by polyatomic molecules is numerical quadrature (see, e.g., Winstead and McKoy 1995, Rescigno et al. 1995). However, in contrast to electronic structure calculations, there are no universal tools, such as standard basis sets, that could be applied to different target molecules. The problem of numerical quadrature is that the radial points corresponding to high momenta cannot be excluded from the integration. These high-momentum radial points require very high angular quadratures, making the calculations cumbersome. We and the others have solved (Polášek et al. 2000, Čársky and Čurík 2006) this problem by a stepwise extension of the size of quadrature, both radial and angular, hoping that the convergence in calculated cross sections will be achieved before the capacity of available computational resources would be exhausted. Success of such an approach depended on the particular molecular target. However, the procedure was time consuming and required considerable human effort. Therefore we developed (Čársky 2010a) a practical and well-defined procedure for obtaining a moderately large quadrature for a particular molecule ensuring sufficient accuracy. Our search for the optimum numerical quadrature was based on the assumption that the integrand in the UGT term behaves similarly as the integrand in the second-Born terms, and that the optimum numerical quadrature found for the latter may also be used for the former. It was also assumed that for the purpose of examination of the numerical quadrature, it is justifiable to calculate the second-Born terms in the static (st) approximation because it requires larger numerical quadratures than the exchange part of the potential. Hence, the task is to find the optimum quadrature of the principal value of the integral

$$P \int dx \, d\Omega \, f(x,\Omega) =$$

$$\sum_{pi} w_p w_i \frac{k_p^2 \langle \mathbf{k}_1 | U_{st} | \mathbf{k}_{pi} \rangle \langle \mathbf{k}_{pi} | U_{st} | \mathbf{k}_2 \rangle - k_0^2 \langle \mathbf{k}_1 | U_{st} | \mathbf{k}_{0i} \rangle \langle \mathbf{k}_{0i} | U_{st} | \mathbf{k}_2 \rangle}{k_0^2 - k_p^2}$$

$$\times \frac{2k_0}{(1-x_p)^2} \tag{8.23}$$

for which we have used the substitution (8.14). First, we concentrated on the radial quadrature. We have used the Lebedev angular quadrature with 5810 points (Lebedev and Laikov 1999). This size should guarantee that the integration over angular coordinates is of sufficient accuracy for any radial point x. We assumed nine values for k_{max}, 10^5, 100, 20, 15, 12, 10, 8, 6, and 5 a.u., and for each integration range (k_{min}, k_{max}) we integrated $f(x)$ function numerically by the Simpson method for

1000 radial points. Integration was performed for selected pairs of $\mathbf{k}_1$ and $\mathbf{k}_2$ vectors for a training set of eight molecules. The k_{max} value was stepwise lowered as far as the error in the squared principal values did not exceed 1%. The values of a (Equation 8.14) thus obtained were in the range from 1.07 to 1.27. In a similar way, also the optimum numbers of radial and angular points were determined (Čársky 2010a). Once this procedure is coded, it gives an optimum quadrature for a particular molecule, which in spite of its modest size guarantees sufficient accuracy for the UGT term.

8.5 EVALUATION OF COULOMB INTEGRALS

With the standard quantum chemical software, Hartree–Fock calculations are done almost exclusively with Gaussian basis sets. Hence, the *ab initio* treatments of electron scattering by polyatomic molecules need evaluation of Coulomb integrals in a mixed Gaussian and plane-wave basis. The formulas for this type of integrals have been available in the literature for decades (Ostlund 1975, Rescigno et al. 1975, Watson and Mckoy 1979, Polášek and Čársky 2002, Füsti-Molnar and Pulay 2002). However, their applications to somewhat larger molecules is becoming too demanding computationally. This problem was overcome in quantum chemistry by using the density fitting and its use contributed greatly to extend applicability of quantum chemical *ab initio* calculations to larger molecules (see, e.g., Weigend et al. 1998, Schütz et al. 2004). We have shown (Čársky 2007, 2009) that density fitting can also be beneficial for evaluation of integrals in a mixed Gaussian and plane-wave basis. The integrals in question are of the type

$$I \equiv \langle \mathbf{k}_1 \mid V_s \mid \mathbf{k}_2 \rangle, \tag{8.24}$$

where $\mathbf{k}_1$ and $\mathbf{k}_2$ are plane-wave functions $\langle \mathbf{r} \mid \mathbf{k}_1 \rangle = \exp(i\mathbf{k}_1 \cdot \mathbf{r})$ and $\langle \mathbf{r} \mid \mathbf{k}_2 \rangle = \exp(i\mathbf{k}_2 \cdot \mathbf{r})$ and V_s is so-called static potential defined by means of the charge density ρ as

$$V_s = \int d\mathbf{r}' \, \frac{\rho(\mathbf{r}')}{\mid \mathbf{r} - \mathbf{r}' \mid}. \tag{8.25}$$

Denote the difference of the $\mathbf{k}$ vectors (called momentum transfer vector) as

$$\mathbf{K} = \mathbf{k}_2 - \mathbf{k}_1 \tag{8.26}$$

to express the integral (8.24) as

$$I = \iint d\mathbf{r} \, d\mathbf{r}' \frac{e^{i\mathbf{K} \cdot \mathbf{r}} \rho(\mathbf{r}')}{\mid \mathbf{r} - \mathbf{r}' \mid}. \tag{8.27}$$

By means of the Gaussian transform, we obtain the known basic working formula

$$I = \frac{4\pi}{K^2} \int d\mathbf{r} \, e^{i\mathbf{K} \cdot \mathbf{r}} \rho(\mathbf{r}). \tag{8.28}$$

The essence of the density fitting, as it was first used in the density functional theory (Baerends et al. 1973), is the expansion of the electron density in an auxiliary basis set

$$\rho(\mathbf{r}) = \sum_{\alpha} c_{\alpha}\alpha(\mathbf{r}),$$ (8.29)

instead of the traditional use of the atomic basis set

$$\rho(\mathbf{r}) = \sum_{\mu\nu} D_{\mu\nu}\mu(\mathbf{r})\nu(\mathbf{r}).$$ (8.30)

If the size of the atomic basic set is N and the size of the auxiliary basis set is n, then the factor of time saving is N^2/n. Use of density fitting for Coulomb integrals in a mixed Gaussian and plane-wave basis set is straightforward by substituting electron density in formula (8.28) by Equation 8.29. However, the integral in Equation 8.28 as a function of K has a singularity at $K = 0$. For small K, say for $K < 0.3$ a.u, it is therefore preferable to calculate the integrals in a rigorous manner. Auxiliary basis set $\alpha(\mathbf{r})$ compatible with the molecular Gaussian basis set can be found in the literature (see e.g., Eichkorn et al. 1995). The expansion coefficients c_{α} are obtained as described in the literature (Eichkorn et al. 1995);

$$\sum_{\beta}^{n} (\alpha \mid \beta) c_{\beta} = \sum_{\mu\nu} (\alpha \mid \mu\nu)D_{\mu\nu},$$ (8.31)

$$c_{\alpha} = \sum_{\beta}^{n} (\alpha \mid \beta)^{-1}\gamma_{\beta},$$ (8.32)

where

$$\gamma_{\alpha} = \sum_{\mu\nu} (\alpha \mid \mu\nu)D_{\mu\nu}.$$ (8.33)

The integrals (8.28) can now be expressed as

$$I = \frac{4\pi}{K^2}\int d\mathbf{r}\, e^{i\mathbf{K}\cdot\mathbf{r}} \sum_{\alpha} c_{\alpha}\alpha(\mathbf{r}).$$ (8.34)

For the derivative of the integral with respect to nuclear coordinate A_x, we have

$$\frac{\partial}{\partial A_x}I = \frac{4\pi}{K^2}\int d\mathbf{r}\, e^{i\mathbf{K}\cdot\mathbf{r}} \sum_{\alpha} \left[\frac{\partial c_{\alpha}}{\partial A_x}\alpha(\mathbf{r}) + c_{\alpha}\frac{\partial\alpha(\mathbf{r})}{\partial A_x}\right].$$ (8.35)

The derivatives of the expansion coefficients, $c_{\alpha}' \equiv \partial c_{\alpha}/\partial A_x$ are obtained from differentiation of Equation 8.31:

$$\sum_{\beta}^{n} \left[(\alpha \mid \beta)' c_{\beta} + (\alpha \mid \beta)c_{\beta}'\right] = \sum_{\mu\nu} \left[(\alpha \mid \mu\nu)' D_{\mu\nu} + (\alpha \mid \mu\nu)D_{\mu\nu}'\right].$$ (8.36)

Define the elements of the A matrix and the B vector as

$$A_{\alpha\beta} = (\alpha \mid \beta)$$ (8.37)

$$B_{\alpha} = \sum_{\mu\nu} \left[(\alpha \mid \mu\nu)' D_{\mu\nu} + (\alpha \mid \mu\nu)D_{\mu\nu}'\right] - \sum_{\beta}^{n} (\alpha \mid \beta)c_{\beta}.$$ (8.38)

The vector of derivatives of expansion coefficients can be expressed in a compact form as

$$\mathbf{C}' = \mathbf{A}^{-1}\mathbf{B}. \tag{8.39}$$

Determination of c coefficients and their derivatives requires additional evaluation of two-electron integrals of the types $(g_1|g_2)$ and $(g_1|g_3g_4)$ and their derivatives, where g_1 and g_2 stand for Gaussians from the auxiliary basis set and g_3 and g_4 are Gaussians from the atomic basis set. Computer time for the evaluation of these integrals is negligible compared to the computer time of the complete electron scattering calculation. In actual scattering calculations, the use of density fitting accelerated the evaluation of Coulomb integrals by one to two orders of magnitude.

8.6 EVALUATION OF EXCHANGE INTEGRALS

Evaluation of exchange integrals is a harder problem than evaluation of Coulomb integrals. In a mixed Gaussian and plane-wave basis, they have the following form:

$$(g_ik' \mid g_jk) = \iint dr_1\, dr_2\, e^{-ik'\cdot r_1}\, (x_1 - A_x)^{m_x}\, (y_1 - A_y)^{m_y}\, (z_1 - A_z)^{m_z}\, e^{-\alpha(r_1 - A)^2}$$

$$\times \left(\frac{1}{r_{12}}\right) e^{-ik\cdot r_2}\, (x_2 - B_x)^{n_x}\, (y_2 - B_y)^{n_y}\, (z_2 - B_z)^{n_z}\, e^{-\beta(r_2 - B)^2}, \tag{8.40}$$

and they can be calculated in different ways: By means of Bessel functions (Watson and McKoy 1979), by means of Shavitt functions with a complex argument (Ostlund 1975, Čársky and Polášek 1998a, Čársky and Reschel 1998), or by means of complex Rys polynomials (Čársky and Polášek 1998b). In either way, their evaluation is time consuming (Winstead and McKoy 1995). If calculation is to be done for a somewhat larger molecule, the only feasible way is to use a supercomputer (Winstead and McKoy 1995) or to evaluate the exchange integrals in a semiempirical manner (Gianturco and Scialla 1987). In the electronic structure theory, a considerable progress was achieved when it was found that the Shavitt functions $F_n(x)$ (Shavitt 1963) need not be evaluated for each two-electron integral but that they can be obtained with a required accuracy by interpolation from a precalculated table. We used the same idea for an efficient evaluation of exchange integrals (8.40) in a mixed Gaussian and plane-wave basis. For elastic scattering, the problem is not so critical because calculations on even large target systems can be done by existing software either on supercomputers (Winstead and McKoy 1995) or by expressing the exchange part of the potential in a semiempirical manner (Gianturco and Scialla 1987). Our ultimate goal was, however, efficient *ab initio* calculations of vibrationally inelastic scattering for which also derivatives of integrals with respect to nuclear coordinates are needed. Moreover, our aim was also to develop a software for calculations of vibrationally inelastic electron scattering that would be feasible on a single-processor desktop computer.

    Shavitt functions used in electronic structure calculations are real monotonous functions defined only for nonnegative real argument x. Their behavior is very favorable for interpolation from precalculated tables of $F_n(x)$ values for a preselected grid (Gill et al. 1991). With exchange integrals 8.40 in a mixed Gaussian and plane-wave basis, interpolation is more complicated because the Shavitt functions $F_n(z)$ needed for their evaluation are complex functions with a complex argument $z = x + iy$. Hence, instead of a one-dimensional table, the precalculated tables with $F_n(z)$ have to be two-dimensional. Plot of $F_0(z)$ in Figure 8.1 reveals the next set of complications.

In contrast to real $F_n(x)$ functions, complex $F_n(z)$ functions are also defined for negative x, they are oscillatory with respect to y, and in absolute value they depend strongly on x going to infinity as x goes to minus infinity. This indicates that a grid used for precalculated tables must be rather dense. Even worse the actual scattering calculations showed that both x and y range from very large negative values to very large positive values. At the first sight, the situation seemed to be hopeless because of memory limitation of computers. Still such precalculated tables for integrals and their derivatives can be done (Čársky 2010b) with the memory requirement of <1GB which is tolerable for desktop computers available nowadays. Even if the interpolation and grids of our tables may seem to be rough, they guarantee that the $F_n(z)$ values are accurate to within 0.01%. For electron scattering calculations, this approximate way of evaluation of exchange integrals introduces an error in the calculated cross sections which is negligible when compared with other sources of errors such as numerical quadrature of the UGT term or density fitting in the evaluation of Coulomb integrals. However, for other than electron scattering calculations, a more accurate evaluation of exchange may be required and therefore a more sophisticated interpolation must be applied.

Acceleration of calculations of exchange integrals was not as dramatic as with Coulomb integrals, but complete calculations for all vibrational modes showed that the total CPU time was reduced by a factor of 5–9. The calculations can be relatively easily parallelized on now commonly available four-core Opteron machines, which make such calculations feasible for any laboratory. With the parallelized code, the computer time for the calculation on benzene with 8126 quadrature points was reduced from 20 to 6 days.

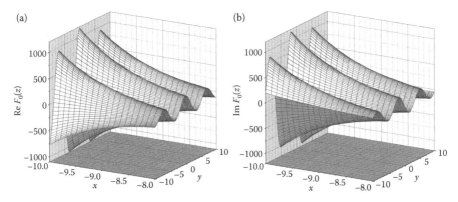

FIGURE 8.1 Plot of the real (a) and imaginary (b) components of the complex Shavitt function $F_0(x)$. (Reprinted with permission from Čársky, P. 2010b. *J. Phys. B: At. Mol. Opt. Phys.* *43*, 175204. Copyright 2010 IOP Publishing Ltd.)

8.7 POLARIZATION EFFECTS

Accounting for polarization–correlation effects for a somewhat larger molecule is difficult, particularly for vibrationally inelastic scattering, and it seems that the only manageable approach is that based on density functional theory. Therefore, we implemented a model based on a short-range DFT correlation potential V_c that is, smoothly connected to asymptotic polarization form

$$V_p(\mathbf{r}) = -\frac{1}{2r^6}\sum_{ij}\alpha_{ij}x_ix_j, \tag{8.41}$$

where α_{ij} is a dipole polarizability tensor of the target molecule. The details about the available correlation potentials are given in Section 2.3 and they shall not be repeated here. It suffices to say we have implemented several DFT correlation functionals widely used in computational quantum chemistry, namely a local functional (LDA) of Perdew and Zunger (1981), and two nonlocal functionals: PBE functional of Perdew et al. (1992) and LYP functional of Lee et al. (1988).

In the remaining part of this section, we shall describe the procedure we use to make a smooth connection between the short-range part V_c and the long-range asymptote V_p in Equation 8.41. As the first step, we expand the general tensor form (8.41) into partial waves

$$V_p(\mathbf{r}) = \sum_{l=0}^{2}\sum_{m=-l}^{l}v_l^m(r)S_l^m(\hat{\mathbf{r}}), \tag{8.42}$$

where $S_l^m(\hat{\mathbf{r}})$ are normalized real spherical harmonics:

$$S_l^m(\vartheta,\varphi) = \left[\frac{2l+1}{2\pi(1+\delta_{0m})}\frac{(l-m)!}{(l+m)!}\right]^{\frac{1}{2}}P_l^m(\cos\vartheta)\begin{cases}\cos m\varphi; & m\geq 0\\ \sin m\varphi; & m<0\end{cases}. \tag{8.43}$$

The radial functions $v_l^m(r)$ in the Equation 8.42 can be obtained analytically as

$$v_l^m(r) = -\frac{\alpha_l^m}{2r^4}, \tag{8.44}$$

where α_l^m are the irreducible components of the Cartesian tensor of rank 2 (Weissbluth 1978, p. 174):

$$\alpha_0^0 = \frac{\sqrt{4\pi}}{3}(\alpha_{xx}+\alpha_{yy}+\alpha_{zz})$$

$$\alpha_1^0 = \alpha_1^1 = \alpha_1^{-1} = 0$$

$$\alpha_2^0 = \sqrt{\frac{4\pi}{5}}\left(\alpha_{zz}-\frac{\alpha_{xx}+\alpha_{yy}+\alpha_{zz}}{3}\right)$$

$$\alpha_2^1 = -2\sqrt{\frac{4\pi}{15}}\alpha_{xz}$$

$$\alpha_2^{-1} = -2\sqrt{\frac{4\pi}{15}}\alpha_{yz}$$

$$\alpha_2^2 = 2\sqrt{\frac{4\pi}{15}}\frac{\alpha_{xx} - \alpha_{yy}}{2} \tag{8.45}$$

$$\alpha_2^{-2} = 2\sqrt{\frac{4\pi}{15}}\alpha_{xy}$$

The short-range part V_c is then separated into two orthogonal angular subspaces:

$$V_c(\mathbf{r}) = \sum_{l=0}^{2}\sum_{m=-l}^{l} w_l^m(r)S_l^m(\hat{\mathbf{r}}) + W_0(\mathbf{r}). \tag{8.46}$$

All the terms on the left and right side of Equation 8.46 decay exponentially. As the DFT form of the short-range correlation potential $V_c(r)$ is evaluated numerically, the radial functions $w_l^m(r)$ are obtained by a numerical angular projection. It follows that $W_0(r)$ contains only partial components with $l > 2$.

Having both (short- and long-range) interactions split into the partial waves, the connection procedure is readily available. We connect each partial wave of $v_l^m(r)$ and $w_l^m(r)$ independently up to $l = 2$ forming six crossing points R_l^m. Three short-range functions $w_l^m(r)$ with $l = 1$ decay exponentially as they do not have the long-range counterparts $v_l^m(r)$ to connect to. It is a consequence of symmetry of the polarizability tensor. Therefore, we can define the connected radial functions

$$u_l^m(r) = \begin{cases} w_l^m(r) & \text{for } r < R_l^m \\ v_l^m(r) & \text{for } r \geq R_l^m. \end{cases} \tag{8.47}$$

The total correlation–polarization potential V_{cp} is then evaluated by the following formula

$$V_{cp}(\mathbf{r}) = \sum_{l=0}^{2}\sum_{m=-l}^{l} u_l^m(r)S_l^m(\hat{\mathbf{r}}) + W_0(\mathbf{r}). \tag{8.48}$$

A visual comparison of the spherical components $u_0^0(r)$ defined by Equation 8.47 is displayed in Figure 8.2 for two polyatomic molecules. In this case, cyclic C_3H_6 exhibits slightly stronger correlation in the core area of the molecule while the open ring C_3H_8 allows stronger correlation in the low-density outskirt regions of the molecule resulting in a stronger asymptotic polarizability. As far as the differences among different functionals are concerned, the LYP gradient correction to the LDA potential results in a stronger correlation especially on the peripheral low-density region. This feature is valid for both presented molecules and it seems to extend even for more unpublished results of polyatomic molecules.

The final step needed to obtain the momentum-space matrix elements (Equations 8.11 through 8.13) is a Fourier transform (FT) integral:

$$\langle \mathbf{k}_1 | V_{cp} | \mathbf{k}_2 \rangle = \frac{1}{(2\pi)^3}\int d\mathbf{r}\, V_{cp}(\mathbf{r})e^{i(\mathbf{k}_2 - \mathbf{k}_1)\cdot\mathbf{r}}. \tag{8.49}$$

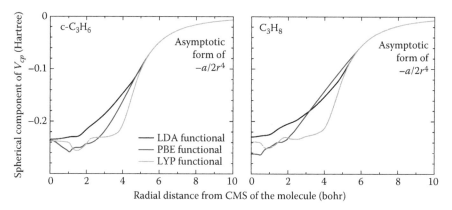

FIGURE 8.2 (**See color insert.**) Visual comparison of different correlation functionals used in the present study. For simplicity, only the dominant spherical components $u_0^0(r)$ of Equation 8.47 are displayed. Left panel shows the data for the cyclopropane molecule, right panel for the propane molecule. (Reprinted with permission from Čurík, R. and M. Šulc 2010. *J. Phys. B: At. Mol. Opt. Phys.* **43**, 175205. Copyright 2010 IOP Publishing Ltd.)

The integral in Equation 8.49 is typically solved by a discrete fast Fourier transform (FFT) method. This is achieved by bounding the integral in a three-dimensional rectangular volume and by evaluating the argument on a three-dimensional rectangular grid. In our implementation, we used a cubic volume. We observed a cube size of several hundred bohr is needed to achieve a sufficient accuracy of the resulting integrals for the small $\mathbf{k}_2 - \mathbf{k}_1$ vectors. The cause of the slow size convergence is the long-range nature (8.41) of the V_{cp} potential. However, with such a large integration volume, it becomes difficult to sample the potential with a sufficient density in the area of the molecule, where the V_{cp} changes rapidly.

In order to solve this dilemma, we subtract and add a well-behaved analytical function that cancels the V_{cp} potential in the long range. We also require this function to have an analytical FT form:

$$V_{cp} = \underbrace{\left[V_{cp} + \frac{1}{2(r^6 + a^6)} \sum_{i,j}^{3} \alpha_{ij} x_i x_j \right]}_{\text{Numerical FFT } V_{cp}^n} - \underbrace{\frac{1}{2(r^6 + a^6)} \sum_{i,j}^{3} \alpha_{ij} x_i x_j}_{\text{Analytical FT } V_{cp}^a} \qquad (8.50)$$

The parameter a is chosen to be sufficiently small in order to achieve the long-range cancellation in the V_{cp}^n term. However, if the cutoff a is chosen too small, it may produce a rapidly changing behavior at the origin. This would lead to higher density of the FFT grid for the numerical integration of the first term in Equation 8.50. To compromise, we have found the procedure very robust and accurate for a being anywhere between 5% and 15% of the FFT cube size. Evaluation of the FT integral for the V_{cp}^a term is described in Čurík and Šulc (2010). Here, we present only the final results:

$$\frac{1}{(2\pi)^3} \sum_{i,j}^{3} \alpha_{ij} \int d\mathbf{r} \, e^{i\mathbf{k}\cdot\mathbf{r}} \frac{x_i x_j}{2(r^6 + a^6)} = \frac{1}{(2\pi)^3} \sum_{ij}^{3} \alpha_{ij} G_{ij}(\mathbf{k}), \qquad (8.51)$$

with

$$G_{ij}(\mathbf{k}) = \frac{F'(k)}{k}\delta_{ij} + \left[\frac{F''(k)}{k^2} - \frac{F'(k)}{k^3}\right]k_i k_j, \qquad (8.52)$$

and

$$F(k) = -\frac{\pi^2}{3a^4 k}e^{-ka/2}\left[e^{-ka/2} + \sqrt{3}\sin\left(ka\frac{\sqrt{3}}{2}\right) - \cos\left(ka\frac{\sqrt{3}}{2}\right)\right]. \qquad (8.53)$$

For practical implementation, one also needs the limit at the origin

$$\lim_{\mathbf{k}\to 0} G_{ij}(\mathbf{k}) = \frac{2\pi^2}{9a}. \qquad (8.54)$$

8.8 COMPUTATIONAL DETAILS

The first step in our approach is a regular Hartree–Fock calculation on a target molecule by any available quantum chemical software. The saved density matrix, derivatives of the density matrix with respect to atomic coordinates, derivatives of the dipole moment matrix elements, and normal modes are then used as an input for the code for scattering. We have found that for static-exchange calculations the calculated cross sections do not depend much on the size of Gaussian basis set used for the target, or at least, they depend on it less than on the number and selection of quadrature points, density fitting, and polarization effects. We are using the valence-shell DZP basis set (Dunning and Hay 1977) as our standard basis, which by its size seems to be sufficient for most applications. As an auxiliary basis, we are using the decontracted SVP (Eichkorn et al. 1995) basis set (8s3p3d1f/6s2p). Additional calculations must be performed if polarization-correlation is to be included in the interaction potential. Our DFT-type approach needs a tensor of polarizabilities and their derivatives with respect to atomic coordinates. Such input data should be obtained by a post-Hartree–Fock method with a large basis set. Integration of the UGT terms is done with the Gauss–Legendre radial and Lebedev angular quadratures, as described in Section 8.4.

The system of Equations 8.20 through 8.22 or 8.7 has to be solved separately for each vibrational mode which means recalculation of integrals for each vibrational mode. For molecules with many vibrational modes this means an enormous increase in the computer time. This problem too can be avoided. A typical highest vibrational frequency is about 3000 cm^{-1} which corresponds to a narrow range of k values for the outgoing electron (about 0.017 a.u.). Test calculations showed that the integrals of the $(k|U|k_{out})$ type change linearly as k_{out} is increased from k_0 to the value corresponding to the highest vibrational frequency. An example of this linear dependence is shown in Figure 8.3. Such a linear dependence may also be expected with the derivatives of integrals, which suggests that it is enough to evaluate integrals and their derivatives only for k_0 and the highest vibrational mode and to obtain integrals and their derivatives for other modes by interpolation (Čársky 2010b). This approximation contributed to a further considerable acceleration of calculations and made them feasible for molecules of the size of benzene and uracil on Opteron-type computer.

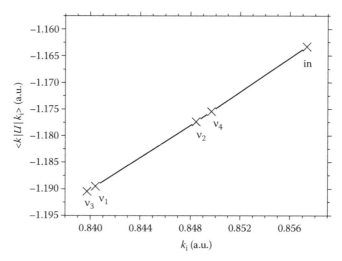

FIGURE 8.3 Plot of $(k_0|U|k_{out})$ integrals for four vibrational modes of methane and the incident electron energy of 10 eV. The k_{out} vector was held parallel to the k_0 vector. (Reprinted with permission from Čársky, P. 2010b. *J. Phys. B: At. Mol. Opt. Phys. 43*, 175204. Copyright 2010 IOP Publishing Ltd.)

For comparison with experimental electron energy loss spectra, the plots were done in the following way. The bands were centered at positions of experimentally observed frequencies (http://webbook.nist.gov/chemistry/). For all bands, the Gaussian form was assumed. The heights of bands were set at the values of calculated differential cross sections and half-widths of bands were adjusted to reproduce best the shapes of the observed bands.

8.9 APPLICATION TO POLYATOMIC MOLECULES

The selected examples concern molecules that by their size cannot be viewed as large, but availability of reliable experimental data makes it profitable to use them for the assessment of performance of the DMR method.

8.9.1 METHANE

Vibrational excitation of methane by electron impact is important in a wide variety of technological and atmospheric applications. As a result of interests from these various fields, there has been considerable amount of experimental and theoretical work devoted to the study of electron–methane collisions in the last two decades. While calculation and measurement of elastic scattering cross sections is a well-established task, attempts at a theoretical analysis of vibrational excitation cross sections are rather rare, and the agreement between experimental and theoretical data is not as satisfactory as could be expected for such a simple molecule. Also the experimental data coming from different laboratories were not in a satisfactory agreement. Therefore, we decided to undertake a joint theoretical and experimental study (Čurík

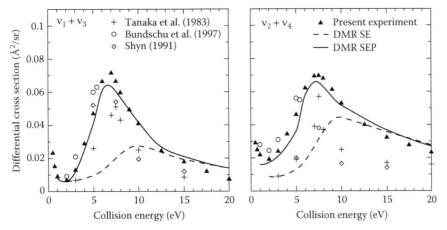

FIGURE 8.4 Rotationally summed $0 \to 1$ differential cross section for methane as a function of the collision energy. The scattering angle is fixed at $\vartheta = 90°$. Our SE and SEP results are compared with the experimental data. The left panel is for the composite stretching mode $v_1 + v_3$ and the right panel displays data for the composite deformation mode $v_2 + v_4$. (Reprinted with permission from Čurík, R., P. Čársky, and M. Allan. 2008. *J. Phys. B: At. Mol. Opt. Phys. 41*, 115203. Copyright 2008 IOP Publishing Ltd.)

et al. 2008) for a better understanding of the problem. We selected from that study two figures. From the first of them (Figure 8.4) it is seen that, in contrast to previous experiments, the latest experimental data are in agreement with the calculations and that at lower energies agreement with experiment cannot be achieved unless polarization effects are included. The second figure (Figure 8.5) shows that with polarization

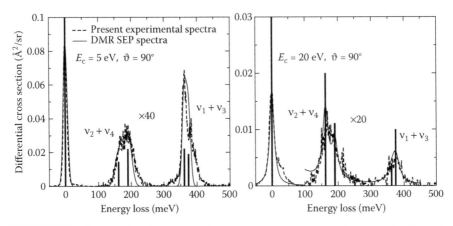

FIGURE 8.5 Computed and measured electron energy-loss spectra of methane for the scattering angle $\vartheta = 90°$. The left panel is for the collision energy $E_c = 5$ eV and the right panel is for $E_c = 10$ eV. Vertical bars represent the actual value of the differential cross section for the particular mode. (Reprinted with permission from Čurík, R., P. Čársky, and M. Allan. 2008. *J. Phys. B: At. Mol. Opt. Phys. 41*, 115203. Copyright 2008 IOP Publishing Ltd.)

effects included, the experimental electron energy loss spectrum can be well reproduced even at low energies.

8.9.2 CYCLOPROPANE

Vibrationally inelastic electron scattering by cyclopropane was studied experimentally (Allan and Andrič 1996, Čurík et al. 2011) and theoretically (Čurík and Gianturco 2002). The most marked feature observed by experiments has to do with the excitation of v_3 vibration, the C–C ring stretching located about 1188 cm^{-1} (147 meV) (Schimanouchi 1972). This finding was confirmed by Čurík and Gianturco (2002) where the authors used SEP interaction in the single-center expansion model. They exploited the method by calculating full-symmetric modes only. With these limitations, they obtained a qualitative agreement of the resonant position but the energy dependence of cross section differed above the resonant energies. In Figure 8.6, we attempt to explain the origin of this discrepancy. Our SEP calculations of vibrational excitation of v_3 mode reproduce the height and width of the 5.5 eV resonance very well. However, the broad shape centered around 9–10 eV visible in the experimental data seems to be due to a vibrational excitation of HCH twisting mode v_{13} that has accidentally identical energy-loss energy of 147 meV (Schimanouchi 1972). Therefore these two vibrational modes appear to be experimentally undistinguishable and the measured cross section reflects a sum of two probabilities of excitation of the two different modes.

In Figure 8.7, we also show an impact of correlation–polarization forces on the cross section at a resonant energy of 5.5 eV. The exact static-exchange model underestimates resonant cross section by a factor of 3. A good agreement with experiment

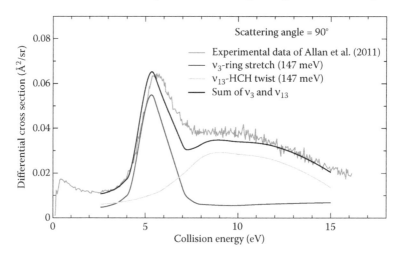

FIGURE 8.6 (**See color insert.**) Rotationally summed $0 \rightarrow 1$ differential cross section of cyclopropane as a function of the collision energy. Scattering angle is fixed at $\vartheta = 90°$. Blue line shows present calculations of the resonant vibrational excitation of v_3 mode while the green line displays the excitation of HCH twisting mode v_{13}. The sum of these two cross sections (black line) is compared to experimental data. (Adapted from Čurík, R., P. Čársky, and M. Allan. 2011. In preparation.)

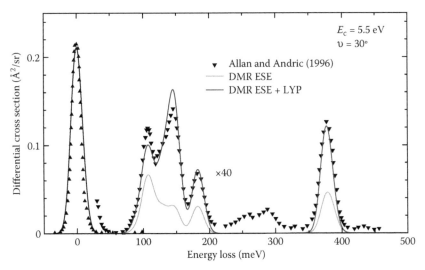

FIGURE 8.7 (**See color insert.**) Computed and measured electron energy-loss spectra of cyclopropane for the scattering angle $\vartheta = 30°$ at the resonant energy of 5.5 eV. The red line displays exact static-exchange approximation while the blue lines include also polarization effects modeled by LYP functional. Bands observed at 250–300 meV are not due to fundamental vibrational transitions. Triangles denote experimental spectra. (Adapted from Allan, M. and L. Andrič. 1996. *J. Chem. Phys. 105*, 3559–3568.)

can only be achieved if polarization effects are included. To be on the safe side, the calculations were performed with a larger plane-wave basis than that given by the procedure described in Section 8.4. Hence, instead of 18 radial points at energy of 5.5 eV and 14 radial points at energy 10 eV, we used 30 and 32 radial points, respectively. We present only data obtained with the larger basis sets, though the differences in data obtained by the two basis sets are small.

8.9.3 DIACETYLENE

As with cyclopropane, we selected energy of 10 eV to show the performance of static-exchange and static-exchange-plus-polarization approaches. The experimental data (Allan et al. 2011) were obtained for the scattering angle of 135°. The experimental and calculated spectra were obtained on absolute scale. For smearing the calculated elastic and differential cross sections, we used half-width of 14 meV. The calculations were performed with a single plane-wave basis with 32 radial points. The resulting electron energy-loss spectra are presented in Figure 8.8. It is clear that the correlation–polarization effects have no impact on the calculated cross sections for excitation of C–H stretching modes and very little effect is visible for C–C stretching modes. However, the cross section is halved for the bending modes on inclusion of the correlation–polarization forces as displayed in Figure 8.8. A comparison of SEP results with experimental data is good for the bending and C–C stretching modes but both computational models overestimate excitation cross sections

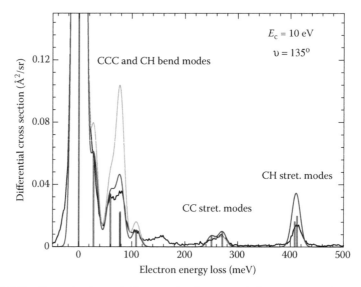

FIGURE 8.8 (**See color insert.**) Computed and measured electron energy-loss spectra of diacetylene for the scattering angle $\vartheta = 135°$ and collision energy of 10 eV. Vertical bars represent the actual calculated value of the differential cross section for the particular mode. Assuming the Gaussian band shape, the calculated SEP results (blue line) are compared with the experimental spectra (black curve). For completeness, we also present static-exchange calculations denoted by the red line. (Adapted from Allan, M. et al. 2011. *Phys. Rev. A, 83*, 052701.)

for the C–H stretching modes. The experimental band around 140–160 meV is attributed to overtone excitation of C–H bend modes (Allan et al. 2011) and therefore it is not visible in the two-channel and harmonic DMR method.

ACKNOWLEDGMENT

This work has been done within the frame of COST Actions CM0601 (Electron Controlled Chemical Lithography) and CM0805 (Chemical Cosmos) and supported also by the Czech Ministry of Education (grants OC09079 and OC10046) and by the Grant Agency pf the Czech Republic (grants 202/08/0631 and P208/11/0452).

REFERENCES

Allan, M. and L. Andric. 1996. σ* Resonances in electron impact-induced vibrational excitation of propane, cyclopropane, ethylene oxide, cyclopentane, and cyclohexane. *J. Chem. Phys. 105*, 3559–3568.

Allan, M., O. May, J. Fedor, B. C. Ibănescu, and L. Andric. 2011. Absolute angle-differential vibrational excitation cross sections for electron collisions with diacetylene. *Phys. Rev. A, 83*, 052701.

Baerends, E. J., D. J. Ellis, and P. Ross. 1973. Self-consistent Hartree-Fock-Slater calculations. I. The computational procedure. *Chem. Phys. Lett. 2*, 41–51.

Čársky, P. 2007. Efficient evalution of Coulomb integrals in a mixed Gaussian and plane-wave basis. *Int. J. Quantum. Chem. 107*, 56–62.

Čársky, P. 2009. Density fitting for derivatives of Coulomb integrals in ab initio calculations using mixed Gaussian and plane-wave basis. *Int. J. Quantum. Chem. 109*, 1237–1242.

Čársky, P. 2010a. Towards efficient ab initio calculations of electron scattering by polyatomic molecules I. Efficient numerical quadrature of the UGT term. *J. Phys. B: At. Mol. Opt. Phys. 43,* 175203.

Čársky, P. 2010b. Towards efficient ab initio calculations of electron scattering by polyatomic molecules II. Efficient evaluation of exchange integrals. *J. Phys. B: At. Mol. Opt. Phys. 43*, 175204.

Čársky, P. and R. Čurík. 2006. Vibrational spectra by electron impact: Theoretical models for intensities. *Computational Chemistry: Reviews of Current Trends 10*, pp. 121–137. Singapore: World Scientific.

Čársky, P. and R. Čurík. 2008. Vibrationally inelastic electron scattering in a two-channel approximation. *J. Phys. B: At. Mol. Opt. Phys. 41*, 055203.

Čársky, P. and M. Polášek. 1998a. Evaluation of molecular integrals in a mixed Gaussian and plane-wave basis by Rys quadrature. *J. Comp. Phys. 143*, 266–277.

Čársky, P. and M. Polášek. 1998b. Incomplete Gamma $F_m(x)$ functions for real negative and complex arguments. *J. Comp. Phys. 143*, 259–265.

Čársky, P. and T. Reschel. 1998. Evaluation of molecular integrals in a mixed Gaussian and plane-wave basis by means of the Fadeeva function and it s derivatives. *Collect. Czech. Chem. Commun. 63*, 1264–1284.

Čurík, R. and P. Čársky. 2003. Vibrationally inelastic electron scattering on polyatomic molecules by the discrete momentum representation (DMR) method. *J. Phys. B: At. Mol. Opt. Phys. 36*, 2165–2177.

Čurík, R., P. Čársky, and M. Allan. 2008. Vibrational excitation of methane by slow electrons revisited: Theoretical and experimental study. *J. Phys. B: At. Mol. Opt. Phys. 41*, 115203.

Čurík, R., P. Čársky, and M. Allan. 2011. Resonant and non-resonant vibrational excitation of cyclopropane by electron impact. *In preparation*.

Čurík, R. and F. Gianturco. 2002. Quantum calculations for resonant vibrational excitations of cyclopropane by electron impact. *J. Phys. B: At. Mol. Opt. Phys. 35*, 1235–1250.

Čurík, R. and M. Šulc. 2010. Towards efficient ab initio calculations of electron scattering by polyatomic molecules: III. Modelling correlation-polarization interactions. *J. Phys. B: At. Mol. Opt. Phys. 43*, 175205.

Dunning Jr, T. H. and P. J. Hay. 1977. Gaussian basis sets for molecular calculations. *Modern Theoretical Chemistry, Vol. 3*, pp. 1–27. New York, NY: Plenum Press.

Eichkorn, K., O. Treutler, H. F. Öhm, M. Häser, and R. Ahlrichs. 1995. Auxiliary basis sets to approximate Coulomb potentials. *Chem. Phys. Lett. 240*, 283–290.

Füsti-Molnar, L. and P. Pulay. 2002. Accurate molecular integrals and energies using combined plane wave and Gaussian basis sets in molecular electronic structure theory. *J. Chem. Phys. 116*, 7795–7805.

Gianturco, F. A. and S. Scialla. 1987. Local approximation of exchange interaction in electron-molecule collisions: The methane molecule. *J. Phys. B: At. Mol. Opt. Phys. 20*, 3171–3189.

Gill, P. M., B. G. Johnson, and J. A. Pople. 1991. Two-electron repulsion integrals over Gaussian *s* functions. *Int. J. Quantum. Chem. 40*, 745–752.

Haxton, D. J., Z. Zhang, H. D. Meyer, T. N. Rescigno, and C. W. McCurdy. 2004. Dynamics dissociative attachment of electrons to water through $^2B_1$ metastable state of the anion. *Phys. Rev. A 69*, 062714.

Itikawa, Y. 2004. Vibrational excitation of polyatomic molecules by electron collisions. *J. Phys. B: At. Mol. Opt. Phys. 37*, R1–24.

Kato, K., M. H. H. Kahawara, C. Makochekanwa, S. J. Buckman, M. J. Brunger, H. Cho, M. Kimura, et al. 2009. Cross sections for electron-induced resonant vibrational excitations in polyatomic molecules. *National Institute for Fusion Science Research Report NIFS-DATA-105*, NIFS, Oroshi–cho, Toki–shi, Japan.

Lane, N. F. 1980. The theory of electron-molecule collisions. *Rev. Mod. Phys. 52*, 29–119.

Lebedev, V. I. and D. N. Laikov. 1999. In Russian. *Dokl. Akad. Nauk 366*, 741–745.

Lee, C., W. Yang, and R. Parr. 1988. Development of the Colle-Salvetti correlation-energy formula into a functional of the electron-density. *Phys. Rev. B 37*, 785–789.

Nishimura, T. and Y. Itikawa. 1995. Electron-impact vibrational excitation of water molecules. *J. Phys. B: At. Mol. Opt. Phys. 28*, 1995–2005.

Nishimura, T. and Y. Itikawa. 1996. Vibrationally elastic and inelastic scattering of electrons by hydrogen sulphide molecules. *J. Phys. B: At. Mol. Opt. Phys. 28*, 4213–4226.

Ostlund, N. S. 1975. Polyatomic scattering integrals with Gaussian orbitals. *Chem. Phys. Lett. 34*, 419–512.

Perdew, J., J. Chevary, S. Vosko, K. Jackson, M. Pederson, D. Singh, and C. Fiolhais. 1992. Atoms, molecules, solids, and surfaces—Applications of the generalized gradient approximation for exchange and correlation. *Phys. Rev. B 46*, 6671–6687.

Perdew, J. and A. Zunger. 1981. Self-interaction correction to density-functional approximations for many-electron systems. *Phys. Rev. B 23*, 5048–5079.

Polášek, M. and P. Čársky. 2002. Efficient evaluation of the matrix elements of the Coulomb potential between plane waves and Gaussians. *J. Comp. Phys. 181*, 1–8.

Polášek, M., M. Juřek, M. Ingr, P. Čársky, and J. Horáček. 2000. Discrete momentum representation of the Lippmann-Schwinger equation and its application to electron-molecule scattering. *Phys. Rev. A 61*, 032701.

Rescigno, T. N., W. A. Isaacs, A. E. Orel, H. D. Meyer, and C. W. McCurdy. 2002. Dynamics dissociative attachment of electrons to water through $^2B_1$ metastable state of the anion. *Phys. Rev. A 65*, 032716.

Rescigno, T. N., B. H. Lengsfield III, and C. W. McCurdy. 1995. The incorporation of modern electronic structure methods in electron-molecules collision problems: Variational calculations using the complex Kohn method. *Modern Electronic Structure Theory*, pp. 501–588. Singapore: World Scientific.

Rescigno, T. N., C. W. McCurdy Jr., and V. McKoy. 1975. Low-energy e^-–H_2 elastic cross sections using discrete basis functions. *Phys. Rev. A 11*, 825–829.

Shimanouchi, T. 1972. *Tables of Molecular Vibrational Frequencies Consolidated*. Vol 1, Washington, DC: National Bureau of Standards.

Schütz, M., H. Werner, R. Lindh, and F. R. Manby. 2004. Analytical energy gradients for local second-order Møller–Plesset perturbation theory using density fitting approximation. *J. Chem. Phys. 121*, 737–750.

Shavitt, I. 1963. The Gaussian function in calculations of statistical mechanics and quantum mechanics. *Methods in Computational Physics, Vol. 2*, pp. 1–45. New York, NY: Academic Press.

Watson, D. K. and V. McKoy. 1979. Discrete-basis-function approach to electron-molecule scattering. *Phys. Rev. A 20*, 1474–1483.

Weigend, F., M. Häser, H. Patzelt, and R. Ahlrichs. 1998. RI-MP2: Optimized auxiliary basis sets and demonstration of efficiency. *Chem. Phys. Lett. 294*, 143–152.

Weissbluth, M. 1978. *Atoms and Molecules*. New York: Academic Press Inc.

Winstead, C. and V. McKoy. 1995. Studies of electron-molecule collisions on massively parallel computers. *Modern Electronic Structure Theory*, pp. 1375–1462. Singapore: World Scientific.

Wong, S. F. and G. J. Schulz. 2003. Vibrational excitation in benzene by electron impact via resonances: Selection rules. *Phys. Rev. Lett. 35*, 1429–1432.

Index

A

Adenine
    C-H and N-H bond breaking, 190
    resonances, 178–181
Adiabatic approximation, 62, 93, 94
    definition, 101
Associative detachment, 3, 109, 146, 147
Auxiliary basis set, 268–270, 275

B

Benzene, 12, 271, 275
Boomerang oscillations
    in CH_2NO_2, 79
    in DBr, 64
    in HCOOH, 81
    in higher vibrational states, 67, 81
    in hydrogen halides, 60
    in H_2, 129, 130, 140
    in N_2, 57, 130
    in OCO bending, 74
    origin, 61–63
    in vibrational excitations, 139–142

C

CCl_4, 11
CF_4, 11
Chlorobenzene, 70–72
CH_4, 11, 51, 276–278
CHF_3, 51
CH_2F_2, 51
CH_2NO_2, 77–79
CH_3F, 51
CH_3I, 66
C_2F_6, 51
C_2H_2, 72, 73, 81
C_2H_5OH, 76, 77
C_2H_5SH, 76, 77
C_3F_8, 51
CO, 58–61
    elastic scattering, 59
    total cross section, 61
    vibrational $0 \rightarrow 1$ excitation, 58
CO_2, 66–69
Correlation-polarization effects, 15–18, 272–275,
    See also Polarization
Cross section
    absolute values, 46, 49
    normalization, 55
    temperature dependence, 153–155
Cyclopropane, 278
Cytosine
    C-H and N-H bond breaking, 188
    resonances, 178–180, 182, 193

D

DBr, 64
DEA, *See* Dissociative electron attachment
Density fitting, 268–270
Density functional theory
    applications to DNA subunits, 167, 177–179,
        181–188, 202–207
    use for electron scattering, 16–18
    use for polarizitation effects, 272–275
DFT, *See* Density functional theory
Diacetylene, 279, 280
Diamond films
    composition analysis, 245–247
    electron energy loss spectra, 242–245
    surface character, 250–253
Dipole
    bound states, 62, 63, 67–71, 77–79, 186–188
    dipolar excitation, 237, 238
    polarizability, 15, 272
    scattering, 253
Dissociative electron attachment, 2, 3
    DNA and its subunits, 180, 188, 202,
        205–207
    hydrogen halides, 132–135
    H_2, 128–131
    measurement of, 48–50
    temperature dependence, 145, 146
Doorway state
    shape resonance, 70–77
    vibrational Feshbach resonance, 69, 70

E

EELS, *See* Electron energy loss spectroscopy and
    HREELS
Electron energy loss spectroscopy
    high resolution, 239, *See also* HREELS
    of CH_4, 276–278
    of cyclopropane, 278, 279
    of diacetylene, 279, 280
    of DNA bases, 177–184
    of HBr, 63

T - #0384 - 071024 - C16 - 234/156/14 - PB - 9780367381806 - Gloss Lamination